Advanced Technologies for Physical Properties Measurement of Food and its Processing

Hao Lin • Lei Zhang • Quansheng Chen •
Man Zhou • Leiqing Pan • Qin Ouyang •
Solomon Yao-Say Selorm Adade • Yanyu Li

Advanced Technologies for Physical Properties Measurement of Food and its Processing

Springer

Hao Lin
School of Food and Biological Engineering
Jiangsu University
Zhenjiang, Jiangsu, China

Lei Zhang
School of Food and Biological Engineering
Jiangsu University
Zhenjiang, Jiangsu, China

Quansheng Chen
School of Food and Biological Engineering
Jimei University
Xiamen, Fujian, China

Man Zhou
School of Food and Biological Engineering
Jiangsu University
Zhenjiang, Jiangsu, China

Leiqing Pan
College of Food Science and Technology
Nanjing Agricultural University
Nanjing, Jiangsu, China

Qin Ouyang
School of Food and Biological Engineering
Jiangsu University
Zhenjiang, Jiangsu, China

Solomon Yao-Say Selorm Adade
School of Food and Biological Engineering
Jimei University
Xiamen, Fujian, China

Yanyu Li
Academy of National Food and Strategic
Reserves Administration
Beijing, Beijing, China

ISBN 978-981-96-2346-4 ISBN 978-981-96-2347-1 (eBook)
https://doi.org/10.1007/978-981-96-2347-1

The original submitted manuscript has been translated into English. The translation was done using artificial intelligence. A subsequent revision was performed by the author(s) to further refine the work and to ensure that the translation is appropriate concerning content and scientific correctness. It may, however, read stylistically different from a conventional translation.

This Springer imprint is published by the registered company Springer Nature Singapore Pte Ltd.
The registered company address is: 152 Beach Road, #21-01/04 Gateway East, Singapore 189721, Singapore

Preface

The study of food's physical properties involves examining the shared physical characteristics across various foods, with a focus on understanding them through the lens of physics. This field explores the composition, structural components, and physical attributes of food, as well as the changes these properties undergo during processing, storage, quality inspection, and transportation. These insights form a critical foundation for food science and are essential to the advancement of related disciplines.

By the late twentieth century, significant progress in food processing technologies, driven by advancements in food chemistry and biology, provided crucial technological support for the expansion of the food industry. As we entered the twenty-first century, considerable strides were made in areas related to the physical properties of food, particularly in detecting and analyzing these properties, as well as in physical processing techniques. Despite these advancements, there remains a shortage of comprehensive and practical guides that keep pace with emerging technologies in the study of food physical properties.

This book is the result of a collaborative effort among experts who have dedicated their careers to researching and understanding the physical properties of food and their applications in food processing. The chapters have been authored as follows: Hao Lin and Quansheng Chen contributed to Chap. 1, "Introduction." Chapter 2, "Basic Physical Characteristics of Food," was written by Man Zhou, Hao Lin, and Selorm Yao-Say Solomon Adade. Lei Zhang authored Chap. 3 on the "Rheological Properties of Food." Chapter 4, focusing on "Texture of Food," was prepared by Hao Lin and Selorm Yao-Say Solomon Adade. Chapter 5, "Optical Characteristics and Color in Foods" was authored by Hao Lin. Man Zhou authored Chap. 6, "Thermal Characteristics of Food." Hao Lin wrote Chap. 7, "Acoustical Characteristics in Foods." Yanyu Li and Leiqin Pan authored Chap. 8, "Electromagnetic Properties in Foods," while Chap. 9, "Advanced Physical Technologies in Foods," was written by Qin Ouyang.

This book provides a systematic review of recent research, along with an outlook on future developments in the study of food physical properties. It offers an in-depth exploration of key topics, including optical, thermal, acoustic, rheological, and textural properties, with a focus on the principles, measurement techniques, data analysis, and their applications in evaluating and monitoring food quality.

Furthermore, the book delves into advanced technologies such as near-infrared spectroscopy, computer vision, spectral imaging, acoustic resonance analysis, and electronic sensory systems, including electronic noses, tongues, and eyes, as well as multi-sensor data fusion. It also introduces modern nonthermal physical food processing techniques, covering methods such as ultrasound, ultra-high pressure, high-pressure pulsed electric fields, ionizing radiation, and pulsed electric and magnetic fields. Additionally, nontraditional heating technologies, including infrared, radio frequency, pulsed intense light, microwave, and ohmic heating, are discussed for their applications in food processing.

The primary objective of this book is to equip readers with a solid understanding of the principles behind the measurement and processing of food's physical properties, enabling them to apply these techniques for real-time assessments and field monitoring. This book serves as both a textbook for undergraduate and graduate students and a reference guide for researchers and professionals in the field of food science and engineering.

We extend our deepest gratitude to Du Shidan and Zhang Kexin for their meticulous revisions and editorial contributions; without their help, this book would not have been completed smoothly. We also wish to express our thanks to Chen Quanzhou, Ji Men, and Huang Baocheng for their invaluable input in revising this manuscript. Special thanks go to Bridget Ama Kwadzokpui for her efforts in refining and polishing the content. Lastly, we acknowledge the financial support from the National Natural Science Foundation of China [Grant No. 32472429].

Zhenjiang, Jiangsu, China Hao Lin

Contents

Chapter 1
Introduction

Hao Lin and Quansheng Chen

1.1 Definition of Food Physical Properties

Food physics is a fundamental discipline that takes food (including food raw materials and intermediate products in processing) as the research object, studying the physical properties like structure, properties, changes of food systems, their mechanisms, and so on. The properties of food are determined by its organizational structure and the composition of its materials. For the properties of food raw materials (agricultural products), physiological activities are also influencing factors. Food physical properties mainly study the physical characteristics common to various foods, with research content primarily belonging to the realm of physics. It studies the composition, structural components, physical properties of food, and their changes and quality control in food processing, storage, quality inspection, and transportation from a physics perspective. It is an important theoretical foundation for food science and all disciplines related to food [1].

Food physical properties focus on exploring and characterizing the physical properties exhibited by foods with different chemical compositions and structures, thereby providing scientific technical data for all aspects of food processing, testing, grading, quality control, storage, transportation, management, and even consumption, and providing direct foundational information for the research and design of food equipment. The study of physical properties involves all disciplinary fields, and materials can be understood as the research objects of all disciplinary fields (broadly defined materials), thus forming the physical properties science (physics) of different disciplinary fields, such as biological material physical properties science (biological material physics), inorganic material physical properties science, engineering

---

H. Lin (✉)
School of Food and Biological Engineering, Jiangsu University, Zhenjiang, Jiangsu, China

Q. Chen
School of Food and Biological Engineering, Jimei University, Xiamen, Fujian, China

H. Lin et al., *Advanced Technologies for Physical Properties Measurement of Food and its Processing*, https://doi.org/10.1007/978-981-96-2347-1_1

material physical properties science, petroleum physical properties science, and geological physics; moreover, physical properties can also be broadly understood as physical characteristics, chemical characteristics, and biological characteristics, etc. (broadly defined physical properties); therefore, it can be collectively referred to as (broadly defined) material physical properties science.

1.2 Characteristics and Content of Food Physical Properties

1.2.1 Characteristics of Food Physical Properties Research

1.2.1.1 The Complexity and Diversity of Food Materials

Food is a very broad concept and a very complex material system. From the perspective of food processing, food includes primary raw materials (such as grains, fruits, and vegetables), intermediate products, semi-finished products, and finished products; from the perspective of food morphology, there are various types of food such as liquid, gel-like, solid, and semi-solid; from the perspective of food composition, most foods are complex mixtures, including organic matter, inorganic matter, and biological entities with cellular structures, and are mostly heterogeneous structures. Therefore, there are significant differences in physical properties between food, biological materials, and general non-biological materials, and their research methods are also different [2].

1.2.1.2 The Uncertainty of Sensory Index Evaluation in Food Physical Properties

The ultimate goal of food is to meet people's material needs. As humans are the main producers and consumers of food, the sensory properties of food constitute an important aspect of food physical properties. Therefore, research on the quantification of sensory properties by instruments, scientific reproduction, information exchange, and sharing also constitutes an important part of food physical properties science research.

1.2.1.3 The Practicality of Food Physical Properties

Physical properties science shows strong practicality in the food industry. When studying, it requires a lot of practical experience in food processing. Research in food physical properties science often does not have ready-made models or testing instruments, requiring the design of testing devices or the establishment of models based on experimental results. Only in this way can one truly master this science and be adept at applying it to solve various problems in food development.

1.2.1.4 The Interdisciplinary Integration of Food Physical Properties

Food physical properties science is an interdisciplinary field, and research should master certain basic knowledge of physics, physical chemistry, food biochemistry, polymer chemistry, and food engineering principles. It also involves content from disciplines such as biology, physiology, and psychology, so attention should be paid to the comprehensive application of this knowledge.

1.2.1.5 The Imperfection of the Theoretical System of Food Physical Properties

Currently, in the field of food physical properties science, many areas of research are still only preliminary experiments, and systematic conclusions need to be refined from long-term research results in the future. Therefore, when studying, one should be good at comprehensive association, bold innovation, drawing inferences, and developing new research ideas, not only truly mastering its research methods but also contributing to the formation of the food physical properties science system [3].

1.2.2 Content of Food Physical Properties Research

Despite the complex nature and distinct characteristics of food physical property research, an important method in scientific research is to discard the dross and select the essential, seeking common ground while reserving differences. The main content of food physical property research includes the following aspects.

1.2.2.1 Basic Physical Characteristics of Food

These include shape, size, particle size, particle size distribution, volume, porosity, shape factor; center of gravity, weight, density; area, surface characteristics; macro and microstructure, etc.

1.2.2.2 Rheological Properties of Food

When materials are subjected to external forces, they undergo deformation, and when deformation continues to expand, it becomes flow. Food rheology studies the laws of deformation, flow, vibration, and fracture of food and raw materials under the action of force. The generation and development of deformation and flow always require a certain time course. Therefore, in rheology, the rheological behavior of materials is expressed by three parameters: force, deformation, and time. The application of rheological characteristics in optimizing production processes, controlling food quality, evaluating food quality, improving eating quality, and enhancing food stability provides a basis for actual production [4].

1.2.2.3 Texture and Bionic Characteristics of Food

The texture characteristics of food (softness, hardness, viscoelasticity, chewability) are important for sensory evaluation of food quality and instrumental evaluation of texture. They are the external manifestation of biochemical changes in food and are an important foundation for food mechanical design and food engineering design. Bionic detection technology refers to simulating the physical information of the object being detected as perceived by a biological entity and presenting the physical information of the biological entity to be simulated in a concrete form. Accurately speaking, bionic detection is a simulation test of the texture characteristics of food, which mainly depends on its internal organizational structure and external morphology. Research on food texture characteristics can optimize the setting of processing equipment or its processing parameters and plays an important role in the evaluation of preparation processes and final products [5].

1.2.2.4 Optical Properties of Food

This involves the absorption and reflection of light by food substances and their properties in sensory response. The research and application fields mainly include two aspects: the determination of food components and the study and utilization of food color through optical properties. Changes in food components can cause changes in their optical properties, and the measurement of optical properties is characterized by being fast, accurate, simple, and non-destructive. Therefore, optical properties are increasingly valued in both instrumental analysis and online detection in production processing. The most representative device in the application of optical properties is the color sorter, which sorts materials (various seeds, processed rice, fruits, etc.) based on color differences and has been widely used [6].

1.2.2.5 Thermal Properties of Food

This involves the specific heat capacity, latent heat, phase change laws, heat transfer laws, and thermal expansion laws related to temperature exhibited by food during thermal processing (such as drying, curing, freezing, etc.). The thermal properties of food are not only the dominant physical properties in cold and hot processing but also an important means of studying the microstructure of food. Currently, cold and hot operations are used not only as basic units of food engineering but also in the improvement of food quality [7].

1.2.2.6 Acoustic Properties of Food

Food acoustic quality detection and acoustic wave processing technology are application technologies based on the acoustic properties of food, especially ultrasonic technology, which is widely used in food property detection (such as non-destructive testing of mango ripeness, non-contact detection of material characteristics inside extruder barrels, etc.). Moreover, in terms of utilizing the mechanical energy of ultrasound, technologies and equipment such as ultrasonic atomization, ultrasonic homogenization, ultrasonic sterilization, ultrasonic-assisted extraction, and ultrasonic food quality detection are also widely applied [8].

1.2.2.7 Electromagnetic Properties of Food

The electromagnetic properties of food refer to the conductive characteristics, dielectric characteristics, and other electromagnetic physical properties of food and its raw materials. The research fields of electromagnetic properties mainly focus on monitoring the quality state of food (especially in non-destructive testing of food) and electromagnetic physical processing. The former uses the interaction rules between the electromagnetic properties of food and its component and state changes, employing electrical sensor methods to detect and evaluate food quality or state; the latter uses electromagnetic physical technology (such as electrostatic field technology, electromagnetic wave technology, electric heating technology, electro-osmotic dehydration technology, magnetic treatment technology, etc.) to process food. In electromagnetic physical processing, it is necessary to study the physical properties of the object being processed on one hand and the electromagnetic physical parameters that change the object's properties on the other [9] (Fig. 1.1).

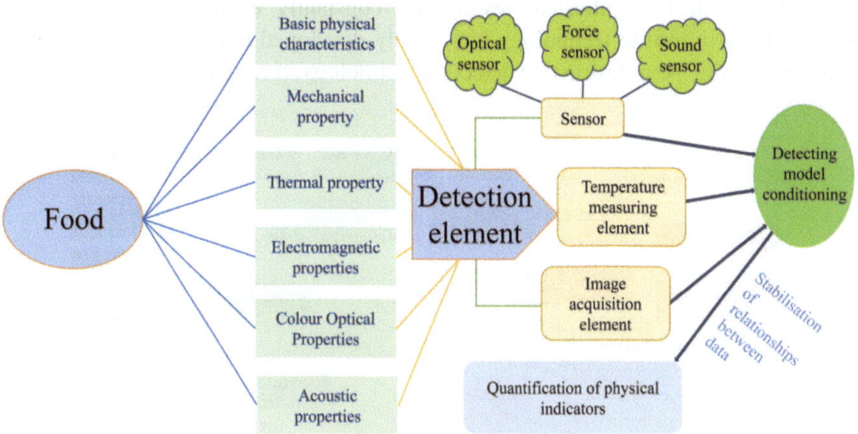

Fig. 1.1 Physical properties of food and advanced technologies

From the above, it can be seen that the research content of food physical properties is extensive and involves many fundamental disciplines. With the continuous emergence of high-tech, the research and utilization of food physical properties will be increasingly strengthened. It can be said that the research content of food physical properties is reflected on two levels: first, objectively understanding and characterizing food objects, scientifically quantifying, managing, and evaluating their information; second, in terms of the analysis, synthesis, and utilization of physical properties, making the external physical effects that food undergoes throughout its life process fully and effectively adapt and utilize the attributes of the targeted objects, providing basic data for food engineering and equipment design.

1.3 Physical Characteristics in Food Measurement

Food quality measurement based on physical characteristics usually refers to the use of specialized instruments and equipment to measure the physical characteristics of food, and then revealing the relationship between the physical characteristics of food and specific physical property indicators through data analysis. Based on this relationship, the physical characteristics of food can be quantitatively evaluated. The execution steps of the instrumented testing method for food physical properties are shown in Fig. 1.2.

The instrumented measurement method for food physical properties is divided into direct and indirect detection methods based on the processing operations of the obtained data. Direct testing methods include general testing methods and empirical testing methods. Indirect testing methods include optical methods, acoustic methods, electrical methods, mechanical methods, etc. With the development of society, new technologies in food physical testing are constantly emerging, among which non-destructive testing technology has rapidly developed in its application in food physical testing. Non-destructive detection technology, through physical means such

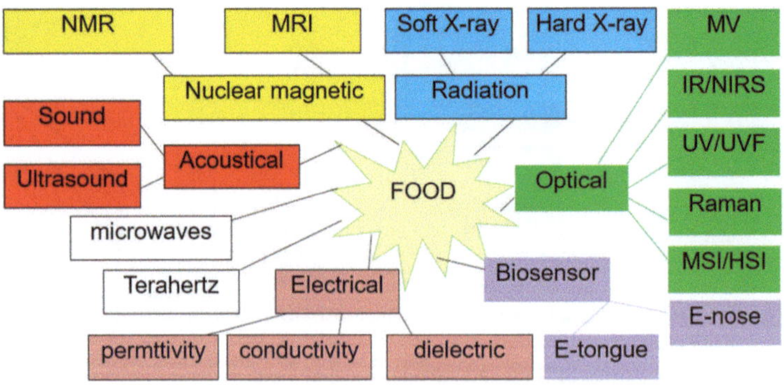

Fig. 1.2 Schematic diagram of the execution steps of the instrumental test method

as spectroscopy, sound waves, and electromagnetism, can detect the internal structure, composition, and characteristics of food without damaging it, providing important technical support for food safety monitoring and processing. Below is a brief introduction to the application of several non-destructive testing technologies in the physical characteristics of food.

1.3.1 Food Detection Technologies Based on Optical Characteristics

Food optical characteristic testing technology is mainly based on the principle of interaction between light and matter, obtaining internal information of food by measuring phenomena such as absorption, emission, reflection, and scattering of light. When light is irradiated onto a material, part of it is reflected by the material's surface, and another part refracts into the material's internal structure. The light entering the material is partially absorbed and converted into heat energy, partially scattered in all directions, and the remaining part passes through the material. The refraction, reflection, absorption, and transmission properties of the material to light are called optical characteristics.

1.3.1.1 Spectroscopy Technologies in Foods

Near-Infrared Spectroscopy

The wavenumber of near-infrared light is approximately $10{,}000 \sim 4000\ cm^{-1}$, which is the electromagnetic spectrum between visible light and mid-infrared. By measuring the absorption of near-infrared light by food, components such as moisture, protein, and fat in food can be detected [10].

Ultraviolet-Visible Spectroscopy

By measuring the absorption of ultraviolet or visible light by molecules or ions of substances, the composition, content, and structure of substances in food, such as vitamins, pigments, and proteins, can be detected.

Raman Spectroscopy

Raman spectroscopy is an analytical method based on analyzing scattered spectra with frequencies different from the incident light to obtain molecular vibration information. In food quality and safety testing, the fingerprint information and high sensitivity of Surface Enhancement of Raman Scattering (SERS) technology can achieve rapid trace detection of substances [11].

1.3.1.2 Imaging Detection Technology

Visible Wavelength Imaging

The range of electromagnetic wave wavelengths that the human eye can directly perceive is 400–780 nm, forming various colors in the brain. Red (R), green (G), and blue (B) are called the three primary colors, which stimulate the photoreceptor cells in the human brain, forming various different colors after analysis and synthesis by the brain. Visible wavelength imaging technology refers to converting the object being captured into image signals using image capture devices within the wavelength range of electromagnetic waves that the human eye can directly perceive, and performing image processing. Currently, visible wavelength imaging technology has been applied to the quality detection of size, shape, texture, color, and defects of foods such as vegetables, fruits, and poultry eggs.

Spectral Imaging

Spectral imaging technology combines advanced technologies from fields such as optics, electronics, information processing, and computer science, organically integrating spectroscopy and imaging technology. It is generally believed that images with a spectral resolution in the range of 10^{-1} λ are called multi-spectral images, and images with a spectral resolution in the range of 10^{-2} λ are called hyper-spectral images. In recent years, the advantages of spectral imaging technology in detecting subtle surface features such as internal structure and organization of food have received more attention and recognition [12].

X-Ray Imaging

X-ray imaging technology is based on the attenuation and refraction of the X-ray wave field as it propagates through materials, and the detection of the resulting intensity distribution. By measuring the scattering pattern of lasers in food, it evaluates the particle size distribution and structural characteristics of the food.

Other Imaging Detection Technologies

Such as Computed Tomography (CT), which is a new imaging technology introduced to overcome the problem of projection overlap in X-ray linear scan imaging; Magnetic Resonance Imaging (MRI) utilizes the phenomenon of nuclear magnetic resonance generated when hydrogen nuclei in a biological body are excited in an external magnetic field. This is spatially encoded, received by detectors, emitted as electromagnetic signals, and processed by computers to form images.

1.3.2 Acoustic Property-Based Detection Technology

The phenomenon of mechanical vibrations of objects propagating in a medium with mass and elasticity is called a wave, and waves that can be perceived by the human ear are called sound waves. The acoustic properties of food are often perceived through the sound waves of vibrations. Food acoustic property detection technology is mainly based on the principle of interaction between sound waves (audible sound waves, ultrasonic waves) and food. By measuring the propagation characteristics of sound waves in food (such as speed, attenuation, reflection, scattering, etc.) and using computer numerical analysis, it evaluates the internal quality of the food.

1.3.3 Electromagnetic Property-Based Detection Technology

Electromagnetic property detection technology evaluates the physical and chemical attributes of food by utilizing the principle of interaction between electromagnetic waves and substances. These technologies include dielectric property analysis, microwave technology, radio frequency identification, nuclear magnetic resonance, electromagnetic spectrum technology, conductivity and impedance measurement, capacitance measurement, electromagnetic wave scattering technology, and terahertz technology. They analyze the moisture content, composition, maturity, freshness, and processing conditions of food by measuring the absorption, reflection, transmission, or scattering characteristics of substances to electromagnetic waves, thus providing important information for food quality control.

1.3.4 Multi-physical Property Information Fusion-Based Detection Technology

Food multi-sensor information fusion detection technology combines data on the physical, chemical, and biological properties of food obtained from various sensors, using advanced data processing algorithms such as chemometrics and machine learning to improve the accuracy and efficiency of food quality, safety, and adulteration detection. This method can integrate spectral data, image analysis, electronic nose and electronic tongue signals, and possibly other physical measurements, to comprehensively evaluate food from different angles and levels. Information fusion technology can overcome the limitations of a single sensor, providing more comprehensive and reliable food attribute information, enhancing the robustness of detection systems, and is of great significance for rapid, non-destructive, high-throughput food detection.

1.4 Physical Properties in Food Processing

1.4.1 Physical Processing of Food

Food physical processing technology refers to the process of changing the form, structure, appearance, or taste characteristics of food through physical means without altering its chemical composition. These processing technologies are widely used in the food industry to improve food quality and safety, increase food diversity, and extend shelf life.

The use of food physical processing technology plays a crucial role in the food industry by providing various methods to improve food safety, quality, and processing efficiency. One way of food processing is non-thermal physical processing, such as ultrasonic technology, which has been used in food and biological processing to improve processes like emulsification, extraction, enzymolysis, fermentation, and microbial inactivation. The potential of non-thermal plasma in microbial inactivation has also been explored, providing insights into new methods for food safety and preservation. Another physical method of food processing involves thermal physical processing, especially in fruits, vegetable juices, and beverages, where heat treatment plays an important role in pasteurization and sterilization to ensure product safety and extend shelf life. Food physical processing technology shows good results in improving moisture loss, fat absorption, texture, color, and nutrient retention in food processing processes such as frying, fermentation, and emulsification by applying mechanical properties, electromagnetic properties, and ultrasonic waves. These physical processing technologies provide innovative solutions to challenges in food processing manufacturing and quality safety [13].

1.4.2 Physical Properties in Non-thermal Food Processing

Non-thermal physical food processing technology utilizes modern physical techniques such as ultrasound, ultra-high pressure, high-pressure pulsed electric fields, ionizing radiation, pulsed electric/magnetic fields, plasma, ultraviolet light, supercritical or subcritical fluids, to perform sterilization, enzyme inactivation, and other processes without increasing the food's temperature. This approach preserves the original nutritional components, texture, color, and freshness of the food. Compared to traditional thermal processing, non-thermal processing has less environmental pollution, lower energy consumption, and reduced pollutant emissions. The application of this technology in the food industry has become a new growth point, especially as consumer demands for freshness, nutrition, safety, and functionality in food continue to rise, leading to increased attention on the research and development of non-thermal processing technology.

Some non-thermal processing technologies, such as ultra-high-pressure technology, high-pressure carbon dioxide technology, ionizing radiation, and pulsed magnetic fields, have been industrially applied in food processing. As research deepens, the application scope of non-thermal processing technology continues to expand, and it can be used for the extraction of functional food components and the modification of food macromolecules [14].

Research involving the application of modern physical technologies such as sound, light, electricity, magnetism, and force in food processing focuses on three areas: food physical processing, food functional properties and functional foods, and food equipment and intelligent manufacturing. The aim is to promote the integration of modern physical technology with food production and manufacturing, explore the manufacturing theory of healthy foods, product development and result transformation, as well as the design theory and product development of food equipment and intelligent manufacturing. Non-thermal physical processing technology, due to its advantages of being efficient, green, environmentally friendly, and intelligent, aligns with the future international development trends of the food industry. Future development requires strengthening basic theoretical research and breakthroughs in key technologies.

1.4.3 Physical Properties in Thermal Food Processing

In recent years, non-traditional heating technologies such as infrared, radio frequency, pulsed intense light, microwave, and ohmic heating have been applied in the field of food physical processing. These technologies are used not only for food drying and maturation but have also expanded into new areas such as sterilization, blanching of fruits and vegetables, pest control, fruit peeling, protein and starch modification, and rapid thawing of frozen foods. The application of these technologies can improve the efficiency and safety of food processing while reducing the destruction of nutritional components, meeting modern consumers' demands for health and nutrition. As research deepens, the application of these physical methods in the field of food processing will become more widespread [15].

Infrared technology has made significant progress in the field of food processing, particularly in the development of catalytic infrared equipment using natural gas as an energy source. This equipment reduces energy consumption to about one-third of traditional electric infrared equipment, becoming a low-carbon heating solution. At the University of California, Davis, after nearly 20 years of research, infrared processing technology has been widely applied in various stages of food processing, such as dehydration of vegetables, drying and sterilization of rice and dried fruits, and peeling of fruits and vegetables, with successful development of industrial equipment.

Radio frequency, as an electromagnetic wave, utilizes the reciprocal rotation of polar molecules and the reciprocal movement of charged ions to heat the entire volume of water-containing materials simultaneously. The United States Department of Agriculture has been committed to researching the application of radio frequency technology in food sterilization. Studies have shown that radio frequency heating can effectively kill pathogenic microorganisms in food while maintaining the nutritional components and sensory characteristics of the food. For example, radio frequency technology has been applied to the sterilization of low-moisture foods such as nuts, spices, and grains. The New Zealand Food Research Institute has used radio frequency technology for the rapid thawing of frozen foods. This technology reduces the partial uneven heating problem that occurs during traditional thawing processes, making the thawing of meat, seafood, and other products more uniform, thereby improving the final quality of the products.

Microwave, as a traditional heating method, is widely used in the food industry due to its high efficiency and energy-saving properties, and can be used for thawing, sterilization, extraction, and more. Microwave technology has also been successfully applied to the enzyme inactivation and stabilization processing of cereal germ, with the development of drum-type microwave processing equipment that solves the problem of uneven microwave heating, successfully applied in industrial production.

Food rheology is an extension of physics in the field of food science and engineering. Currently, food rapid non-destructive detection technology and food physical processing technology, which are based on food rheology, are research hotspots in the food sector. Food rapid non-destructive testing technology comprehensively applies the basic theories of food rheology, introducing fusion technology from the field of information science on the basis of various single detection technologies such as computer vision, electronic nose, spectral detection technology, and acoustic characteristics, to organically integrate multiple detection information for a comprehensive assessment of food quality. The main research directions of food physical processing include the influence of its own physical properties and environmental factors on various properties of food, the development of physical methods and equipment for food processing, the development of physical methods and instruments for food inspection, and the impact of food rheology on processing. It also includes the physical characteristics of food materials under extreme environmental factors such as ultra-high temperature, ultra-high pressure, ultra-fine grinding, and ultrasound, which are of great significance for process optimization and product development in food processing.

1.5 Conclusion

In recent years, the rapid development of some high-tech technologies, such as artificial intelligence, nanotechnology, and 4D printing technology, has brought challenges and opportunities to the development of food rheology. For example, artificial neural networks (ANN) have shown great potential in process control and simulation, providing an innovative solution for simulating complex tasks and enhancing food safety and quality control [16]. The significant progress in nanotechnology offers opportunities for advancements in fields such as food packaging, nanoemulsions, nanosensors, and nanoencapsulation. 4D printed food may play an important role in improving appearance and texture. In the future, with the development of bionic technology, sensing technology, microelectronics technology, and artificial intelligence technology, multi-technology integration based on the physical properties of food can be further applied in the food sector. This can better meet the needs of diversified and high-value food processing, holding significant practical and academic significance for the research and development of modern food engineering, and will have a tremendous impact and far-reaching influence on the development of the modern food industry.

References

1. Pakulska A, Kawecka L, Galus S. Physical properties of selected fruit fibre and pomace in the context of their sustainable use for food applications. Appl Sci. 2024;14:9051.
2. Ma H, Tang J, Ramaswamy HS, Ashokkumar M, et al. Interdisciplinary innovation involving food science and physics: food physics. Food Phys. 2024;1:100001.
3. Oppen D, Attig T, Weiss J, Krupitzer C. Anticipating food structure of meat products from mastication physics applying machine learning. Food Res Int. 2023;174:113576.
4. Vilgis TA. Gels: model systems for soft matter food physics. Curr Opin Food Sci. 2015;3:71–84.
5. Tokifuji A, Matsushima Y, Hachisuka K, Yoshioka K. Texture, sensory and swallowing characteristics of high-pressure-heat-treated pork meat gel as a dysphagia diet. Meat Sci. 2013;93(4):843–8.
6. Ji L, Gravelle AJ. Modulating the behavior of ethyl cellulose-based oleogels: the impact food-grade amphiphilic small molecules on structural, mechanical, and rheological properties. Food Hydrocoll. 2025;159:110703.
7. Wang Z, Pan J, Liu A, Zhang L, et al. Microstructure and properties controlling of Al-xGd alloys for thermal neutron absorbing. J Nucl Mater. 2025;603:155447.
8. Zhu X, Das RS, Bhavya ML, Garcia-Vaquero M, et al. Acoustic cavitation for agri-food applications: mechanism of action, design of new systems, challenges and strategies for scale-up. Ultrason Sonochem. 2024;105:106850.
9. Peng Y, Gong K, Liu A, Yan H, et al. Ultralight and rigid PBO nanofiber aerogel with superior electromagnetic wave absorption properties. J Mater Sci Technol. 2025;211:320–9.
10. Ichiba K, Kato T, Nakauchi D, Kawaguchi N, et al. Ultraviolet–near infrared luminescence characteristics of Nd: La2Be2O5 single crystals for near-infrared emitting scintillators. J Lumin. 2025;277:120938.
11. Tucker ZE, Abedin NM, Wincheski AR. Raman characterization of lunar highlands simulants for in-situ resource utilization in a lunar setting. Acta Astronaut. 2024;225:1039–48.
12. Siwen B. Research on the high-resolution infrared quantum spectral detection imaging technology. In: Quantum Sensing and Nano Electronics and Photonics XVII, vol. 11288. SPIE; 2020.
13. Neng J, Wang J, Wang Y, Zhang Y, et al. Trace analysis of food by surface-enhanced Raman spectroscopy combined with molecular imprinting technology: principle, application, challenges, and prospects. Food Chem. 2023;429:136883.

14. Liu Q, Hou S, Zhang Y, et al. Applications of novel non-thermal physical field technologies in enhancing the quality and storage stability of grains. J Stored Prod Res. 2024;108:102398.
15. Lei W, Jing Y, Lihua Y, et al. Effect of hydrostatic pressure on the mechanical, thermodynamic, and optical properties of orthorhombic WAlB. J Mater Res. 2022;37(23):4232.
16. Quansheng C, Hao L, Jiewen Z. Advanced nondestructive detection technologies in food. Springer Nature; 2021.

Chapter 2
Basic Physical Characteristics of Food

Man Zhou, Hao Lin, and Solomon Yao-Say Selorm Adade

2.1 Introduction

The basic physical characteristics of food include size, shape, surface area, volume, density, porosity, and more. These properties are essential for designing and developing agricultural machinery used in various unit processes such as cleaning, grading, drying, milling, handling, shipping, and storage. Additionally, these characteristics help monitor the safety and quality of foods. Understanding the physical properties of food is crucial for engineers, food scientists, and processors, as it enables the creation of effective processing techniques and equipment.

2.1.1 Basic Physical Properties of Food Materials

The basic physical properties of food materials include size, size distribution, shape, surface area, volume, density, and porosity. Figure 2.1 illustrates various types of food materials along with their corresponding physical characteristics.

These basic physical characteristics are widely used in the food industry:

M. Zhou · H. Lin (✉)
School of Food and Biological Engineering, Jiangsu University, Zhenjiang, Jiangsu, China
e-mail: manzhou@ujs.edu.cn; linhao@ujs.edu.cn

S. Y. S. S. Adade
Food Science and Biological Engineering, Jimei University, Xiamen, Fujian, China

H. Lin et al., *Advanced Technologies for Physical Properties Measurement of Food and its Processing*, https://doi.org/10.1007/978-981-96-2347-1_2

Fig. 2.1 Various food materials and their basic physical characteristics

2.1.1.1 Size and Shape

Size and shape are critical parameters in determining the quality of various agro-based commodities. For instance, in chocolate production, the particle size of slurry after fine grinding significantly influences the texture and taste of the final product. Additionally, size and shape play important roles in many unit processes, including screening, grading, separation, and dehydration/drying. When determining the sieve diameter for seed grading, it is essential to consider the size distribution of the seeds. Similarly, ensuring consistent appearance features such as the shape index is necessary for the automated pitting of dates or peaches and the automatic grading of fruits.

2.1.1.2 Volume and Surface Area

The volume of food materials, such as baked goods and puffed snacks, indicates their degree of expansion and overall quality. In the fruit and vegetable industries, measuring surface area is crucial. For example, surface area plays a vital role during the wax coating of fruits. During drying processes, surface area also affects water loss from grains, seeds, and other materials. Moreover, the leaf area of crops is indicative of their productivity. Surface area measurements are essential for calculating spray coverage, heat and mass transfer, and respiration rates in various heating and cooling processes. To accurately model heat and mass transfer during cooling and drying, it is necessary to consider the volume and surface area of solids, as well as the influence of porosity on airflow resistance through a large number of solids.

2.1.1.3 Density

Density is a fundamental physical property important for equipment design, process control, and quality characterization. It can be used to assess the quality of grains and seeds. For example, in the dairy industry, milk fat and skimmed milk are separated based on density. Additionally, density directly correlates with the foaming degree of ice cream and whipped dairy toppings. In grain processing, materials containing stones, glass, metal fragments, or lighter grains can be separated using gravity platforms according to their density and weight.

2.1.1.4 Particle Size

The size of food particles, such as milk powder particles, must be carefully controlled. Particles that are too large may dissolve too slowly, affecting the product's texture and usability, while particles that are too small can lead to caking, reducing product flowability and shelf life. The particle size of a material is related to its specific surface area, nutritional value, adsorption capacity, and functional properties in various food products.

2.1.2 Organizational Structure of Food Materials

According to the organizational structure of food, food materials can be roughly divided into granular materials, porous food, and powdered food (Fig. 2.1).

2.1.2.1 Granular Materials

Granular materials refer to food products composed of small, distinct particles or solids. These particles are typically small, discrete, and can be either spherical or irregularly shaped. Examples of granular food materials include grains such as wheat, rice, and soybeans; fruits like apples, oranges, and pears; vegetables such as potatoes and tomatoes; nuts including pine nuts and pecans; spices like pepper and cumin; and other materials like coffee beans and cocoa beans. Additionally, processing units such as grinding, crushing, chopping, pelletizing, and agglomeration produce granular solids or particles.

In the food industry and engineering applications, granular materials play a crucial role. The size, shape, and density of granular materials are important quality indicators that significantly impact processing, storage, nutritional value, market value, and consumer acceptance. For instance, the size and shape of fruits directly reflect their maturity, variety characteristics, and growth conditions. Different shapes and sizes influence market attractiveness and consumer preferences. Understanding the basic physical properties of granular food materials is essential for food manufacturers and consumers, as these properties affect the market value, processing adaptability, and overall quality of products.

2.1.2.2 Porous Food

Porous foods, also known as solid foam foods, are characterized by a dispersed system where gas is entrapped within a solid or semi-solid matrix. This structure can be viewed from the perspective of colloidal systems, where the solid or semi-solid acts as the continuous phase and the gas forms the dispersed phase [1, 2]. Common examples of porous foods include steamed bread, sponge cake, biscuits, puffed snacks, ice cream, and whipped cream. Porous foods can be categorized into soft, porous foods like steamed bread and sponge cake, and hard porous foods like biscuits and puffed snacks. Some relatively hard foams like ice cream and whipped cream are also considered porous foods.

The volume and distribution of bubbles in porous foods significantly influence their physical properties, including texture, density, and structural integrity. These characteristics affect the sensory attributes and shelf life of the products, thereby influencing consumer acceptance. The formation of pores and bubbles is influenced by various factors, including internal factors like food composition, initial moisture content, and porosity, as well as external factors such as application technology, processing methods, and storage conditions [3]. These factors collectively determine the quantity, size, distribution, and orientation of pores and bubbles within the food matrix, shaping the overall quality and structural attributes of processed foods.

In food processing, the volume of bubbles in porous foods impacts texture, density, and structural integrity, which in turn affect hardness, adhesion, chewiness, and water retention capacity. Surface area calculations are essential for modelling heat and mass transfer during cooling and drying processes. In foam-based products, such as foam creams and beverages, the stability of foam and the size distribution of bubbles are critical for taste and appearance. For example, in products like cheese and chocolate, pore formation can enhance nutrient digestion rates but may also reduce shelf life by promoting fat oxidation and microbial growth.

2.1.2.3 Powdered Food

Powdered foods consist of tiny solid particles that exhibit friction between particles, unlike liquids where pressure is directly proportional to the height of the filling. This friction affects properties such as flowability and compressibility. Common powdered food substances include flour, soy flour, various starches, instant milk powder, coffee, and powdered seasonings. Grinding or pulverization is a key process in producing powdered foods. Even when using the same pulverization equipment to grind different grains, the resulting powder may vary significantly in particle size and shape. Similarly, different crushing methods can alter the physical properties of the resulting powder.

The size and size distribution of powdered foods have a significant impact on their solubility, flowability, stability, specific surface area, and mixing, dispersion, and reaction characteristics during processing. In the food industry, controlling particle size and distribution allows for the optimization of the taste, appearance, and shelf life of products. For example, the particle size of flour affects the texture and volume of bread, while the particle size distribution of milk powder influences its solubility and resilience in liquids.

Understanding and controlling the physical properties of powdered foods are essential for ensuring product quality and consistency. Proper management of particle size and distribution can enhance processing efficiency, improve product performance, and meet consumer expectations regarding taste and texture.

2.1.3 Concept of Physical Properties of Food Materials

2.1.3.1 Size and Shape

The size of an object is typically described in terms of its dimensions, with each dimension contributing to the object's overall shape. Consequently, size and shape are two inseparable parameters when characterizing an object. In the context of food and agricultural materials, such as various fruits and grain seeds, size and shape can be highly complex. While some items have regular shapes, most are irregular and cannot be accurately described by a single-size measurement. Generally, objects of regular shapes—such as spheres, cubes, and cones—can be characterized by specific dimensions. Under normal conditions, an object can be represented by three mutually perpendicular axial dimensions: length (l), width (w), and thickness (t). As illustrated in Fig. 2.2, l refers to the maximum size in the plane projection, w refers to the maximum size perpendicular to the length direction, and t refers to the linear size perpendicular to both the length and width directions.

For one certain, most fruits are nearly spherical in shape. In addition, within the kind of spheroid, there are various shape definitions, such as an oblate sphere, ellipsoid, oval, and so on. In general, the shape is described in terms of sphericity and roundness [4]. Sphericity is defined as the ratio of the volume of an irregular

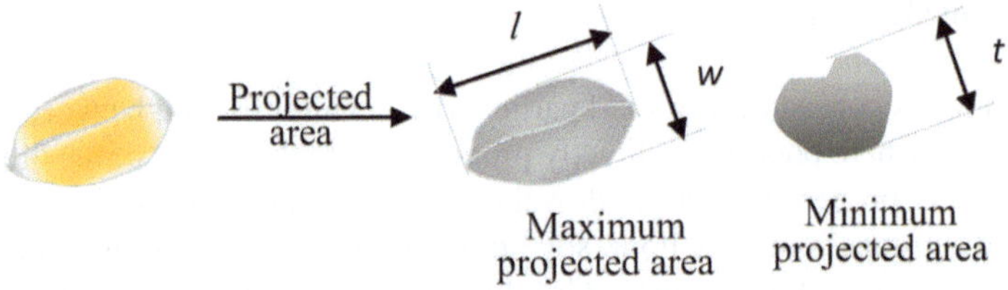

Fig. 2.2 Dimensions of food material based on projected area method

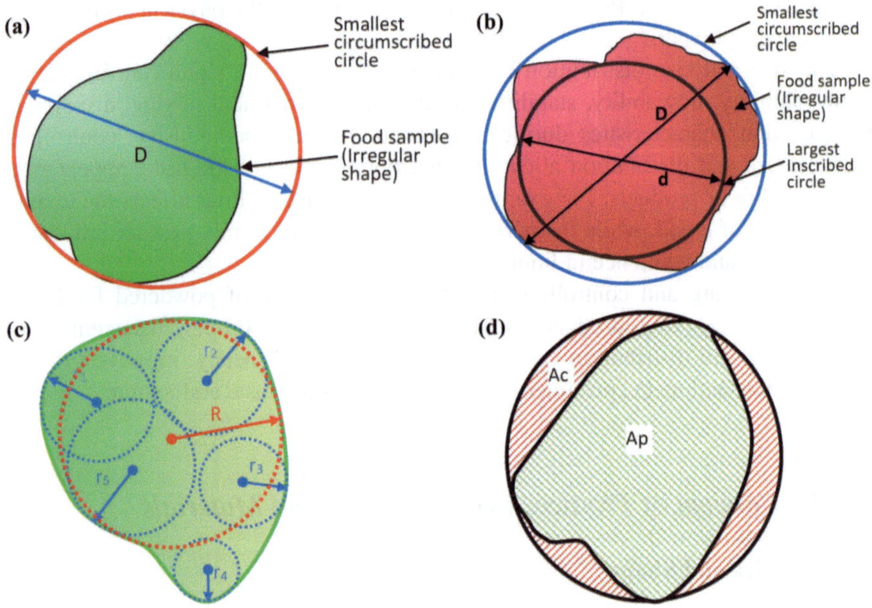

Fig. 2.3 Sphericity and roundness of irregularly shaped food. Sphericity is based on volume (**a**) and diameter (**b**); roundness is based on the ROC of corners (**c**) and projected area (**d**)

solid to the volume of the smallest circumscribed sphere that can encompass the solid, with the sphere's diameter equal to the object's major diameter (Fig. 2.3a). Mathematically, sphericity (∅) can be expressed as:

$$\text{Sphericity}\ (\varnothing) = \left(\frac{\text{Volume of irregular solid}}{\text{Volume of smallest circumscribed sphere}}\right)^{\frac{1}{3}} \quad (2.1)$$

Sphericity measures the degree to which a particle resembles a perfect sphere. Calculating the volume of the smallest circumscribed sphere can be simplified using mathematical equations. In contrast, the volume of an irregular solid can be determined using the liquid displacement method, as detailed in the volume section of this chapter.

Another method to determine sphericity involves the ratio of the largest inscribed circle diameter (d) to the smallest circumscribed circle diameter (D) (Fig. 2.3b):

$$\text{Sphericity } (\varnothing) = \left(\frac{d}{D}\right) \tag{2.2}$$

Roundness is defined as the ratio of the average radius of curvature (ROC) of the corners to the radius of the largest inscribed circle (Fig. 2.3c). ROC indicates how easily an object will roll. The mathematical expressions for roundness are as follows:

$$\text{Roundness2} = \frac{\sum_{i=1}^{n} r_i}{nR} \tag{2.3}$$

$$\text{Rate of roundness} = \frac{r_{m \cdot \text{in}}}{Rp} \tag{2.4}$$

Where r_i is the radius of curvature of edges on the graph of the maximum projection area of the spheroid; R is the maximum radius of the inscribed circle of the graph with the maximum projection area of the spheroid; N is the total number of edges and corners; r_{min} is the minimum radius of curvature (radius of curvature of maximum acute Angle) of spheroid on the maximum projection area diagram; R_{p} is the mean radius of a spheroid on the maximum projected area diagram.

Additionally, roundness can be described using the ratio of the object's largest projected area at rest to its smallest circumscribing circle (Fig. 2.3d):

$$\text{Roundness1} = A_p / A_C \tag{2.5}$$

Where A_{p} is the maximum projection area of the spheroid under the stable state of natural placement, obtainable via the projection or tracing method; A_C is the area of the smallest circumscribing circle around A_p.

2.1.3.2 Area and Volume

Area refers to the size of a planar shape or figure in two-dimensional space. Projected area is the area of an object projected onto a plane from a specific direction. Surface area denotes the total area of a three-dimensional object's outer or inner surfaces. Specific surface area is the ratio of surface area to the mass or volume of an object, typically used to describe pore characteristics or particle fineness of a material. The SI unit for area is the square meter (m^2), while the SI unit for volume is the cubic meter (m^3).

Volume is the amount of three-dimensional space that an object occupies. There are three types of volume measurements: solid volume, apparent volume, and bulk volume. Solid volume refers to the volume of solid material, excluding any air-filled

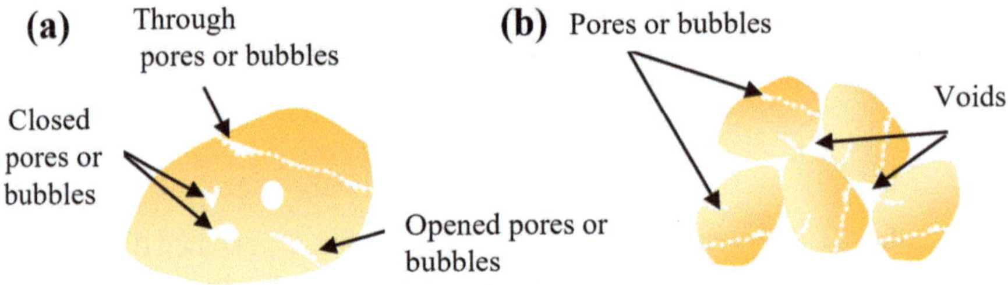

Fig. 2.4 Food materials with pores/bubbles (**a**) and voids (**b**)

internal (closed) pores or bubbles (Fig. 2.4a). Apparent volume includes all interior pores and bubbles, while bulk volume is the volume of a substance when stacked in bulk, encompassing both the material and the voids between particles (Fig. 2.4b).

Porous foods are divided into soft porous foods (such as bread and ice cream) and hard porous foods (like popcorn and puffed grains). Overrun, bubble volume, bubble size distribution, and specific volume are important indicators for soft porous foods, while expansion rate is a key indicator for hard porous foods [1]. Overrun is calculated as the percentage increase in volume attributed to air:

$$\text{Overrun} = \frac{V_f - V_l}{V_l} \times 100\% \tag{2.6}$$

Where V_f is the volume of the sample after aeration, and V_l is the volume of the original sample.

Specific volume is represented as volume per unit mass and is used to evaluate the degree of expansion of food. A higher expansion rate correlates with the palatability and softness of popcorn. The expansion rate is defined as the ratio between the final and initial volume of food materials:

$$\text{Expansion rate} = \frac{V_f}{V_i} \times 100\% \tag{2.7}$$

Where V_e is the volume of the sample after expansion, and V_i is the volume of the original sample.

2.1.3.3 Density

The density of a substance is expressed as mass per unit volume and is the reciprocal of specific volume. Depending on the physical state of the food materials, density can be categorized into three types: solid density, liquid density, and gas density. Bulk solid food materials are typically granular, porous, or powdered and contain hollow spaces or voids. Therefore, solid density can be further divided into true, particle, apparent, and bulk.

1. *True density*: The ratio of a material's mass to its actual volume, excluding the volume of space between particles.
2. *Particle density*. The mass-to-volume ratio of a single particle when it is not packed or compressed with other particles. It focuses solely on the material of the particles.
3. *Apparent density*. The mass per unit volume of a material after it has been packed or settled under certain conditions, typically excluding the volume of the voids between particles.
4. *Bulk density*. The mass per bulk volume, including both the material and the voids or spaces between particles. It measures the amount of mass contained in a certain volume of material.

2.1.3.4 Porosity

Many food materials contain pores and void spaces, as shown in Fig. 2.4. For example, the high loaf volume and consistent crumb structure of bread dough are largely due to the entrainment and evolution of gas bubbles within the dough. Foods with dried and intermediate moisture content are characterized by their porosity, which is a crucial physical component in several processes, including baking, frying, drying, and extrusion. Porosity data are essential for calculating mass and heat transfer during various drying and dehydration procedures.

Porosity, or void fraction (ϵ), is the ratio of the volume of voids between particles to the overall volume of the void-containing material:

$$\varepsilon = \frac{\text{Volume of void space}}{\text{Total volume}} \tag{2.8}$$

If the real density ρ_s or heavy γ_s and the virtual surface density ρ_a or volume heavy γ_a of the material are known, the porosity can be calculated as follows:

$$e = 1 - \rho_d/\rho_s = 1 - \gamma_d/\gamma_s \tag{2.9}$$

Where ρ_s is the true density, ρ_a is the apparent density, and γ_s and γ_a are the true and apparent specific gravities, respectively.

Interparticle porosity refers to the distribution and size of gaps between grain particles [5]. It can be estimated using the following equation:

$$\varepsilon' = \left(1 - \frac{\rho_b}{\rho_p}\right) \times 100\% \tag{2.10}$$

Where ρ_b is the bulk density, and ρ_p is the particle density.

2.1.3.5 Particle Size and Distribution

Particle size is a comprehensive index that represents the dimensions of an object. The particle size of a material is divided into single-particle size and average particle size. The former represents a single particle, while the latter represents a group of particles of different sizes within the system. As listed in Table 2.1, there are many terminologies to define particle size. For single particles, particle size can be represented by arithmetic mean diameter, circular equivalent diameter, directional diameter, and Stokes diameter. For a swarm of particles, particle size can be represented by arithmetic mean diameter, geometric mean diameter, harmonic mean diameter, area-length average diameter, average diameter of body area, weight

Table 2.1 Definition of particle size for single particles and average particle size for particle swarm

Name	Definition
Particle size for single particles	
Arithmetic mean diameter	The arithmetic mean diameter calculated from the three principal axes (long axis diameter, short axis diameter, and thickness) of a single particle.
Circular equivalent diameter	The diameter of a circle whose area matches the projected area of the particle.
Directional diameter	The maximum distance in any direction on the particle projection graph
Stokes diameter	The equivalent diameter exhibited by particles when settling in a fluid under Stokes' law
Particle size for particle swarm	
Arithmetic mean diameter	The average diameter of particles in a swarm, calculated by summing all particle diameters and dividing by the total number of particles.
Geometric mean diameter	The *n*th root of the product of N particle diameters, where N is the total number of particles.
Harmonic mean diameter	The harmonic mean of the particle diameters, calculated as the reciprocal of the average of the reciprocals of the diameters.
Area length average diameter	The total surface area of all particles divided by the total projected perimeter (length).
Average diameter of body area	The total volume of all particles divided by the total surface area.
Weight average diameter	An average diameter calculated by weighting each particle's size by its mass, giving more importance to larger particles.
Average surface area diameter	The square root of the total surface area divided by the number of particles.
Average volume diameter	The cube root of the total volume divided by the number of particles.
Specific surface area diameter	The particle size calculated from the specific surface area S, which is the total surface area per unit volume of the material layer.
Medium diameter	The particle size at the 50th percentile of the particle size distribution, indicating that half the particles are smaller and half are larger.
Most diameter	The particle size with the highest frequency in the particle size distribution, representing the most frequently occurring size.

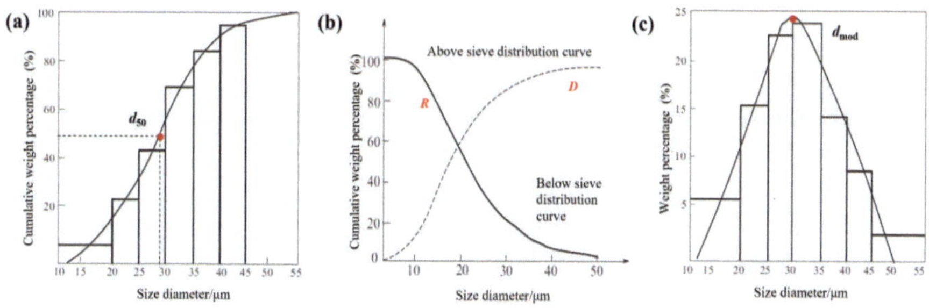

Fig. 2.5 Cumulative distribution curve (**a**, **b**) and size frequency distribution curve (**c**)

average diameter, average surface area diameter, average volume diameter, specific surface area diameter, medium diameter, and most probable diameter.

Particle size distribution can be described as a cumulative distribution curve calculated based on particle mass or particle number percentage (Fig. 2.5a, b) or as a particle size frequency distribution curve (Fig. 2.5c). The frequency distribution curve typically conforms to a normal distribution. As shown in Fig. 2.5a, the particle size at the 50% mark of the cumulative distribution curve is called the median diameter (d_{50}). The particle size at the highest point of the frequency distribution is referred to as the mode diameter (d_mode) (Fig. 2.5c). If the particle size distribution follows a normal distribution, it can be expressed by the probability density function. To test whether the particle size distribution is normally distributed, normal probability paper can be used to determine if the data points lie on a straight line.

Sometimes, the cumulative distribution curve is divided into the above-sieve distribution curve and the below-sieve distribution curve (Fig. 2.5b). If the percentage of the weight of particles larger than a given particle size to the total weight is R %, and the percentage of the weight of particles smaller than this particle size to the total weight is $D\% = 100\% - R\%$, then the R curve is called the above-sieve distribution curve of the material, and the D curve is called the below-sieve distribution curve.

2.2 Measurement of Physical Properties of Foods

The physical properties of foods, such as size and volume, can be directly measured using mechanical measurements and calculations based on their definitions. Additionally, these properties can be obtained through various instrumental methods. Recently, image analysis techniques, including machine vision, laser topography, and hyperspectral imaging [6], have gained popularity for measuring the geometric features of various foods. As illustrated in Fig. 2.6, machine vision acquires the physical properties of foods through the following steps: image capture, image

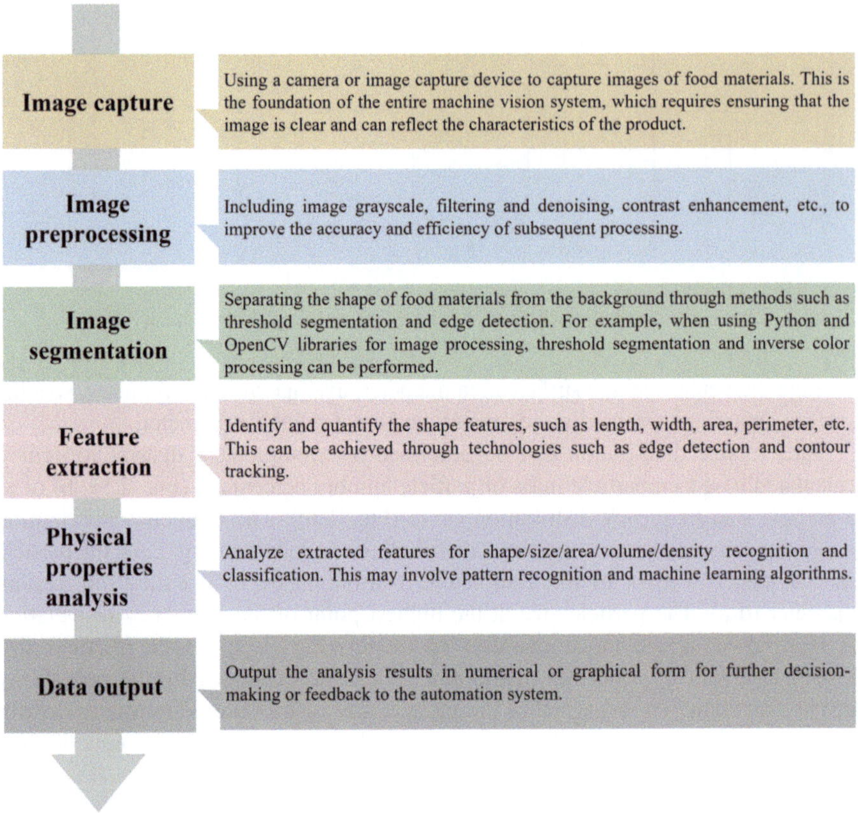

Fig. 2.6 Process diagram for machine vision to obtain the physical properties of foods

preprocessing, image segmentation, geometric feature extraction, physical properties analysis, data output, and system integration.

2.2.1 Measurement Technologies of Shape, Size, Area, and Volume

2.2.1.1 Shape and Size

For objects with irregular shapes, the standard shape is typically defined by consulting reference tables. The specific process involves drawing longitudinal and transverse sections of the object's outline and then defining the shape by comparing the physical Figure with the standard figures. The shape of foods can be directly measured using mechanical measurement tools and calculated based on definitions such as sphericity and roundness. For example, the principal dimensions (width, depth, and length) of coffee beans were measured using digital calipers [7]. Additionally, machine vision technology offers a direct, feasible, economical, and non-destructive method to measure the shape of food materials such as apples [8, 9].

For food materials with large sizes that can be individually selected, the size can be calculated by averaging the diameters. First, n particles (preferably $n = 200$ or more for better accuracy) are collected randomly from the sample. Then, the total weight W_{sn} is measured using an ordinary balance. Assuming that the true weight of the particles is γ_s, the average particle size d_s can be calculated using the following formula:

$$d_s = \sqrt[3]{\frac{6W_{sn}}{\pi\gamma_s n}} \tag{2.11}$$

The particle size d_s calculated from Eq. (2.11) represents the average diameter of all particles, assuming they are spherical particles of equal volume, or powdery materials, such as flour, where particles are too fine to be counted individually, the harmonic mean diameter d_s is often used. It is assumed that the proportion of each component in a certain amount of powder can be expressed as follows:

$$d_s = \sum_{i=1}^{m} (x_i d_i) \tag{2.12}$$

where x_i is the percentage of the total weight of particles with diameter d_i. The percentage of each component can be determined through sieving.

There are numerous methods for particle size measurement, including sieving, microscopy, electro-induction, optical microscopy, electron microscopy, sedimentation, light scattering, laser holography, mass spectrometry, and image analysis. Table 2.2 lists the main methods of particle size measurement. The appropriate measurement method should be selected based on the type of material, particle size, particle shape, and physical properties. In production practice, it is often preferable to use simple methods to determine particle size to improve design speed, rather than relying on expensive instruments that require significant time.

In addition to the above methods, particle size measurement also includes determination methods based on Brownian motion, electrophoresis, mass spectrometry, and others. In general, particle size depends both on the numerical size measured directly or indirectly and on the measurement method. To summarize, due to the different physical bases of various particle size measurement methods, the physical significance and even the particle size of the same sample obtained by different measurement methods also differ. For example, the sieving method determines the sieving diameter; statistical diameters are obtained using microscopy and light scattering; and equivalent diameters are obtained through sedimentation, electro-induction, and mass spectrometry. Additionally, the scope of application for different particle size measurement methods varies. It is very important and necessary to

Table 2.2 Measurement methods for particle size

Methods		Equipment	Pros and cons	Results
Direct observation		Magnifying projector, image analyzer	Rapid, low accuracy	Size distribution and shape index
Screening		Electromagnetic vibration type and acoustic vibration type	Rapid, laborious	Size distribution histogram
Sedimentation	Specific gravity meter, specific gravity balance, settlement sky level light transmission type, X-ray transmission type	Specific gravity meter, specific gravity balance, settlement sky level light transmission type, X-ray transmission type	High resolution	Size distribution
	Light transmission type X-ray transmission type	Light transmission type, X-ray transmission type		Size distribution
Laser	Laser particle size analyzer	Laser particle size analyzer	Online detection	Size distribution
	Photon coherent particle size analyzer	Photon coherent particle size analyzer		Size distribution
Electro-induction		Kurt particle size analyzer	Low accuracy for non-spherical particles	Size distribution and quantity measurement
Fluid projection		Gas permeation particle size analyzer	Non-destructive and suitable for nanoparticles	Specific surface area and average particle size
Adsorption		Brunauer-Emmett-Teller (BET) adsorption instrument	Indirect method	Specific surface area and average particle size
Machine vision		Image capture device	Visibility and high reliability	Size and shape

select the appropriate measurement method according to the measured object and the required measurement accuracy and precision.

2.2.1.2 Area and Surface Area

For regular geometric objects such as spheres, cubes, and cylinders, the surface area can be directly calculated using corresponding mathematical formulas. However, for irregular objects, the calculation of surface area typically requires experimental measurements or numerical simulation methods. For example, the surface area of an object with a hemi-ellipsoidal geometry can be calculated as follows [7]:

$$A_{s,b} \approx 2\pi\left(\frac{(0.5ab)^{1.6} + (0.25ac)^{1.6} + (0.5bc)^{1.6}}{3}\right)^{\frac{1}{1.6}} + \frac{\pi ac}{4} \tag{2.13}$$

The Brunauer-Emmett-Teller (BET) theory can be used to calculate a solid's surface area.

$$\frac{(P_0 - P) \cdot V_m \cdot C}{P} = \frac{1}{(C-1) \cdot P} + \frac{C-1}{C} \cdot \frac{P}{P_0} \tag{2.14}$$

Where P is the relative pressure, which is the ratio of the actual pressure to the saturated vapor pressure, P_0 is the saturated vapor pressure, V_{m} is the adsorption capacity of a single layer, representing the volume of gas required to cover the solid surface with a single layer, C is the BET constant, correlated with the adsorption energy and representing the strength of the interaction between the adsorbed molecules and the solid surface. By fitting experimental data to the BET equation, V_{m} and the constant C can be obtained, allowing the calculation of the specific surface area of the material. Additionally, surface area can be measured using image analysis tools, a planimeter, the coating method, and peeling techniques. Commercial instruments such as the SA-3100 (Beckman Coulter, USA) and ASAP 2020 Plus (Micromeritics, USA) are commonly used to characterize the surface area of materials.

Agricultural product surface area determination methods

1. *Measurement of Fruit Surface Area*

   Measuring the surface area of fruits is challenging due to their complex shapes. Generally, fruits are peeled into narrow strips, which are then laid out flat. After tracing the strips, the area can be calculated from the graph. Alternatively, statistical methods can relate the fruit's surface area to its size or weight, allowing for a quick estimation of surface area. In some cases, the surface area can be calculated from images of the rotating fruit.

   The surface area S can be calculated using the following integral formula:

$$S = \int_a^b 2\pi y\sqrt{1 + \left(\frac{dy}{dx}\right)^2}\,\mathrm{d}x \tag{2.15}$$

2. *Measurement of Egg Surface Area*

   The surface area (S) of fresh eggs can be calculated by the weighing method with an empirical formula:

$$S = KW^m \tag{2.16}$$

Where K and m are constants (typically $K = 4.56\sim5.07$ and $m = 0.66$), S is the egg surface area in cm^2 and W is the weight of the egg in g.

Another method involves drawing the outline of the egg using projection or light shading, then differentiating and calculating the total surface area. The Romanov research group utilized Hügelschäffer's model and simulation modeling to non-destructively measure the surface area of eggs [10]. The surface area is calculated using the following formulas:

$$S = \pi BL\left(\left(0.043\frac{\omega}{L} + 0.292\right)\frac{B}{L} - 0.061\frac{\omega}{L} + 0.704\right) \tag{2.17}$$

$$S_s = 0.933B(B + 2.343L) \tag{2.18}$$

$$s_V = 4.944V^{\frac{2}{3}} \tag{2.19}$$

Where B is the greatest breadth of the egg, L is the length of the egg, w is a parameter that represents the separation between two vertical axes, one of which crosses the egg at $L/2$ and coincides with B, S_v is the egg S recalculated using the meaning of the egg volume V. S_s is the egg S recalculated using the simplified formula under the measurements of the egg geometrical parameters (B and L).

2.2.1.3 Volume

Regularly shaped samples have typical dimensions that can be used to calculate their volume. For samples with irregular shapes, the volume can be determined experimentally using methods such as solid, liquid, or gas displacement. For example, the primary measurement method for baked goods is the rapeseed displacement method, which is an AACC International approved method (10-05.01). The main operational steps of the seed displacement method include calibrating the measuring instrument, placing the bread sample, pouring the seeds, and reading the volume from the calibration column. In addition to the seed displacement method, other methods such as the paraffin wax method and the liquid displacement method are also used, where the sample is wrapped in appropriate film and immersed in oil or water for measurement.

Another measurement method for bread volume involves digitally modeling the dimensions of a test sample using laser topography. The primary measurement processes are as follows: position the sample on a revolving support frame, and the laser sensor will rotate 180° around it. During the rotation process, the sample's length/height, volume, density, and weight can be determined, and a full 3D model can be created. Several commercial products are available for this purpose, such as the VolScan Profiler (Stable Micro Systems, UK) and PV-X (Perkone Scientific, China). These devices comply with AACC International approved methods (10-16.01 and 10-14.01). Compared to the seed displacement method, the laser topography-based method offers several advantages: it is fast (20–30 seconds), requires no pre-processing, is easy to operate, provides accurate measurements, and is suitable for samples of various sizes.

To calculate the volumes of a chicken egg's shell and interior, Narushin and colleagues developed a mathematical assay that takes inputs such as the egg's thickness and its primary external geometric dimensions, including length, maximum breadth, and the value of the shift from the horizontal axis [10, 11]. The shell volume can be calculated by multiplying the surface area estimated along the shell section's midline by the average thickness. Additionally, using X-ray 3D reconstruction, the edible volume of thick-skinned citrus fruits was detected based on the basic physical parameters [12].

2.2.2 *Measurement Technologies of Density and Porosity*

2.2.2.1 Density

The density of a material can be categorized into several types, namely true density, particle density, apparent density, and bulk density. The methods for measuring these different densities vary. Figure 2.7 shows some laboratory methods and devices for determining density [13]. These methods include the buoyancy method (or gravity balance method), pycnometer method, gradient column method, X-ray method [14], and resonator method [15]. Typically, the density of a food material is measured using multiple methods. For example, the intrinsic, bulk, packing, and

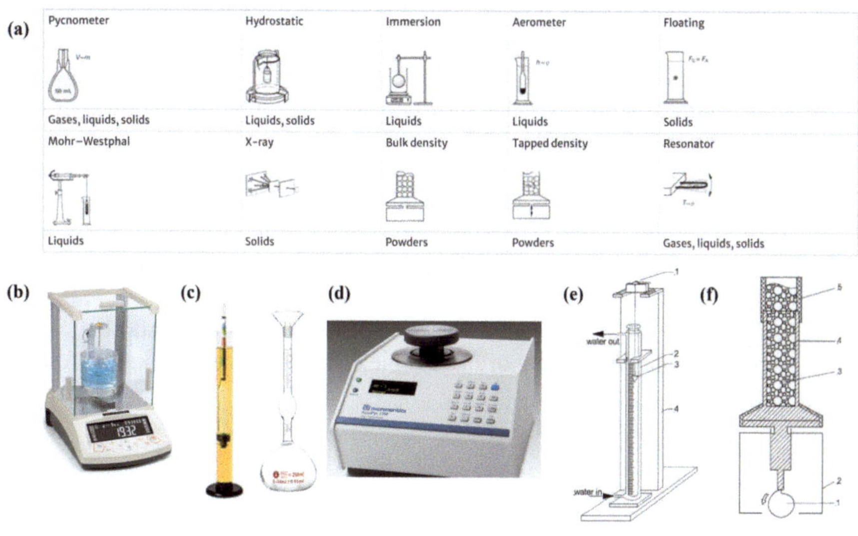

Fig. 2.7 Laboratory methods and devices for determining density

pycnometric density of coffee beans were measured using digital calipers and pycnometry devices [7].

The following section provides a detailed introduction to some of these methods.

1. *Buoyancy Method (or Gravity Balance Method)*

   For smaller materials, a gravity balance can be used to measure the volume of the material. During measurement, the material is weighed in air and liquid, respectively. For fruits and other large materials, the density can be measured by weighing the platform (Fig. 2.7b).

   First, the material to be tested is weighed on the platform scale, and the weight is denoted as m_s. Then, a cup with a certain capacity of water is placed on the scale and weighed, with the total weight denoted as m_1. Next, the material is submerged in water (if the density of the material is smaller than water, the material must be forcibly immersed in water using a glass rod), and the immersed material container is weighed on the platform scale, denoted as m_2. The density of the material ρ_S is calculated as: $\rho_s = \rho m_s/(m_2 - m_1)$.
2. *Pycnometer Method*

   The gravity flask method (Fig. 2.7c) is suitable for measuring the density of fine particles or powder materials. Sometimes, measurements are made using a flask with a volume scale. A commercial instrument, the Helium Pycnometer (Accupyc II, Micromeritics) (Fig. 2.7d), can be used to measure the true density of powders, solids, and slurries. It utilizes the helium gas displacement method and is known for fast, fully automatic, and high-precision determination.
3. *Gradient Column Method*

   The gradient column method is a rapid and accurate way to calculate the density of small solid particles such as agricultural seeds and grains [13]. Liquids of different densities are loaded into the gradient tube of specific gravity (Fig. 2.7e), creating a density gradient along the tube's height. When the sample is placed into the tube, its density can be determined based on its equilibrium position. A key condition for this method is that the liquid in the gradient tube must not permeate the sample.

   To perform the test, place the sample slowly into the gradient tube. When the sample is stably suspended in the liquid, the height at which it rests in the tube is recorded. According to the calibration curve, the sample's density is determined. If the sample does not reach equilibrium in the tube, it becomes saturated with liquid.

   If there is no calibration curve, the balance height of a standard float can be measured first, and the density of the sample can be calculated using the following formula:

$$\rho_s = \frac{(x-y)(b-a)}{z-y} + a \tag{2.20}$$

Where a and b are the densities of two standard floats; y and z are the equilibrium heights of two standard floats *a* and *b, respectively*, *x* is the equilibrium height of the sample.

Bulk density is important in various applications, including food processing and storage, where understanding packing efficiency and the amount of space a material will occupy is crucial. An example of a device for measuring both bulk density and tapped bulk density is illustrated in Fig. 2.7f. The process involves filling a 1000 cm^3 cylinder with the bulk material until it overflows, using a consistent method. Afterwards, a flat spatula is used to remove the excess material from the top of the cylinder, leveling the sample precisely. The 1000 cm^3 sample is then weighed, providing the bulk density measurement.

The sample can also be loaded into a container (preferably using a sample filling thickness and container diameter ratio of six containers) so that it falls from a height to the floor by impact shock. At this time, in the virtual table, the volume V_a and mass m_s are measured, allowing the calculation of the virtual table density ρ_a.

$$\rho_a = m_s / V_a \tag{2.21}$$

If the filled container is continuously knocked until the material reaches its most compact state, the virtual surface density at this time is called the final virtual surface density or the filling density ρ_t.

$$\rho_t = m_s / V_t \tag{2.22}$$

Where V_t is the final filling volume of the material. Table 2.3 shows the bulk density of some foods.

2.2.2.2 Porosity

Although porosity is straightforward to conceptualize, quantifying it presents challenges due to the wide range of void sizes found in food materials, spanning from nanometers to larger dimensions. The measurement process is further complicated by factors such as deformation and hydrothermal changes, which can create new forms of porosity during testing [16]. Additionally, the shape and connectivity of pores significantly influence the outcomes of porosity assessments, varying with the testing approach.

There are two major types of techniques to measure porosity. The imaging method stands out as a modern and accessible technique, facilitated by advancements in digital imagery and software. This method allows for the precise determination of pore characteristics, including size, distribution, overall porosity, and the microstructure of food materials. Imaging techniques include optical microscopy (such as transmitted light microscopy, polarized light microscopy, and differential interference contrast microscopy), scanning electron microscopy [17], X-ray

Table 2.3 Bulk densities of some food materials

Materials	Bulk density /(Kg/m^3)	Materials	Bulk density /(Kg/m^3)	Materials	Bulk density /(Kg/m^3)	Materials	Bulk density /(Kg/m^3)
Pepper	200~300	Corn grain	680~770	Cocoa	400	Sugar	800
Eggplant	330~430	Peanut meat	500~630	Instant coffee	330	Powdered sugar	480
Tomato	580~630	Soybean	700~770	Corn flour	660	Wheat flour	480
Onion	490~520	Broad bean	670~800	Corn starch	560	Whey powder	560
Carrot	560~590	Potato	650~750	Gelatin powder	680	Soy protein (precipitated)	280
Peach	560~590	Sweet potato	640	Oat flour	430	Microcrystalline cellulose	680
Mushroom	450~500	Beet tubers	600~770	Powdered salt	950	Milk powder	610
Pea seeds	700	Egg	340				

Table 2.4 Lists of porosity and density of several food powders

Materials	Average particle size (μm)	Bulk density (kg/m^3)		Volume reduction rate (%)	Porosity e_{max}	Porosity e_{min}	Real density (kg/m^3)
		Loose	Compaction				
Flour	120	484	606	20	0.652	0.567	1400
Lactose powder	100	586	812	28	0.614	0.465	1520
Sugar	400	660	880	25	0.580	0.440	1578
Whole milk powder	110	589	710	17	0.435	0.316	1039
Skimmed milk powder	105	589	746	21	0.517	0.389	1224
Instant skimmed milk powder	300	284	312	9	0.768	0.744	1224
Soluble starch	55	810	966	16	0.420	0.310	1440

computed tomography (CT) [7], and nuclear magnetic resonance. For instance, the 3D pore structure of whole apples was visualized using high-resolution X-ray CT [18].

The second type comprises other methods, including mercury intrusion porosimetry, thermoporometry, physisorption, nitrogen adsorption and BET analysis, dynamic vapor sorption, pycnometry, and envelope density analysis. These methods differ in their physical principles and are selected based on the specific requirements of the measurement, such as the nature of the material and the desired accuracy. Table 2.4 illustrates the relationship between porosity and density for several powders, highlighting how different measurement methods correlate with various density parameters.

The porosity of spherical particles of the same size can be calculated geometrically. For food materials with irregular shapes, porosity can be measured using a device. Figure 2.8 shows a porosity measurement device consisting of two tanks of equal volume. To operate the device, fill tank 2 with the material. Close valve 2 and vacuum tank 2. When the pressure gauge pointer reaches an appropriate value, close valve 1 and record the pressure value P_1. Then, open valve 2 to allow air from tank 1 to fill the material space in tank 2, balancing the pressure in the tank, and measure the pressure P_3 It is noted that $P_3 < P_1$.

The porosity e is calculated as:

$$e = \frac{V_2}{V_1} = \frac{P_1 - P_3}{P_3} \tag{2.23}$$

In the formula, P_1 and P_3 are the absolute pressures measured before and after, respectively. V_1 and V_2 are the volumes of air in tank 1 and tank 2, respectively.

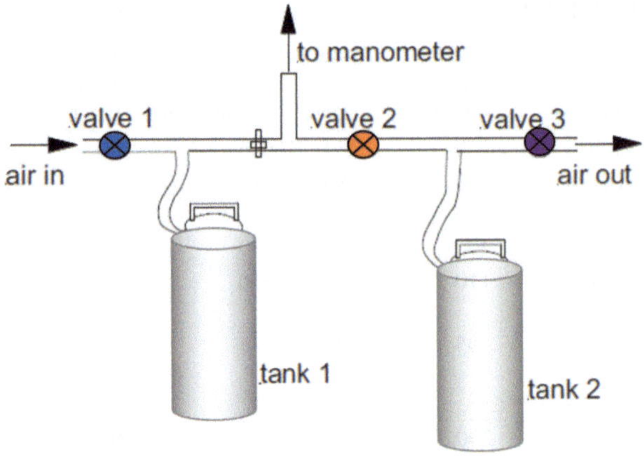

Fig. 2.8 Porosity measurement device

2.3 Application of Basic Physical Properties in Food Engineering

The basic physical characteristics of food and agricultural products are widely used in quality evaluation, sorting, grading, and hydraulic or pneumatic conveying systems. They affect multiple aspects of food processing, including quality control, product development, and food safety. Below are some application examples in food engineering.

2.3.1 Hierarchical Classification of Food Materials Based on Shape and Size

2.3.1.1 Sorting and Grading of Fruits

Basic physical parameters, such as the shape and size of fruits, play a vital role in quality detection and classification. Several techniques are employed to sort and grade various fruits, including machine vision [8], spectroscopy, X-ray CT [19–21], magnetic resonance imaging (MRI) [22], vibro-acoustic signals [23], and infrared thermal imaging [24]. Additionally, these methods can detect internal disorders in fruits, such as bruises, watercore disorder [25], internal browning [26], and cavity formation, allowing for the sorting of defective fruits from healthy ones. Recent studies have shown that the combination of X-ray CT and machine learning is a novel approach to sorting and grading various fruits [19–21]. Table 2.5 presents a comparative analysis of the benefits and drawbacks of different detection approaches in relevant situations.

Particularly, machine vision can assess fruits quality in an efficient, effective, economical, and accurate manner [8]. Automatic fruit grading systems that rely on

Table 2.5 Comparison of detection technologies for measuring fruit quality

Detection technology	Detection object	Benefits	Drawbacks
Machine vision	Fruit size, shape, external damage, and surface defects	Low cost and enables online inspection	Unable to distinguish interior flaws and components; difficult to detect the entire fruit's surface
Spectroscopy	Surface and internal defects, composition	Provides richer detection data than machine vision	Requires filtering of multi-band spectral information; generates large amounts of data and has long processing times.
X-ray Computed Tomography (CT)	Internal defects, composition	Highly sensitive to bulk defects; can simultaneously detect internal defects and components, especially distinguishing between different densities	Radiation risk; costly radiation source; insensitive to surface imperfections.
Magnetic Resonance Imaging (MRI)	Surface and internal defects, composition	Strong penetrating capability; non-toxic, non-radiating; unaffected by peel thickness	Expensive equipment; time-consuming procedures.

machine vision typically encompass a control module, an imaging apparatus, a conveyance system, a sorting apparatus, and a human-machine interface. The imaging apparatus is usually equipped with a charge-coupled device (CCD) and a lighting system to capture images of the fruits' exteriors. The conveyance and sorting apparatus are designed to move the fruits and classify them accurately. In terms of shape, machine vision is used to select cucumbers based on their length and curvature [27]. For the automatic classification of apples, size characteristics (including transverse diameter, longitudinal diameter, transverse diameter ratio, area, circumference, roundness, equivalent radius, etc.), color characteristics, and defect characteristics can be extracted to achieve accurate classification.

The shape index is one of the important indicators of food appearance quality. For example, a fruit shape index of 0.8–0.9 is usually round or nearly round, 0.6–0.8 is oblate, 0.9–1.0 is oval or conical, and 1.0 is oblong. Using computer vision, Kirchgessner and co-workers developed a morphometric device named FruitPhenoBox for automated fruit weighing and appearance description [28]. A comparison of fruit photos from 15 apple varieties revealed substantial variances in shape index, width, height, and weight. FruitPhenoBox determined the optimal fruit sample size to characterize the recorded attributes by comparing the relative margin of error for fruit width, height, and weight descriptions to alternative sample sizes. Lu and co-workers designed a machine vision-based method to detect peel defects and fruit morphology of citrus [9]. The symmetry index, circularity, roundness, and tilt angle of the citrus were used as input parameters. The proposed method achieved accurate and real-time fruit classification.

2.3.1.2 Sorting and Grading of Eggs

In current industrial practices, the sorting and grading of table eggs are performed based on external and internal quality. External quality refers to the shape, dimensions, color, surface defects of the eggshell (such as cracks and stain areas), purity, and weight of the eggshell. Internal quality refers to the freshness of the eggs, blood spots, and the rate of meat spots. For example, in the USDA egg grading standard, qualified eggs are categorized into three grades: AA, A, and B, whereas unqualified eggs are categorized as Class C.

Machine vision technology detects and grades eggs based on appearance quality, whereas vibration analysis methods detect and grade structural stiffness such as cracks and eggshell hardness [29]. Near-infrared spectrometers (NIRS) are used to quickly detect internal and external quality indicators such as egg freshness and to establish optical, acoustic, and electrical transmission characteristics [30]. NIRS can achieve fast sorting and grading of eggs. Lin and co-workers developed a novel approach for detecting egg cracks by analyzing numerous vibration sensors in a convenient, effective, and non-destructive manner [31]. Recently, Egg Industry 4.0 (EI 4.0) has gained popularity in the egg industry. EI 4.0 is powered by several cutting-edge techniques, including various non-destructive optical sensing technologies, the Internet of Things (IoT), robotics, artificial intelligence (AI), big data, and cloud computing [32]. EI 4.0 can be used in quality monitoring, process control, shelf-life prediction, and intelligent logistics.

2.3.2 Granularity of Food Nanoparticles

Various types of food nanoparticles can be prepared using different methods and have wide applications in the food industry. Microemulsions, nanoemulsions, liposomes, nano microcapsules, starch nanoparticles, protein nanoparticles, nano minerals, nano vitamins, nano antioxidants, and other nano foods have been successfully developed. The granularity of food nanoparticles is an important factor affecting their dispersibility, stability, bioavailability, and final application effectiveness in food systems. These food nanoparticles may undergo positive changes in physical and chemical properties, taste, nutritional value, and other aspects [33].

2.3.3 Control of Functional Food Base Material Processing

The development of nanotechnology has enabled various food active ingredients to be prepared in nano forms through different methods. Some trace active substances, such as selenium, are often added to foods—especially functional foods—to achieve health benefits. To avoid the toxic and side effects of additives, the amounts of additives are generally strictly controlled, necessitating the control of particle size through micro-grinding (10–1000 μm) or superfine grinding (0.1–10 μm). Converting functional ingredients into nanoparticles can improve the absorption rate of these minerals by the human body, enhance the absorption rate and bioavailability of fat-soluble vitamins, and improve the stability and bioavailability of antioxidants such as vitamins E and C.

Nanosized modified starch has great potential for applications in food coating packaging, food texture adjustment, and the stabilization and transportation of food functional factors [34]. Additionally, there is a class of fat substitutes based on starch or protein nanoparticles, which are pulverized to a specific size by superfine grinding (micro-milling). Because there is a threshold for the perception of particles of certain sizes and shapes in the human mouth, particles below this threshold will not be felt, resulting in a creamy mouthfeel and smooth texture characteristics.

Liu and co-workers explored how different soy protein particle sizes affected the physical, rheological, tribological, and sensory attributes of fat-free ice cream [35]. The results showed that the size of the protein particles influenced their functionality in ice cream. Sensory evaluation revealed that samples containing protein particles around 4 μm had a denser texture, lower melting rate, superior lubrication (lower friction), and a sensory profile similar to full-fat ice cream.

Controlling the processing particle size is essential for achieving the unique functions of food base materials. Particle size analysis of food nanoparticles is crucial for ensuring product quality and functionality. Various particle size analysis technologies have been successfully applied in the food industry to analyze food in various forms, including powders, lotions, suspensions, and particles. These technologies provide important information about the particle size distribution of food nanoparticles, which can help optimize product formulations and production processes, as well as improve the stability, taste, and nutritional value of food.

Particle size detection technology based on computer vision (image analysis technology) can achieve this goal. The use of a microscopic image analysis system for particle size detection of food ultrafine powders offers many advantages, such as ease of use, fast determination speed, and stable and reliable results. Additionally, it is resistant to significant deviations caused by changes in measuring instruments, operators, and other factors. This represents the development trend in particle detection [36].

2.3.4 *Freshness Identification of Liquid Food Based on Density*

2.3.4.1 Freshness Identification of Milk

The freshness of milk can be determined by measuring its density. Normal milk has a density of approximately 1.030 g/cm^3 at 20 °C. The relative density, which is the ratio of the mass of milk at a specific temperature to the mass of water at another temperature, is typically maintained between 1.028 and 1.032. Additionally, the specific gravity of milk is usually kept between 1.030 and 1.034. Deviations from these normal values may indicate spoilage or adulteration [37]. When fresh milk is stored, its relative density increases by 0.001 within 2~3 h. Therefore, monitoring milk density can be an effective method to estimate the storage time and assess the degree of freshness.

2.3.4.2 Freshness Identification of Eggs

Egg freshness is a significant factor in determining the quality of eggs and their products. Two primary indicators used to assess egg freshness are the Yolk Index (YI) and the Haugh Unit (HU). The Yolk Index is the ratio of the height of the yolk to its diameter. For fresh eggs, the YI is greater than 0.4. Ordinary eggs have a YI ranging from 0.35 to 0.4, while qualified eggs have a YI ranging from 0.3 to 0.35. The Haugh Unit measures the quality of the egg white (albumen) based on its thickness. It is calculated using the formula:

$$\mathrm{HU} = 100 \times \log\left(H - 1.7W^{0.37} + 7.6\right) \tag{2.24}$$

Where W is the egg weight, and H is the height of the thick protein.

High-quality eggs have an HU above 72, medium-quality eggs range from 60 to 70, and inferior eggs range from 31 to 60. The HU value decreases rapidly with storage time, making it a sensitive indicator for early detection of egg degradation. Commercial egg testers, such as the VisionAI Detector (SANOVO, Denmark) and DET6500 (NABEL, Japan), utilize these metrics to grade egg quality. In addition to YI and HU, other indicators have been proposed to reflect egg quality. The Egg Quality Index (EQI) combines measurements of egg weight, thick albumen height, yolk height, and yolk diameter to provide a comprehensive assessment of egg quality [38]. Furthermore, the Density of Egg Interior (Di) serves as a general indicator of the overall quality of the egg's interior and can be used as an alternative to the Haugh Unit for pre-incubation sorting of eggs. Furthermore, the Density of Egg Interior (Di) serves as a general indicator of the overall quality of the egg's interior and can be used as an alternative to the Haugh Unit for pre-incubation sorting of eggs [39].

Harnsoongnoen and Jaroensuk (2021) developed a novel, non-invasive, non-destructive, low-cost, and real-time method for evaluating egg freshness based on density. Their system integrates machine vision and weighing sensors to assess egg quality [40]. The machine vision system measures the external dimensions (breadth and length) of eggs to calculate their volume. Concurrently, the weighing mechanism measures the weight of each egg. By using the measured weight and calculated volume, the system calculates the density of the egg, which is then used to evaluate its freshness. This approach was the first to utilize density measurements for classifying egg freshness into specific intervals. The system is depicted in Fig. 2.9.

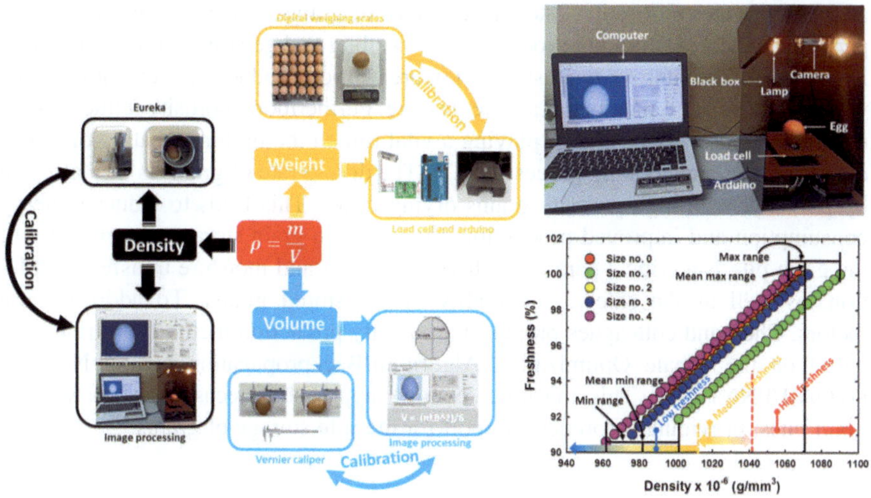

Fig. 2.9 System for density assessment of eggs

2.3.5 *Porosity of Granular Mixtures in Grain*

2.3.5.1 Evaluation and Control of Grain Quality

The quality evaluation of grains such as rice often relies on parameters like thousand-grain weight, whole grain rate, shape, and overall quality. Grains with high volume density typically exhibit high real density, a high whole grain rate, and superior shape and quality. Uniform and consistent grain sizes enhance the processing efficiency and ensure the consistency of finished products. However, variations in grain size distribution can result in a significant number of secondary grains being mixed into the batch. These secondary grains can be separated based on their triaxial dimensions through screening or air sorting. Nevertheless, specific gravity cleaning is more effective for assessing the germination rate of seeds. For instance, rice seeds selected using a salt water solution with a relative density of 1.05 demonstrate much higher precision in seed selection compared to mechanical methods [36]. One of the key indicators of rice quality is the content of broken rice. Since the density of broken rice is identical to that of whole rice grains, density sorting is ineffective for separating broken grains. Instead, broken rice can be effectively removed using a pit eye grain selection tube, which differentiates based on the varying long-axis diameters of rice grains.

Porosity is another essential indicator of grain quality, significantly impacting processing quality, storage stability, and edible quality. During grain processing activities such as grinding or food production, porosity influences the uniformity of crushing and mixing, thereby affecting the quality of the final product. Grains with high porosity generally exhibit better air circulation, which helps reduce mold growth and decay during storage, enhancing storage stability. Additionally, porosity affects the water absorption rate and capacity of grains, which is crucial for their performance during cooking processes like swelling and gelatinization. High porosity in grains can improve the taste and texture of food products; for example, bread with high porosity typically has a softer texture. Furthermore, porosity influences the thermal conductivity of grains, playing a vital role in the uniform heating of grains during baking and cooking processes [17]. The drying rate of grains is also affected by porosity, with high-porosity grains drying faster. This leads to reduced energy consumption and improved production efficiency. In grain storage, the porosity of the grain pile is a critical parameter that affects heat and moisture transfer, ventilation, as well as the safety and quality of the stored grains. To address these factors, Chen and colleagues proposed a porosity prediction method for grain cells based on the Whale Optimization Algorithm-Backpropagation Neural Network (WOA-BPNN) [41]. This machine learning approach forecasts the porosity of grain piles, providing theoretical support for the safe storage of grains.

2.3.5.2 Evaluation of Bread and Steamed Bread Quality

Porosity is one of the key factors affecting the quality of bread. It significantly affects various attributes such as taste, texture, volume, moisture retention, flavor, appearance, processing characteristics, and nutritional value. Extensive research has established a solid theoretical foundation demonstrating how factors like flour type [42], water contents [43], salt concentration [44], emulsifier presence [45], other additives [46], and mixing conditions [47] influence the integration, development, and stabilization of gas bubbles within the dough. Various microscopy techniques, including confocal laser scanning microscopy and X-ray microtomography, are employed as qualitative or quantitative methods to characterize the porosity and bubble size distribution (BSD) of dough at different phases of the bread-making process [2].

For instance, Jha and colleagues discovered through instrumental measurement, light microscopy, and image analysis that longer resting times increase dough porosity, resulting in fewer tiny bubbles and a larger median pore area [47]. As resting time increases, the specific volume and porosity fraction of bread also increase, while the median pore area within the bread crumb decreases. Image analysis of gas bubbles under a microscope revealed that the addition of emulsifiers enhances dough volume during proofing and increases the number of smaller gas cells within the bread crumb [45]. Additionally, confocal laser scanning microscopy, combined with image processing, was utilized to determine the size and shape of bubbles. Ishwarya and colleagues found that increasing bran addition reduces the average bubble size and prevents bubble coarsening in the dough [46].

X-ray microtomography studies showed that clusters of partially coalesced bubbles form within doughs during proofing, with one cluster expanding into a large, interconnected, closed cell. Microscopy results indicated that the open cell structure in bread consists of these partially coalesced bubbles. Comparative studies between Canadian and Australian flour doughs revealed that Canadian doughs have thicker layers between bubbles, smaller bubble sizes, and a slower formation of continuous structures [42]. Furthermore, Sun and colleagues used X-ray microtomography to examine the effects of reduced sodium on gas phase parameters in non-yeasted doughs with various formulations. They found that decreasing salt content led to a reduced gaseous volume within the dough by the end of the mixing process. Doughs made from stronger wheat cultivars, with higher moisture levels, or subjected to shorter mixing durations contained fewer gas bubbles. Additionally, reduced sodium content accelerated the rate of disproportionation in the bubble size distribution [44].

References

1. Li L. Physical properties of foods. China Agriculture Press; 2010. p. 148–64.
2. Sun X, Scanlon MG, Nickerson MT, Koksel F. The entrainment and evolution of gas bubbles in bread dough—a review. Cereal Chem. 2023;100(5):1031–47.
3. Aghajanzadeh S, Sultana A, Mohammad Ziaiifar A, Khalloufi S. Formation of pores and bubbles and their impacts on the quality attributes of processed foods: a review. Food Res Int. 2024;188:114494.
4. Kumar V, Sharma HK, Kumar N. Engineering properties of foods. In: Sharma HK, Kumar N, editors. Agro-processing and food engineering: operational and application aspects. Cham: Springer Singapore; 2022. p. 23–80.
5. Vargas-Solórzano JW, Ascheri JLR, Carvalho CWP, Takeiti CY, Galdeano MC. Impact of the pretreatment of grains on the interparticle porosity of feed material and the torque supplied during the extrusion of Brown Rice. Food Bioprocess Technol. 2019;13(1):88–100.
6. Xiong C, She Y, Jiao X, Shao Y, et al. Application of hyperspectral imaging technology in nondestructive testing of agricultural products. Sci Technol Cereals Oils Foods. 2023;31(1): 109–22.
7. Al-Shemmeri M, Fryer P, Farr R, Lopez-Quiroga E. Development of coffee bean porosity and thermophysical properties during roasting. J Food Eng. 2024;378:112096.
8. Yang Z, Li Z, Hu N, Zhang M, Zhang W, et al. Multi-index grading method for pear appearance quality based on machine vision. Agriculture. 2023;13(2)
9. Lu J, Chen W, Lan Y, Qiu X, et al. Design of citrus peel defect and fruit morphology detection method based on machine vision. Comput Electron Agric. 2024;219:108721.
10. Narushin VG, Romanov MN, Griffin DK. Non-destructive measurement of chicken egg characteristics: improved formulae for calculating egg volume and surface area. Biosyst Eng. 2021;201:42–9.
11. Narushin VG, Romanov MN, Griffin DK. Non-destructive evaluation of the volumes of egg shell and interior: theoretical approach. J Food Eng. 2021;300:110536.
12. Cai J, Liang X, Xu Q, Xia Z, et al. Detecting volumetric edible rate of thick-skinned citrus using X-ray threedimensional reconstruction. Trans Chinese Soc Agric Eng. 2024;40(1):293–300.
13. Figura LO, Teixeira AA. Food physics-physical properties measurement and applications, 2nd ed. 2023, ISBN: 978-3-031-27397-1.
14. Kelkar S, Boushey CJ, Okos M. A method to determine the density of foods using X-ray imaging. J Food Eng. 2015;159:36–41.
15. Voglhuber-Brunnmaier T, Jakoby B. Electromechanical resonators for sensing fluid density and viscosity—a review. Meas Sci Technol. 2021;33(1):012001.
16. Lawrence M, Jiang Y. Porosity, pore size distribution, micro-structure. In: Amziane S, Collet F, editors. Bio-aggregates based building materials; 2017. p. 23.

17. Oliveros NO, Hernández JA, Sierra-Espinosa FZ, Guardián-Tapia R, et al. Experimental study of dynamic porosity and its effects on simulation of the coffee beans roasting. J Food Eng. 2017;199:100–12.
18. Janssen S, Verboven P, Nugraha B, Wang Z, et al. 3D pore structure analysis of intact 'Braeburn' apples using X-ray micro-CT. Postharvest Biol Technol. 2020;159:111014.
19. Wang J, Lin Y, Li Q, Lu Z, et al. Non-destructive detection and grading of chilling injury-induced lignification of kiwifruit using X-ray computer tomography and machine learning. Comput Electron Agric. 2024;218:108658.
20. Van De Looverbosch T, Rahman Bhuiyan MH, Verboven P, Dierick M, et al. Nondestructive internal quality inspection of pear fruit by X-ray CT using machine learning. Food Control. 2020;113:107170.
21. Wang N, Yu S-K, Qi Z-P, Ding X-Y, et al. Pears classification by identifying internal defects based on X-ray images and neural networks. Adv Manuf. 2024;
22. Yakatpure S, Rasane K. In development of an efficient nondestructive grading method for pomegranate using magnetic resonance imaging and neural network. Cham: Springer Singapore; 2020. p. 135–42.
23. Zhang H, Zha Z, Kulasiri D, Wu J. Detection of early core browning in pears based on statistical features in vibro-acoustic signals. Food Bioprocess Technol. 2021;14(5):887–97.
24. Mohd Ali M, Hashim N, Abd Aziz S, Lasekan O. Characterisation of pineapple cultivars under different storage conditions using infrared thermal imaging coupled with machine learning algorithms. Agriculture. 2022;12(7)
25. Herremans E, Melado-Herreros A, Defraeye T, Verlinden B, et al. Comparison of X-ray CT and MRI of watercore disorder of different apple cultivars. Postharvest Biol Technol. 2014;87:42–50.
26. Wood RM, Schut DE, Trull AK, Marcelis LFM, et al. Detecting internal browning in apple tissue as determined by a single CT slice in intact fruit. Postharvest Biol Technol. 2024;211: 112802.
27. Liu FH, Zhang YQ, Du CT, Ren X, Huang B, et al. Design and experimentation of a machine vision-based cucumber quality grader. Food Secur. 2024;13(4)
28. Kirchgessner N, Hodel M, Studer B, Patocchi A, et al. FruitPhenoBox - a device for rapid and automated fruit phenotyping of small sample sizes. Plant Methods. 2024;20(1):74.
29. Chen Q, Lin H, Zhao J. Nondestructive detection technologies for real-time monitoring food quality during processing. In: Advanced nondestructive detection technologies in food. Cham: Springer; 2021.
30. Lin H, Zhao J, Sun L, Chen Q, et al. Freshness measurement of eggs using near infrared (NIR) spectroscopy and multivariate data analysis. Innovative Food Sci Emerg Technol. 2011;12(2): 182–6.
31. Lin H, Xu PT, Sun L, Bi XK, et al. Identification of eggshell crack using multiple vibration sensors and correlative information analysis. J Food Process Eng. 2018;41(8)
32. Ahmed MW, Hossainy SJ, Khaliduzzaman A, Emmert JL, et al. Non-destructive optical sensing technologies for advancing the egg industry toward industry 4.0: a review. Compr Rev Food Sci Food Saf. 2023;22(6):4378–403.
33. Qian X-L, Tao N-P, Wang X-C. Research progress on the preparation, characterization and application of nanoparticles in food. Sci Technol Food Ind. 2018;29(16)
34. Chavan P, Sinhmar A, Nehra M, Thory R, et al. Impact on various properties of native starch after synthesis of starch nanoparticles: a review. Food Chem. 2021;364:130416.
35. Liu X, Sala G, Scholten E. Impact of soy protein dispersibility on the structural and sensory properties of fat-free ice cream. Food Hydrocoll. 2024;147:109340.
36. Jiang S, Zhao J Food basic physical properties. Food Phys. 2015.
37. Fu X, You C. Research progress on evaluation and detection methods of milk freshness. J Food Saf Qual. 2021;12(11):4319–25.
38. Narushin VG, Romanov MN, Griffin DK. A novel egg quality index as an alternative to Haugh unit score. J Food Eng. 2021;289:110176.

39. Narushin VG, Kent JP, Salamon A, Romanov MN, et al. Density of egg interior: looking inside an egg while keeping it intact. Innovative Food Sci Emerg Technol. 2023;87:103387.
40. Harnsoongnoen S, Jaroensuk N. The grades and freshness assessment of eggs based on density detection using machine vision and weighing sensor. Sci Rep. 2021;11(1):16640.
41. Chen J, Li J, Zheng D, Zheng Q, et al. Prediction of grain porosity based on WOA–BPNN and grain compression experiment. Appl Sci. 2024;14(7)
42. Chakrabarti-Bell S, Wang S, Siddique KHM. Flour quality and disproportionation of bubbles in bread doughs. Food Res Int. 2014;64:587–97.
43. Cafarelli B, Spada A, Laverse J, Lampignano V, Del Nobile MA. An insight into the bread bubble structure: an X-ray microtomography approach. Food Res Int. 2014;66:180–5.
44. Sun X, Scanlon MG, Guillermic RM, Belev GS, et al. The effects of sodium reduction on the gas phase of bread doughs using synchrotron X-ray microtomography. Food Res Int. 2020;130: 108919.
45. Garzon R, Hernando I, Llorca E, Rosell CM. Understanding the effect of emulsifiers on bread aeration during breadmaking. J Sci Food Agric. 2018;98(14):5494–502.
46. Ishwarya SP, Desai KM, Naladala S, Anandharamakrishnan C. Bran-induced effects on the evolution of bubbles and rheological properties in bread dough. J Texture Stud. 2017;48(5): 415–26.
47. Jha PK, Chevallier S, Cheio J, Rawson A, et al. Impact of resting time between mixing and shaping on the dough porosity and final cell distribution in sandwich bread. J Food Eng. 2017;194:15–23.

Chapter 3
Rheological Properties of Food

Lei Zhang

3.1 Introduction

This chapter discusses the rheological properties of foods, focusing on the principles of flow behavior and deformation of food systems. Firstly, the definition of rheological properties and models used to understand the rheology of food materials are presented. Next, the principles of viscosity and its measurement methods and the corresponding devices used in these methods are detailed. Finally, the application of rheological properties in food engineering is described.

Food rheology, a key area of study in food physics, is closely tied to the mechanical properties of food; moreover, the rheological properties of food are closely related to their chemical composition, molecular structure, intramolecular and intermolecular binding, and tissue structure [1]. For example, rheology plays a crucial role in three-dimensional (3D) printing, particularly in the development of printable biopolymer inks, which rely on reversible interactions and chain entanglements within the microstructure [2]. Large amplitude oscillatory shear rheology is used to investigate the structure of pizza cheese, allowing for the distinction of different networks at various processing stages [3]. A quantifiable correlation has been reported between the protein foam microstructure of egg whites and their rheological properties, these properties are expressed through the relationship between stiffness (G', Pa) and oscillation strain (%), which is based on foams with varying amounts of sugar, and this relationship is useful for research on food foams to enhance quality control [4].

Thus, food rheological properties play an important role in the transportation, processing, and design of food products, even influencing the satisfaction of consumers during chewing. Understanding changes in food tissue structure through

L. Zhang
School of Food and Biological Engineering, Jiangsu University, Zhenjiang, Jiangsu, China
e-mail: zhangleifd@ujs.edu.cn

H. Lin et al., *Advanced Technologies for Physical Properties Measurement of Food and its Processing*, https://doi.org/10.1007/978-981-96-2347-1_3

rheological research reveals mechanical property changes during processing. Consequently, product quality can be identified and controlled, providing detailed data for technologies, devices, and product development [5].

3.1.1 Basic Conception of Rheological Properties

When materials are subjected to external forces, deformation occurs, gradually leading to flow. This process involves time as the deformation and flow develop. In essence, rheological properties are mechanical properties that result in deformation and flow under stress. The rheological behavior of materials is characterized by three parameters: force, deformation, and time. The rheological properties of materials are represented by the relationship among stress, strain, and time; additionally, rheology can be categorized into different groups with hierarchical levels [1, 6], and is summarized in Table 3.1.

For liquid materials, flow behavior is in the first level, including viscous and plastic fluids in the second level. Newtonian (e.g., honey) and non-Newtonian (e.g., mayonnaise) fluids belong to viscous ones, and Bingham (e.g., toothpaste) and non-Bingham (e.g., cream) fluids belong to plastic ones (the third level). The relationship between shear stress and shear strain (rate) will be introduced later. For solid materials, deformation is in the first level, including elastic and inelastic materials in the second level. Elastic materials contain Hookean (e.g., steel) and non-Hookean (e.g., rubber) ones, and inelastic materials include viscoelastic and viscoplastic ones, such as fruit.

3.1.2 Mechanical Properties and Models of Ideal Materials

3.1.2.1 Definition of Viscoelastic Body

When one end of a cylindrical dough is fixed, and a constant tensile load is applied to the other, the dough can flow slowly like a viscous liquid. However, when the applied load is removed, some parts of the dough may shrink slightly, exhibiting behavior like an elastic solid. Although the dough does not return to its original length due to the permanent deformation characteristic of viscous flow, it simultaneously displays the properties of liquid viscosity and solid elasticity. This combination of elasticity and viscosity is called viscoelasticity, and materials exhibiting this behavior are called viscoelastic bodies. Foods showing both elastic and viscous components are known as viscoelastic foods. Research has shown that zein, the major prolamin storage protein, plays an important role in forming viscoelastic dough, and this viscoelastic property is crucial to the quality of baked goods, including their mechanical properties [7].

Table 3.1 Grouping of rheology

Level 1	Level 2	Level 3	Stress-strain	Material
Flow (For liquid)	Viscous	Newtonian	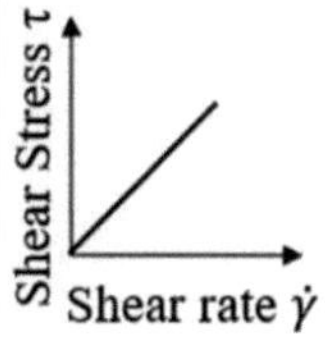	Honey
		Non-Newtonian	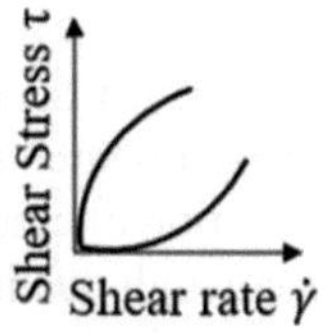	Mayonnaise
	Plastic	Bingham	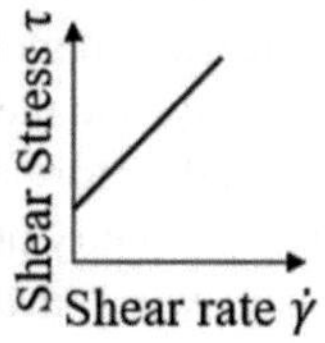	Toothpaste
		Non-Bingham	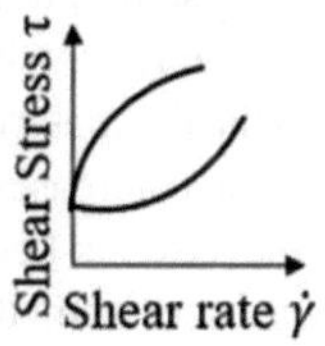	Cream
Deformation (for solid)	Elastic	Hookean	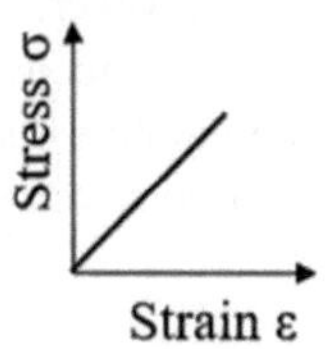	Steel
		Non-Hookean	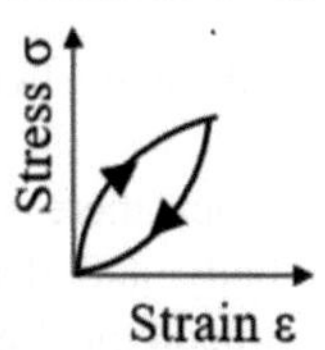	Rubber
	Inelastic	Viscoelastic	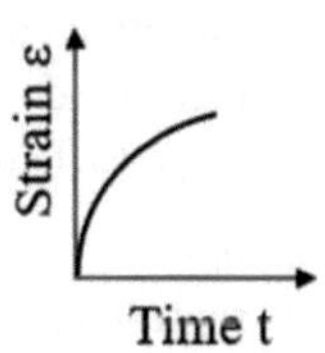	Fruit

(continued)

Table 3.1 (continued)

Level 1	Level 2	Level 3	Stress-strain	Material
		Viscoplastic	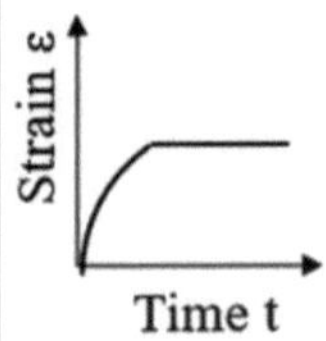	Fruit

The dough is a typical viscoelastic material, and it is necessary to research its rheology. The primary research tool for this purpose is the rheometer. The study is shown that a combination of empirical and fundamental rheology can be effectively used to characterize dough quality. Empirical rheology is assessed using Brabender farinograph analysis and Chopin alveograph analysis; the former records dough resistance with standard parameters such as water absorption, dough development time, and dough stability, and the latter measures dough tenacity, extensibility, curve configuration ratio and deformation energy. Modular rheometers equipped with 20 mm parallel plate geometry probes are used for fundamental rheology to conduct oscillatory frequency sweep measurements and creep and recovery tests [8].

Furthermore, it is essential to understand the relationship among stress, strain and time in viscoelastic materials, which defines their rheological behavior. When a load is applied to viscoelastic materials, elastic deformation occurs immediately. And then, the deformation increases, but the rate of deformation decreases, which corresponding to the viscous behavior. Upon removal of the applied load, some of the elastic deformation is immediately recovered, while other viscous parts recover gradually over time. However, given an infinite amount of time, a small portion of the deformation remains unrecovered and is permanently retained because of viscous characteristic. Representative viscoelastic materials include apple pulp, potato pulp, keratin endosperm of corn kernels, starch endosperm of wheat kernels, and wheat straw [1, 6].

3.1.2.2 Mechanical Properties and Models of Ideal Materials

In the field of mechanics, there are models of ideal materials, such as the ideal elastic body, ideal plastic body, and ideal viscous body; these basic models, also known as model elements or rheological elements, can be combined to simulate rheological behavior through constitutive equations or rheological equations [1, 6, 9]. Table 3.2 summarizes three typical model elements. Meanwhile, their constitutive equations are also shown from Eqs. (3.1) to (3.3) [1, 6]:

1. *Hook's body (spring)*

   Hook's body represents an ideal elastic body, which is also can be called spring. The symbol of the rheological element and its mechanical properties are shown in Table 3.2. The model symbol is spring. The relationship between stress

Table 3.2 Rheological elements

Name	Symbol	Mechanical property
Hook's body (spring)		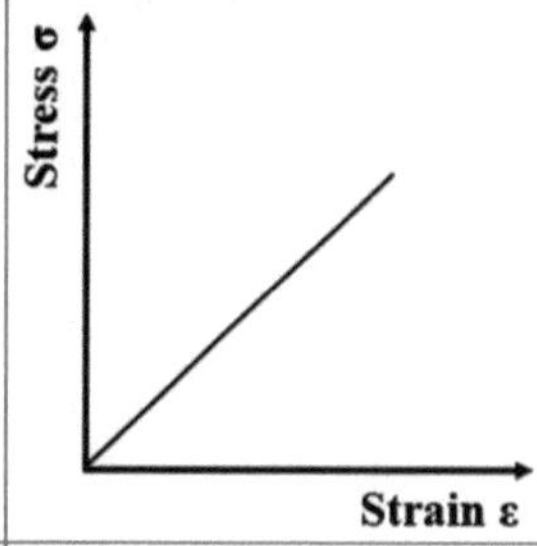
St. Venant (friction block)		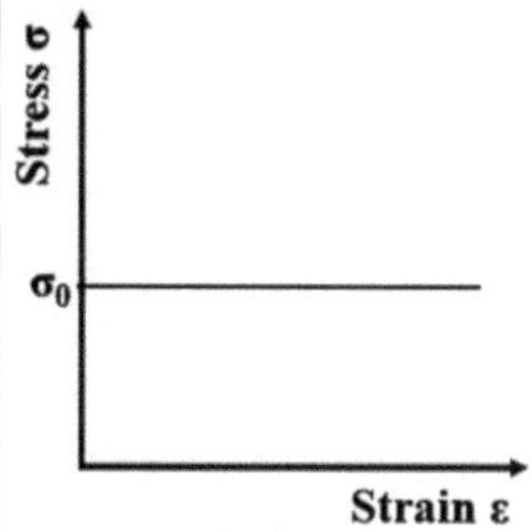
Newton's body (dashpot)	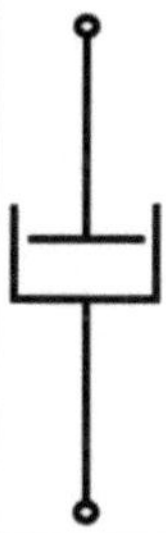	

and strain is linear from origin of coordinate axis. The constitutive equation is given in Eq. (3.1), where E is the mechanical constant known as the elasticity modulus [1, 6].

$$\sigma = E\varepsilon \tag{3.1}$$

2. *St. Venant (friction block)*

   St. Venant that can be named friction block represents an ideal plastic body. The symbol of the rheological element and its mechanical properties are shown in Table 3.2. When the applied force exceeds the corresponding yield stress (σ_0), the strain is induced. The constitutive equation is given in Eq. (3.2) [1].

$$\begin{cases} \upsilon \leq \upsilon_0 & \varepsilon = 0 \\ \sigma > \sigma_0 & \varepsilon > 0 \end{cases} \tag{3.2}$$

3. *Newton's body (dashpot)*

Newton's body represents an ideal viscous body and can be named as dashpot. The symbol of the rheological element and its mechanical properties are shown in Table 3.2. With the increment of shear stress, the shear strain rate increased linearly. The constitutive equation is given in Eq. (3.3), where η is the mechanical constant [1, 6].

$$\tau = \eta\dot{\gamma} \tag{3.3}$$

3.1.2.3 Combination Models

The previously mentioned rheological models, comprising mechanical elements such as springs and dashpots, are instrumental in interpreting the rheological behavior of viscoelastic materials. A significant advantage of these models is their ability to qualitatively mimic the behavior of real materials. Mechanical behavior can be described through force-deformation or stress-strain relationships over time, leading to mathematical equations. These equations not only elucidate but also, in some cases, predict how materials respond under various loading conditions, making these models exceptionally useful. Most rheological models are designed to represent idealized behavior. By combining these fundamental mechanical elements in various configurations, more sophisticated models can be developed to accurately depict the behavior of realistic viscoelastic materials.

Based on the models of ideal materials, the most fundamental combination model is the two-element model. When one spring and one dashpot are arranged in series, the Maxwell model is established [1, 6], as seen in Fig. 3.1a. This model effectively

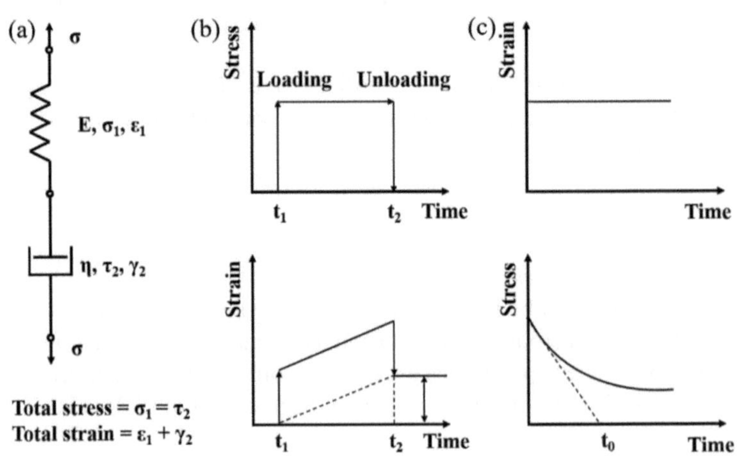

Fig. 3.1 Maxwell model. (**a**) Symbol combination, (**b**) Relationship between stress and strain, (**c**) Stress relaxation

interprets the stress relaxation behavior of viscoelastic liquids, particularly polymeric liquids. The total stress is equal to the stress of Hook's body (spring) and that of Newton's body (dashpot), and the total strain is the sum of the strain of Hook's body (spring) and that of Newton's body (dashpot) [1, 6]. The relationship between stress and strain is illustrated in Fig. 3.1b, when applied stress is loaded, the elastic strain of spring is induced promptly, and then the viscous strain of dashpot is generated together. As applied stress is unloaded, the elastic strain of spring is recovered firstly, while the viscous strain of dashpot is caused permanently and cannot be recovered. The corresponding rheological property, namely the stress relaxation of the Maxwell model, is shown in Fig. 3.1c; when the strain of Maxwell model is constant, the total stress decreases in the materials [1, 6].

Correspondingly, when one spring and one dashpot are arranged in parallel, the Kelvin-Voigt model is established [1, 6], as seen in Fig. 3.2a, where the spring and dashpot undergo the same displacement and the total strain is equal to the strain of Hook's body (spring) and that of Newton's body (dashpot), and the total stress is the sum of the stress of Hook's body (spring) and that of Newton's body (dashpot) [1, 6]. This model is used to interpret the creep behavior of materials. The relationship between stress and strain for the Kelvin-Voigt model is depicted in Fig. 3.2b, when the applied stress is loaded, the elastic strain of spring and the viscous strain of dashpot is generated at the same time. When the applied stress is unloaded, the viscous strain of dashpot is forced to recover by the spring elastic recovery synchronously. The corresponding rheological property of the Kelvin-Voigt model is presented in Fig. 3.2c, as the applied stress of Kelvin-Voigt model is constant, the generated strain increases gradually [1, 6].

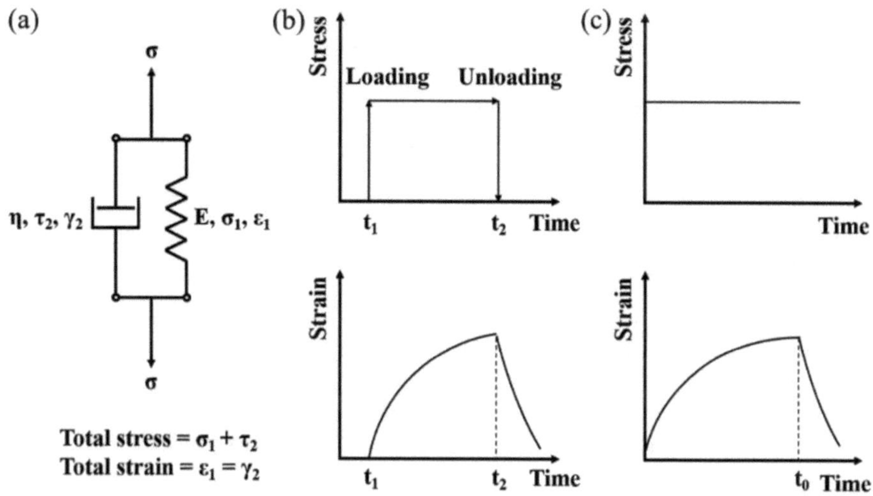

Fig. 3.2 Kelvin-Voigt model. (**a**) Symbol combination, (**b**) Relationship between stress and strain, (**c**) Creep behavior

3.1.2.4 Application on Combination Models of the Viscoelasticity

Rheological models are essential for representing material structure. The rheological elements, their combinations, and the corresponding constitutive relations form the foundational theory for material development. Additionally, rheological models, such as viscoelastic models, can be derived from isochoric tensile tests [10]. To characterize the rheological behavior of dough, the four-element Burgers model, along with several equivalent models, is established Fig. 3.3 [8]. In the Burgers model, the creep deformation stage comprises instantaneous elastic strain G_1, retarded elasticity strain G_2 and viscous fluid η_1; during the recovery stage, the model includes instantaneous elastic recovery G_1, retarded elastic recovery G_2 and permanent deformation (non-recovery) [1, 6, 8]. Furthermore, the Burgers model incorporates one retarded time and two relaxation times [1]. In summary, the Burgers model is constructed by connecting one two-element Maxwell model and one two-element Kelvin-Voigt model in series. This configuration effectively describes the time-dependent creep deformation behavior under shear stress and the subsequent recovery phases [11].

The viscoelastic mechanical model of a polymer gel is studied, the creep behavior curve which can be measured by a dynamic mechanical analyzer and combination models, that is, a Maxwell model and multiple Kelvin model in series, are given in Fig. 3.4 [12]. When the applied stress works on the polymer gel, the instantaneous

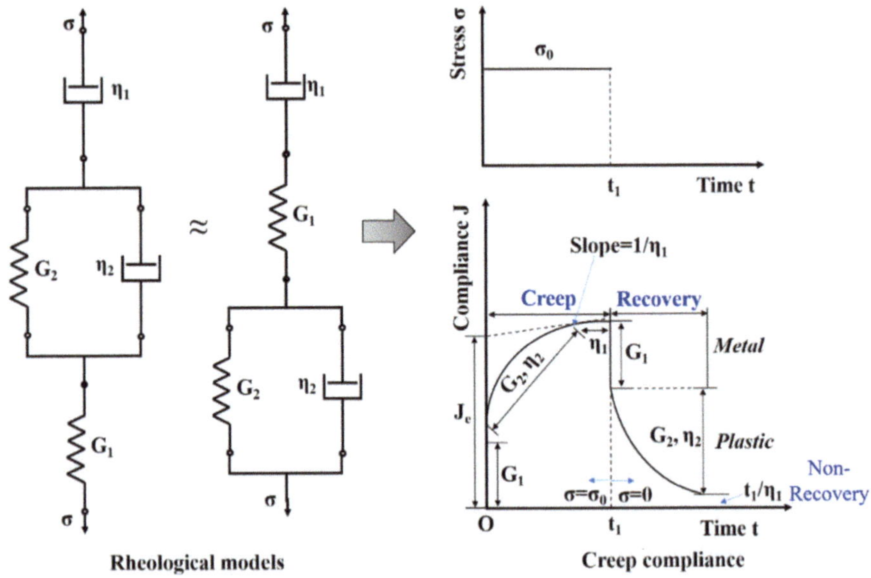

Fig. 3.3 Four-element model (Burgers model) [8]

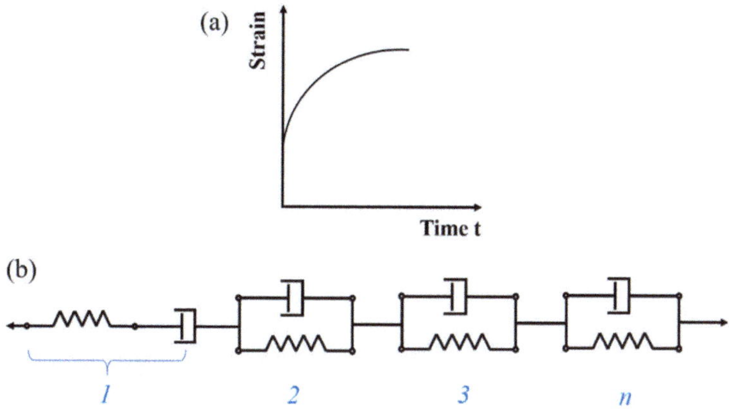

Fig. 3.4 Viscoelastic mechanical model of a polymer gel. (**a**) Creep behavior curve, (**b**) combination models [12]

elastic deformation is induced immediately because of the independent Hook's body (spring), and then creep stage begins with creep acceleration stage firstly and deceleration stage subsequently, which is related to Kelvin model in series [12]. The corresponding stress and stain expressions are given in Eqs. (3.4)~(3.8) [12]. This study can be used to simulate the creep deformation of polymer gel, predict creep fracture and optimize material structure.

$$\varepsilon = \varepsilon_1 + \varepsilon_2 + \varepsilon_3 + \ldots + \varepsilon_n \tag{3.4}$$

$$\varepsilon_1 = \frac{\sigma}{E_1} + \frac{\sigma}{\mu_1} t \tag{3.5}$$

$$\varepsilon_2 = \frac{\sigma}{E_2}\left(1 - e^{-\frac{E_2}{\mu_2}t}\right) \tag{3.6}$$

$$\varepsilon_n = \frac{\sigma}{E_n}\left(1 - e^{-\frac{E_n}{\mu_n}t}\right) \tag{3.7}$$

$$\varepsilon = \frac{\sigma}{E_1} + \frac{\sigma}{\mu_1} t + \frac{\sigma}{E_2}\left(1 - e^{-\frac{E_2}{\mu_2}t}\right) + \frac{\sigma}{E_3}\left(1 - e^{-\frac{E_3}{\mu_3}t}\right) + \ldots + \frac{\sigma}{E_n}\left(1 - e^{-\frac{E_n}{\mu_n}t}\right) \tag{3.8}$$

In the study of the viscoelasticity of polymer nanocomposites, viscoelastic behavior is evaluated through tensile and relaxation tests that analyze the stress-strain relationship. Additionally, a finite-element-based micromechanical model, specifically the generalized Maxwell model, is established based on a representative constitutive viscoelastic law. The relaxation modulus is calculated using a Prony series, as illustrated in Fig. 3.5 [13].

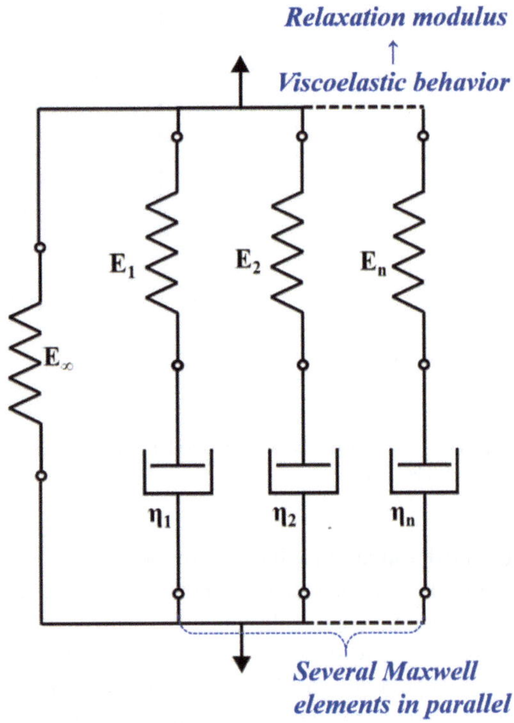

Fig. 3.5 Viscoelasticity of polymer and combination model [13]

3.2 Rheological Properties of Liquid Foods and Its Measurement Technology

3.2.1 Concept of Rheological Properties of Liquid Foods

3.2.1.1 Viscosity and Its Theory

Liquid foods exhibit flow behavior, which includes both viscous and plastic characteristics. The simplest viscous liquid is water, while oil is more viscous than water. This variation is directly related to viscosity, which is defined as the resistance of a fluid to flow [6]. Viscosity is a critical parameter in determining the quality of liquid foods, influencing their texture, mouthfeel, and overall consumer acceptance. Viscosity can be classified into four primary types: absolute viscosity, kinematic viscosity, conditional viscosity, and relative viscosity. Absolute viscosity, measured in pascal-seconds (Pa·s), quantifies the tangential force required per unit area for a liquid to flow at a speed of 1 cm/s. This measure provides an intrinsic property of the fluid's internal resistance to flow, independent of its density. Kinematic viscosity, expressed in square meters per second (m^2/s), is the ratio of absolute viscosity to the fluid's density under identical temperature conditions. This classification accounts for both the fluid's resistance to flow and its density, offering a more comprehensive understanding of its flow behavior in practical applications. Conditional viscosity refers to the ratio of the flow time of one liquid to that of water with the same volume under specified temperature conditions within the same viscometer. This comparative measure allows for the assessment of a liquid's flow characteristics relative to a standard reference (water), facilitating easier comparisons across different fluids. Lastly, relative viscosity is defined as the ratio of the absolute viscosity of one liquid to that of another liquid (e.g., water) under specified temperature conditions. This type of viscosity provides a straightforward comparison, indicating how much more or less viscous a particular liquid is compared to a reference fluid [1].

Viscosity is influenced by several factors, including temperature, stability, the quality of dry matter, and the concentration of liquids. These factors are intrinsically linked to the molecular structure of the liquids. Among these, temperature has the most pronounced effect. Generally, the viscosity of liquids decreases as temperature increases. This relationship can be explained by Eyring's theory, which posits that vacancies exist within liquids, as molecules move into these vacancies continuously, the flow is permitted but the needed energy should be required; when temperatures get higher, the activation energy becomes more readily available, allowing the fluid to flow more easily. The effect of temperature on the viscosity can be quantitatively described by the Arrhenius-type equation as follows [6, 14]:

$$\eta = \eta_{\infty} \exp\left(\frac{E_a}{\mathrm{RT}}\right) \tag{3.9}$$

Where, E_a is the activation energy (J/kg mol), R is gas constant (8314.34 J/kg mol K), T is absolute temperature (K), η_{∞} is a constant (Pa·s).

Cohesive force theory explains that liquid molecules are closely spaced with strong cohesive forces between them, and these intermolecular or cohesive forces play a critical role in determining viscosity [15]. When the temperature increases, the cohesive forces between molecules diminish, resulting in freer flow. Notably, the viscosity of liquids exhibits minimal dependence on factors such as density, molecular velocity, or mean free path. Consequently, the viscosity of liquids decreases as temperature rises [6, 15].

3.2.1.2 Newton's Law of Viscosity

When liquids flow, the motion can be characterized by numerous laminar flows. The velocity of these laminar flows varies due to the friction between liquid molecules, known as viscous resistance. Consider the example of laminar flow between solid plates (Fig. 3.6), the adhesive force between the plate walls and the liquid molecules is stronger than the intermolecular cohesive forces among the liquid molecules near the walls [5]. As a result, the velocity of the liquid is zero at the plate walls, creating a stationary laminar flow. Moving away from the plate walls, the velocity of the laminar flow increases. This variation in velocity induces a velocity gradient perpendicular to the flow direction within the fluid [5].

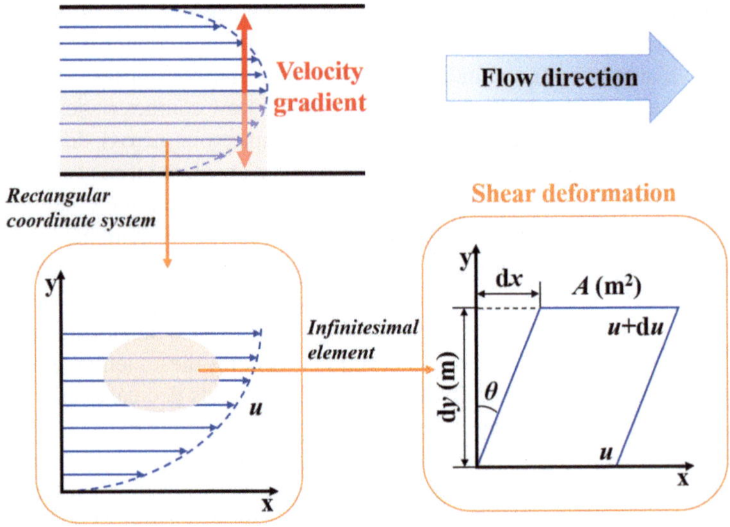

Fig. 3.6 Calculation model of Newton's law of viscosity

To analyze this phenomenon, a rectangular coordinate system is established. An infinitesimal flow element parallel to the flow direction is isolated, forming a parallelogram-shaped body due to shear deformation. The contact area between the upper and lower layers of this parallelogram is denoted as A (m^2), the distance between the two layers is dy (m), and the viscous resistance between them is represented by F (N). Because of shear deformation of the liquid material, the velocity of the lower layer is u, while that of the upper layer is u + du (m/s). During a short interval of time, the shear deformation of the infinitesimal flow element can be represented by the angle θ (in radians) under the action of shear stress, and this relationship is expressed as [5]:

$$\gamma = \theta = \frac{\mathrm{d}x}{\mathrm{d}y} \tag{3.10}$$

Consequently, the shear rate, which is the velocity gradient perpendicular to the flow direction (s^{-1}), is given by [5]:

$$\dot{\gamma} = \frac{\theta}{\mathrm{d}t} = \frac{\frac{\mathrm{d}x}{\mathrm{d}y}}{\mathrm{d}t} = \frac{\frac{\mathrm{d}x}{\mathrm{d}t}}{\mathrm{d}y} = \frac{\mathrm{d}u}{\mathrm{d}y} \tag{3.11}$$

Furthermore, the viscous resistance F is directly proportional to the contact area A and the shear rate γ. This relationship is mathematically described as [5]:

$$F = \eta A \dot{\gamma} \tag{3.12}$$

Here, η represents the viscosity of the liquid (Pa·s). If τ denotes the shear stress, which is the viscous resistance per unit area of the liquid, then [5]:

$$\tau = \frac{F}{A} \tag{3.13}$$

Substituting the expression for F into this above-mentioned equation yields Newton's law of viscosity, the fundamental law of viscosity [6]:

$$\tau = \eta \dot{\gamma} \tag{3.14}$$

Where shear stress τ (Pa) refers to the stress component parallel to the tangential direction of the cross-section. The proportionality constant η (Pa·s), known as viscosity, quantifies the internal friction between two laminar layers when the velocity gradient γ is 1 s^{-1} and the contact area A is 1 cm^2.

3.2.1.3 Classification of Viscous Flow and Its Characteristics

Viscous fluids tend to deform continuously under the influence of applied stress and can be categorized into Newtonian and non-Newtonian fluids. Fluids that adhere to Newton's law of viscosity are termed Newtonian fluids. In these fluids, the relationship between shear stress and shear rate is linear, meaning that viscosity—the slope of the shear stress versus shear rate—is constant and independent of the shear rate [6], as given in Table.3.3.

Conversely, fluids that do not adhere to Newton's law of viscosity are known as non-Newtonian fluids. These fluids follow the power-law model, also referred to as the Ostwald-de Waele equation, which is defined as [6]:

$$\tau = k\left(\dot{\gamma}_{yz}\right)^{n} \tag{3.15}$$

where k is the consistency coefficient (Pa·s^n), and n is the flow behavior index. Unlike Newtonian fluids, the slope of shear stress versus shear rate varies for non-Newtonian fluids, as illustrated in Table 3.3. By defining the apparent viscosity η_a *as* [6]:

$$\eta_a = k\left(\dot{\gamma}_{yz}\right)^{n-1} \tag{3.16}$$

Table 3.3 Characteristics of viscous flow

Shear stress versus shear rate	Viscous flow	Flow behavior
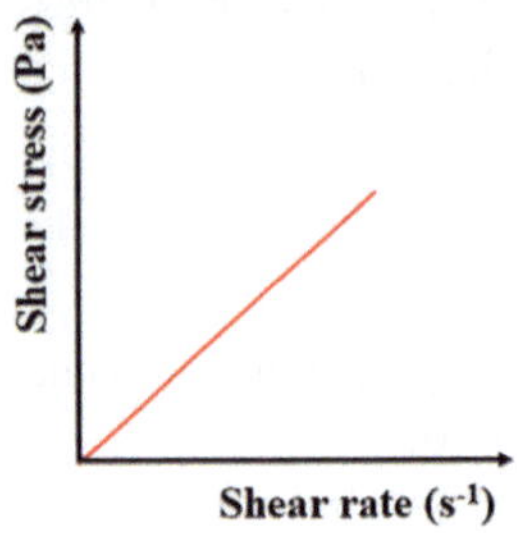	Newtonian	Linear
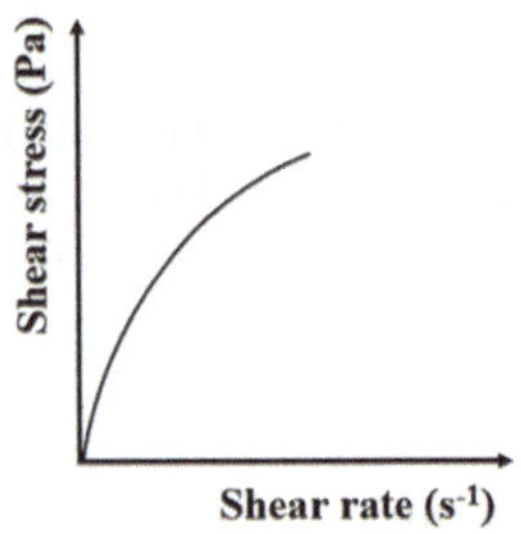	Non-Newtonian	Shear-Thinning
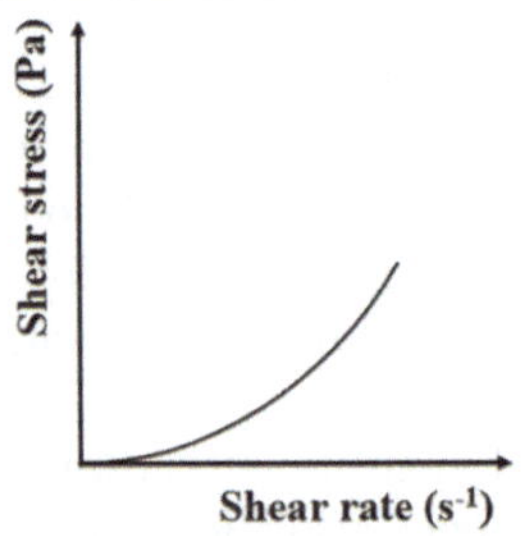	Non-Newtonian	Shear-Thickening

The empirical equation can be transformed into:

$$\tau = \eta_a \dot{\gamma} \tag{3.17}$$

This transformation indicates that the apparent viscosity η_a changes with the shear rate γ for non-Newtonian fluids, resulting in different viscosity values at varying shear rates [6]. Specifically, when the flow behavior index n satisfies $0 < n < 1$, the non-Newtonian fluids are classified as shear-thinning (pseudoplastic) fluids [6]. It means the shear rate decreases with the increment of shear stress, such as stirring force. Conversely, when $1 < n < \infty$, the fluids are classified as thickening fluids [6]. The shear strain rate increases as the shear stress increases, which also can be explained by the slope of shear stress versus shear rate curve, as shown in Table 3.3.

As the shear rate increases, the friction between layers of shear-thinning (pseudoplastic) fluids decreases. This behavior is characteristic of one type of non-Newtonian fluids known as shear-thinning fluids. Shearing causes entangled,

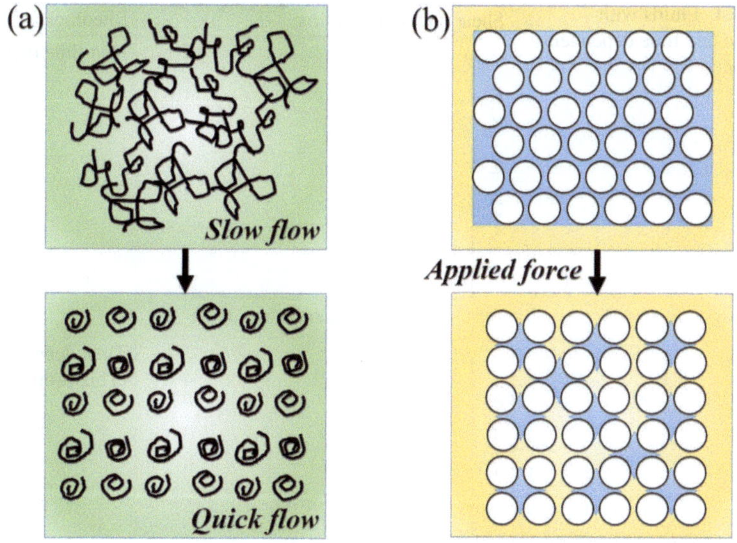

Fig. 3.7 Flow mechanism of non-Newtonian fluids. (**a**) Shear thinning, (**b**) Shear thickening

long-chain molecules to straighten and align with the flow, thereby reducing viscosity, as illustrated in Fig. 3.7a. Common examples of pseudoplastic fluids in food systems include fruit and vegetable products, including applesauce, banana puree, and concentrated fruit juices [6].

On the contrary, when the shear rate increases, the internal friction and apparent viscosity in shear-thickening (dilatant) fluids also increase. Shear thickening is another category of non-Newtonian fluids. This behavior can be explained by the dense packing of colloidal particles within a paste-like liquid, where the dispersion medium (e.g., water) fills the minimal voids between particles. Applying shear disrupts this close packing, increasing the void fraction. As a result, water can no longer adequately fill the spaces between colloidal particles, leading to reduced lubrication and increased resistance to flow, as depicted in Fig. 3.7b. An example of shear-thickening fluids in food systems is corn starch suspension. Additionally, when the increment of the viscosity is accompanied by the volume expansion, these fluids are specifically referred to as dilatant fluids [6].

Plastic fluids can be further categorized into Bingham plastic fluids and non-Bingham plastic fluids, each exhibiting distinct flow behaviors under applied shear stress. When the magnitude of shear stress is below a certain threshold known as the yield stress (τ_0), the flow phenomenon of Bingham plastic fluids cannot be generated. Unless the applied shear stress exceeds τ_0, these fluids begin to flow and their flow behaviors are similarly to Newtonian fluids. The representative example of the Bingham plastic fluids is toothpaste, and the toothpaste can flow when the big squeezed force acting on the tube. As for food materials, ketchup, tomato paste, and mayonnaise also exhibit Bingham plastic behavior [6].

Table 3.4 Fluids with non-time and time dependent behavior

Shear stress versus time	Rheological behavior
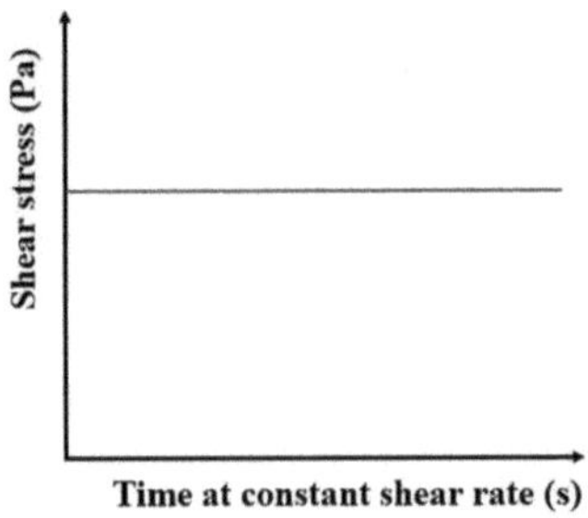	Non-time dependent
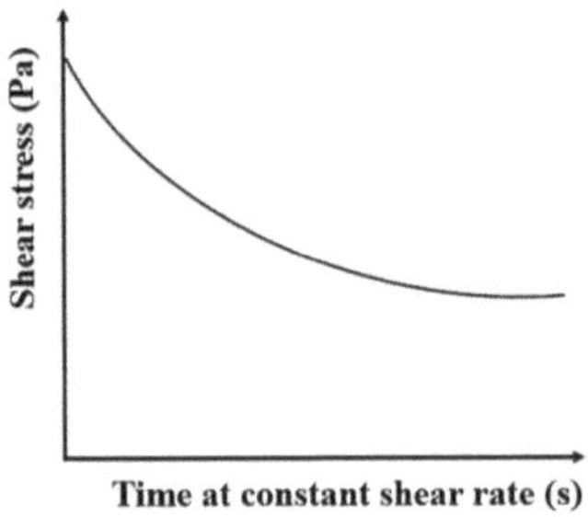	Time dependent (Thixotropic)
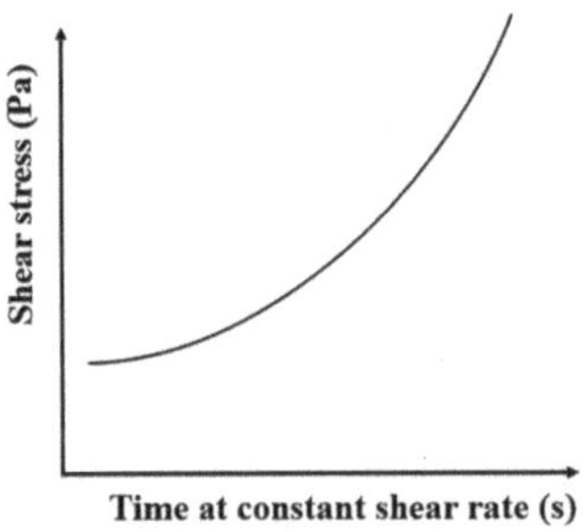	Time dependent (Rheopectic)

Unlike Bingham plastics, although non-Bingham plastic fluids also require the shear stress to exceed a minimum yield stress (τ_0) to initiate flow; however, the relationship between shear stress and shear rate in these fluids is nonlinear. Non-Bingham plastic fluids can exhibit either shear-thinning (pseudoplastic) or shear-thickening (dilatant) behavior while maintaining a yield stress. Examples of non-Bingham plastic fluids in food systems include minced fish paste, raisin paste, rice flour-based batter used in fried products, and molten milk chocolate [6].

The aforementioned pseudoplastic and thickening fluids, along with plastic fluids, are classified as non-Newtonian fluids with non-time dependence. This means their viscosity changes with shear rate but not with time under a constant shear rate. In contrast, some non-Newtonian fluids exhibit time dependent behavior, where their viscosity changes with time even when subjected to a constant shear rate, and these time dependent non-Newtonian fluids either become thinner (shear-thinning) or thicker (shear-thickening) over time [6], as depicted in Table 3.4.

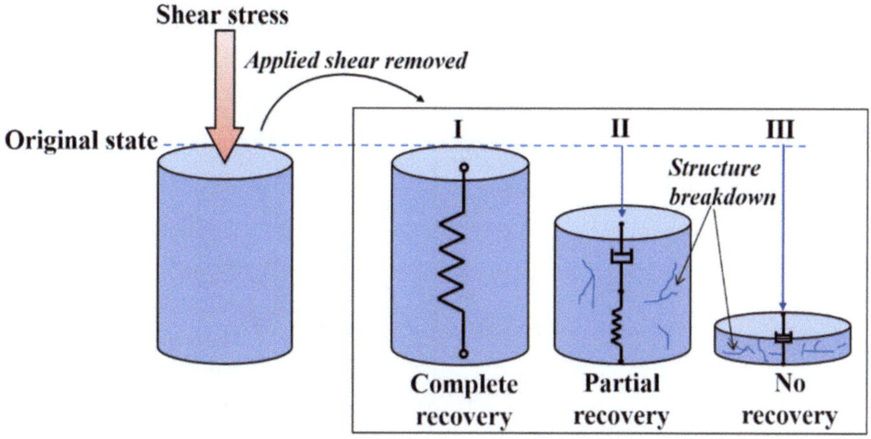

Fig. 3.8 Schematic diagram of thixotropic behavior

As the fluids are characterized by shear thinning with time, that is, the shear stress decreases with the increasing time at the fixed shear strain rate, and the fluids are known as thixotropic fluids. This rheologic behavior is related to the breakdown of the material's internal structure when continuing shearing is affected. When the applied shear is removed and the fluid is allowed to rest, thixotropic behavior can present three states, including complete recovery, partial recovery, and no recovery, while thixotropy with no recovery is referred to as rheomalaxia or rheodestruction [6], and the schematic diagram is given in Fig. 3.8. The common examples of thixotropic fluids include gelatin, egg white, and shortening.

Conversely, when the fluids are characterized by shear thickening with time, that is, the shear stress increases with the increasing time at the constant shear strain rate, and the fluids are named rheopectic fluids. This kind of behavior occurs because the structure within the fluid builds up when continuing shearing is affected. Rheopectic behavior is rarely observed in food systems; however, examples include fenugreek paste, a traditional food in Turkey [16], and bentonite-clay suspensions.

Furthermore, the special behaviors of thixotropic and rheopectic fluids warrant detailed study. When shear stress is measured as a function of shear rate as shown in Fig. 3.9, and the shear rate is increased firstly and then decreased, a hysteresis loop is observed in the curves of shear stress versus shear rate [6]. For thixotropic fluids, the viscosity decreases and the mobility increases with the time when the fluids are vibrated, stirred, or shaken; after rest for a while, the mobility of the fluids become worse and the viscosity may recover also with the time. In the shear stress-strain rate curves, the curve as the shear rate decreases is under the curve as the previous shear rate increases, and the hysteresis loop is formed. The food with thixotropic phenomenon has a soft and refreshing taste. For rheopectic fluids, the viscosity increases with the fluids flow. After the flow velocity increases to the maximum value, the flow velocity begins to decrease, and the curve as flow velocity decreases is above the

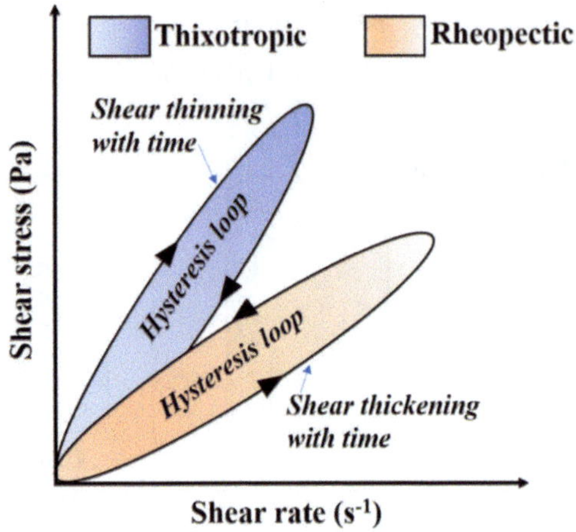

Fig. 3.9 Hysteresis in shear stress versus shear rate curve

curve as flow velocity increases. The food with rheopectic characteristic, such as whipped cream, often provides a sticky taste [5, 6].

In conclusion, compared with Newtonian fluids with constant viscosity, that is, the straight line through the origin in the flow curve, non-Newtonian fluids own complicated apparent viscosity and the corresponding flow curve is not the straight line also through the origin. Non-Newtonian fluids can be broadly classified into two categories, one is non-Newtonian fluids without time dependence, and the other is non-Newtonian fluids with time dependence [9]. Non-Newtonian fluids without time dependence have an apparent viscosity that remains constant over time when shear stress is applied, meaning their viscosity is solely dependent on the shear rate and not on the duration of applied shear stress [6, 9]. Pseudoplastic and dilatant fluids belong to non-Newtonian fluids without time dependence, the viscosity of the former decreases with the increment of shear rate, such as tomato paste; but the viscosity of the latter increases with the increment of shear rate, which can be seen in highly concentrated suspensions [6, 9]. On the other hand, non-Newtonian fluids with time dependence exhibit changes in apparent viscosity over time under constant shear stress, including thixotropic and rheopectic fluids; hereinto, the viscosity of thixotropic fluid decreases, such as ketchup, but that of rheopectic fluid increases over time at constant shear rate, which is some paste and in fact can hardly be found in foods [5, 6, 9].

3.2.2 Measurement Technology of Viscosity

Viscosity is a critical parameter for evaluating the quality of liquid foods, making accurate measurement technologies essential. The most commonly used viscosity measurement devices include capillary flow viscometers, rotational viscometers, and rheometers.

3.2.2.1 Capillary Viscometers

Capillary flow viscometers are simple, inexpensive, and particularly suitable for measuring low-viscosity fluids. A typical design of a capillary viscometer is the U-tube Ostwald viscometer [1], as shown in Fig. 3.10. It primarily consists of a main tube, wide tube, bent tube, measuring ball, capillary, starting line (S1), and ending line (S2). The fundamental principle of measurement is based on the laminar flow behavior of Newtonian fluids within the tube, which can be described by the Hagen-Poiseuille equation as follows [5]:

$$\eta = \frac{\pi \Delta P \mathrm{R}^4 t}{8 Q_t \mathrm{L}} \tag{3.18}$$

Where ΔP is the pressure drop causing flow between the two ends of the capillary (Pa), R is the inner radius of the capillary (m), Q_t is the volumetric flow rate through the capillary at time t (m^3/s), L is the length of the horizontal cylindrical tube of the capillary (m).

Pure water serves as the standard liquid with a known viscosity η_0. The relative viscosity can be obtained using the following equation [5]:

$$\frac{\eta}{\eta_0} = \frac{\frac{\pi \Delta P \mathrm{R}^4 t}{8 Q_t \mathrm{L}}}{\frac{\pi \Delta P_0 \mathrm{R}^4 t_0}{8 Q_t \mathrm{L}}} = \frac{\Delta P t}{\Delta P_0 t_0} = \frac{\rho t}{\rho_0 t_0} \tag{3.19}$$

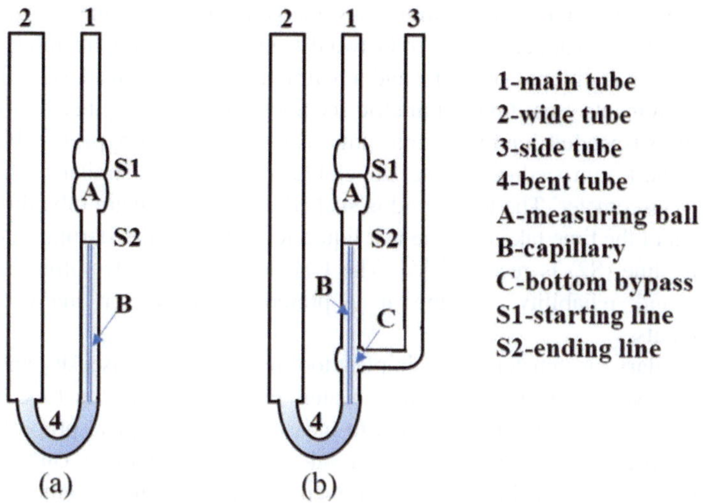

Fig. 3.10 Capillary flow viscometers. (**a**) Ostwald's viscometer, (**b**) Ubbelohde viscometer

Where ρ and ρ_0 are the densities of the test liquid and standard liquid, respectively (kg/m^3), t and t_0 are the transit times of the test liquid and standard liquid in the capillary flow viscometer, respectively (s).

In capillary flow viscometers, the pressure drop is caused by the height difference of a liquid column. Changes in this height difference can affect the shear rate. For non-Newtonian fluids, viscosity relates to the flow rate. Based on the above-mentioned classification of viscosity, the kinematic viscosity (v), measured in m^2/s, is defined as:

$$v = \frac{\eta}{\rho} \tag{3.20}$$

Subsequently Eq. (3.19) can be transformed into:

$$v = \frac{v_0 t}{t_0} \tag{3.21}$$

where v_0 is kinematic viscosity of the standard liquid.

The test procedure for Ostwald's viscometer involves the following steps. Firstly, the viscometer is accurately filled by the test liquid with a precisely known volume and the measuring temperature should be constant. The test liquid is then sucked up from the main tube through the capillary until it reaches above the starting line (S1). Subsequently, the suction is removed, allowing the test liquid to flow through the main tube of the capillary under the influence of gravity or an induced pressure head. The time t it takes for the test liquid to flow from the starting line (S1) to the ending line (S2) is recorded. This transit time t is a direct measure of the kinematic viscosity, as it depends on both the viscosity and density of the fluid [6].

Despite its simplicity, the Ostwald viscometer has certain disadvantages that limit its application. To address these limitations, the Ubbelohde viscometer is developed as shown in Fig. 3.10b; compared to Ostwald's viscometer, the Ubbelohde viscometer features an additional side tube and a bottom bypass [5]. This design allows for lower requirements regarding volumetric accuracy and verticality because the presence of the suspended liquid column induced by the bottom bypass stabilizes the flow [5]. The test procedure for the Ubbelohde viscometer is similar to that of the Ostwald's viscometer. The test liquid is sucked up from the main tube through the capillary, and the time taken for the test liquid to flow from the starting line (S1) to the ending line (S2) is recorded [5]. The Ubbelohde viscometer offers improved accuracy and reliability, making it a preferred choice for many viscosity measurements.

The capillary viscometer is the simplest tool for measuring viscosity and is most suitable for Newtonian fluids, such as water. Furthermore, in the food industry, capillary viscometers can also be used to measure the rheological properties (e.g., power-law parameters, thixotropy) of food materials, including carboxymethyl cellulose solutions, pregelatinized starch, locust bean gum, and enzymes hydrolyzing complex polysaccharides [17]. When a capillary viscometer is used to measure the viscosity of water, the flow time of water is typically recorded as it freely flows through capillary tubes. If the viscosity of the measured fluid increases, free flow through the capillary tubes becomes difficult, and an applied force must be introduced while maintaining a stationary temperature. According to Poiseuille's flow principle, a newly developed capillary viscometer is capable of measuring the viscosities of methanol-water and acetonitrile-water mixtures at ultrahigh pressures, ranging from atmospheric pressure to 3500 bar, at 25 °C [18]. Capillary viscometers can be combined with other instruments for enhanced functionality. For instance, combining a capillary viscometer with a dynamic surface tension detector allows for the measurement of both viscosity and surface tension changes in bovine serum albumin (BSA), and this combination can reveal alterations in composition or conformation, including unfolding and aggregation, thereby providing methods for quality control [19]. With advancements in computer technology, capillary viscometers have been integrated with computer vision systems based on deep convolutional neural networks for laser speckle contrast imaging processing. A high-speed camera records the inertial flow in a transparent capillary illuminated by a laser, aiding in the study of the rheology of physiological fluids [20].

3.2.2.2 Rotational Viscometers

Rotational viscometers are commonly used in most laboratories, and their basic structure is illustrated in Fig. 3.11. A rotational viscometer consists of two annular cylinders—the cylinder rotor and the cup—with a narrow gap between them. The fluid to be measured is placed in this gap. Either the inner cylinder (Searle system, which is more common) or the outer cylinder (Couette system) is rotated. This type of viscometer can be used for both Newtonian and non-Newtonian food materials [6].

The fundamental measurement principles are established with the test liquid considered in a Searle-type viscometer (Fig. 3.11); based on the torsion system,

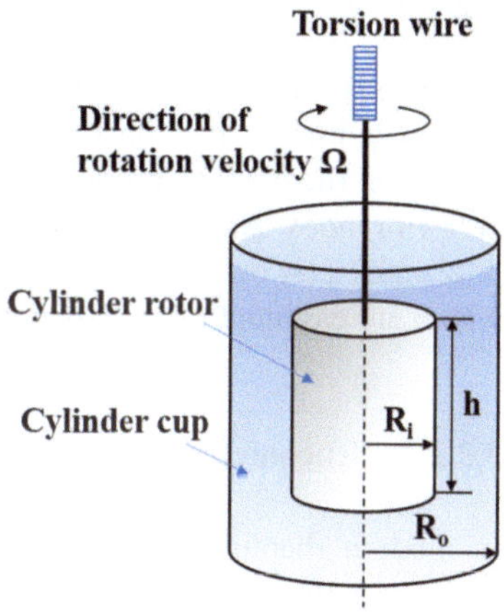

Fig. 3.11 Rotational viscometers

the cylinder rotor rotates at a constant angular velocity (Ω), while the cylinder cup remains stationary. Meanwhile, the torque (M) that is required to maintain this constant angular velocity Ω of the cylinder rotor is measured in the viscometer, and the opposing torque is generated due to the shear stress exerted on the cylinder rotor by the liquid; the force balance is performed as [6]:

$$M = 2\pi rhr\tau = 2\pi hr^2\tau \tag{3.22}$$

Where r is the radial position within the liquid and h is the height of the cylinder rotor. Furthermore, Eq. (3.22) indicates that the shear stress (τ) is not constant across the gap between the two concentric cylinders, but τ increases from the radius of outer cylinder cup R_o to that of the inner cylinder rotor R_i.

For Newtonian fluids, the relationship between the viscosity (η) and angular velocity (Ω) is described by Margules equation [6]:

$$\Omega = \frac{M}{4\pi\eta h}\left(\frac{1}{R_i^2} - \frac{1}{R_o^2}\right) \tag{3.23}$$

Where R_i is the radius of inner cylinder rotor and R_o is the radius of outer cylinder cup.

For non-Newtonian fluids or power law fluids, the equation is expressed as [6]:

$$\Omega = \frac{n}{2k^{1/n}}\left(\frac{M}{2\pi hR_i^2}\right)^{1/n}\left[1 - \left(\frac{R_i}{R_o}\right)^{2/n}\right] \tag{3.24}$$

Where n is the flow behavior index, k is the consistency coefficient ($Pa{\cdot}s^n$).

Rotational viscometers are widely used in laboratories due to their ease of use, cost-effectiveness, and ability to characterize both Newtonian and non-Newtonian fluids. Although they require relatively larger sample volumes, their versatility makes them suitable for a broad range of applications. For example, rotational viscometers have been employed to measure the viscosity of carbon dioxide-expanded fish oil fatty acid ethyl esters, enabling the establishment of rheological models [21]. In such applications, the cylinder cup is filled with the liquid sample, and the cylinder rotor must be fully submerged. This setup allows for the determination of rheological properties such as shear-thickening behavior and the flow behavior index. Additionally, the viscosity data obtained are valuable for developing new processes, optimizing existing processes, and modeling processing equipment, which are essential for enhancing industrial operations [22].

3.2.2.3 Rheometers

In modern rheological measurement technologies, the rheometer stands out as a high-precision, large-scale instrument. It analyzes the flow and deformation of food materials by examining the stress and strain resulting from applied forces. The stress measured by a rheometer is related to torque, similar to rotational viscometers, while strain corresponds to the displacement. It has been reported that a simple, purely mechanical rheometer can be assembled using a syringe and driven by a steel spring, and this setup allows for the measurement of viscosities related to spring displacement for common Newtonian and non-Newtonian liquids, as well as very high-viscosity fluids, with good reproducibility [23]. Ultimately, essential characteristics such as viscosity and modulus are established based on rheological definitions and research conducted using the rheometer.

Furthermore, rheometers enable comprehensive and systematic research into the rheological behaviors of food materials. They surpass the measurement limitations of capillary and rotational viscometers by accurately presenting both Newtonian and non-Newtonian behaviors, including shear thinning, shear thickening, and yield stress. Rheometers are widely applied to a variety of materials, including most food products such as emulsions, food inks, sludge, and other fluids. For example, in the characterization of emulsions, viscosity is a fundamental rheological parameter crucial for understanding the emulsification stage. The viscosity of oil-in-water (O/W) protein-based emulsions is measured using a rheometer with parallel plates based on the power-law model [24]. In the case of sludge, measuring yield stress requires a rheometer equipped with a 40 mm-diameter cone and a 20 mm-diameter plate geometry, along with a cylinder geometry; this configuration allows for various measurements, including transient shear, steady shear, and dynamic oscillatory shear; additionally, stress growth and creep are analyzed during transient shear, and the Herschel-Bulkley model is fitted during steady shear measurements; as a result, parameters such as storage modulus (G'), loss modulus (G''), complex modulus (G^*), and shear stress-shear strain curves are obtained [25]. In the rapidly advancing field of food 3D printing, rheological behaviors are critical for ensuring the smoothness and formability of food inks. Rheometers with stainless steel parallel plates are employed to measure viscosity, perform stress sweeps for viscoelastic properties, and determine yield stress [26]. Studying the rheology of solutions containing complex polymers with chain structures, such as polysaccharides, is essential. Oscillatory rheology (frequency sweep) is a fundamental research method, allowing the recording of parameters like G', G'', and Tan Delta (tan δ) for complex fluids ranging from low-viscosity liquids to stiff solids. Extensional behavior of polysaccharides can also be investigated using capillary-breakup extensional rheometers, illustrating shear and extensional rheology related to intermolecular hydrogen bonds between neighboring chains [27]. To study the dynamic viscoelastic behavior of polymer films, dynamic mechanical analysis (DMA) is employed. DMA provides measurements of G', G'', and the damping factor $\tan \delta = G''/G'$. Additionally, tensile mode isothermal creep analysis based on time-temperature superposition theory reveals that a reduction in G' indicates enhanced polymer chain mobility [28].

Rheometers offer several advantages, including precise rheological data provided by computer-controlled systems and ease of use with a variety of versatile platforms and geometries. These features make rheometers highly valuable for characterizing both Newtonian and non-Newtonian fluids, despite requiring relatively larger sample volumes.

3.3 Rheological Properties of Solid Foods and Its Measurement Technology

3.3.1 Concept of Rheological Properties of Solid Foods

The deformation (strain) of solid foods is induced under the action of an applied force (stress) [1, 6]. Thoroughly discussing stress and strain is crucial for understanding the rheology of solid foods, much like it is for liquid foods. Stress is defined as force per unit area and is generally expressed in pascals (Pa) or newtons per square meter (N/m^2) [1, 6]. Strain refers to the relative change in size or shape of a material when a force is applied. It is calculated as the extension divided by the original length of the material, that is [1, 6]:

$$\text{Strain} = \frac{\text{Extension}}{\text{Original length}} \tag{3.25}$$

Understanding the relationship between stress and strain is fundamental in rheological studies, as it allows for the characterization of how solid food materials respond to applied forces. This knowledge is essential for optimizing processing methods, improving texture, and ensuring the quality and consistency of food products.

3.3.1.1 Deformation of Solid Foods

As materials are subjected to external forces, deformation occurs until fracture happens. For typical elastic solid materials, when an external force is applied, deformation is generated, eventually leading to plastic fracture. Examples of such materials include noodles, rice, fruits, and vegetables [5]. The corresponding relationship between force and deformation is illustrated [1, 9], as shown in Fig. 3.12. Line (OL) represents the linear region, and Point (L) is the elastic limit point. Within the elastic limit range, deformation is proportional to the applied force, with the proportionality constant being the elastic modulus. If the applied force is removed, the deformation of the solid material is reversible [5]. Point (Y) is the bioyield point, defined as the point at which an increase in deformation is observed with a decrease or no change in force, and the force reaches its maximum value [5]. Not all materials exhibit a bioyield point. At the bioyield point, some internal structures of the solid material break, inducing yield and resulting in microstrain. The stress at the bioyield

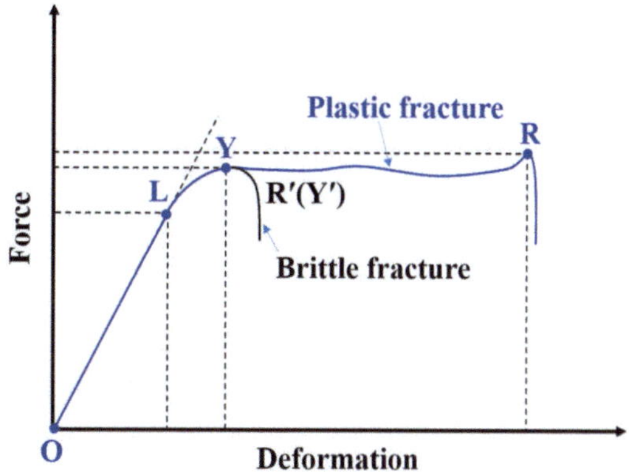

Fig. 3.12 Rheological relationship between force and deformation for solid foods

point is referred to as bioyield strength [1, 5, 6]. Beyond the bioyield point, even though the force remains almost unchanged, deformation continues to increase, a phenomenon known as plastic deformation [9]. Eventually, the deformation cannot sustain the balance with the applied force, leading to fracture at the rupture point (R). The rupture point is located on the stress-strain or force-deformation curve and represents the moment when the axially loaded solid material ruptures under the load [9]. The stress at the rupture point is termed rupture strength or rupture limit, and the rupture point corresponds to macrostrain or failure in the material's macrostructure; in contrast, the bioyield point corresponds to failure in the material's microstructure [5, 6].

Generally, for elastic solid materials, the rupture point (R) occurs well after the plastic deformation stage [9]. However, for brittle solid materials, the rupture point (R′) may appear at the initial stage of the force-deformation curve; therefore, fracture patterns can be categorized into plastic fracture and brittle fracture [1, 9], as illustrated in Fig. 3.12. Plastic fracture occurs after an extended stage of plastic deformation and is characteristic of solid foods such as bread, noodles, rice, fruits, and vegetables [5]. In contrast, brittle fracture occurs when the bioyield point (Y′) and rupture point (R′) appear almost simultaneously [5, 9]. Solid foods that typically exhibit brittle fracture include biscuits, agar gel, chocolate, and peanuts; additionally, some candies may undergo plastic fracture when stretched slowly and brittle fracture when stretched quickly [5, 6].

3.3.1.2 Load-to-Unload Curve of Solid Foods

If the applied force is removed quickly before the fracture of solid foods occurs, a load-to-unload curve is formed [1, 6], as shown in Fig. 3.13; this curve, also known as the stress-strain curve, represents the functional relationship between stress and strain during the loading and unloading phases. Strain is categorized into two types: elastic strain (ε_e) and plastic strain (ε_p). Elastic strain (ε_e) is recoverable during the unloading stage, whereas plastic strain (ε_p) is irreversible and cannot be recovered

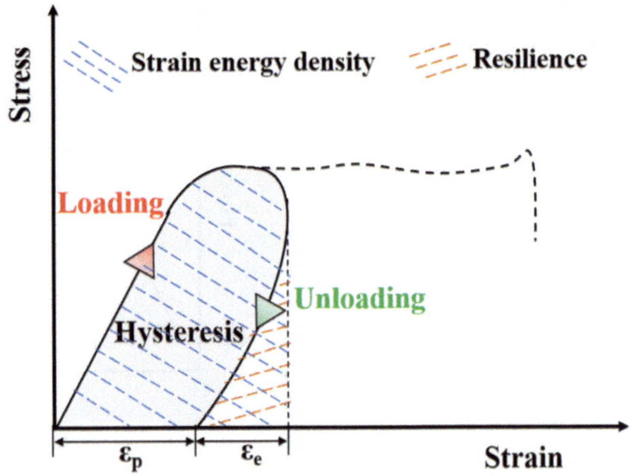

Fig. 3.13 Load-to-unload curve of foods

[6]. The degree of plasticity is defined as the ratio of plastic strain (ε_p) to the total strain, which is the sum of plastic strain (ε_p) and elastic strain (ε_e), when a material is loaded to a certain force and then unloaded; similarly, the degree of elasticity is defined as the ratio of elastic strain to the total strain [6].

For a purely elastic solid, applying stress results in finite deformation, and the material returns to its original position once the stress is removed [6]. Such materials are referred to as Hookean solids due to their adherence to Hooke's law, which describes elastic behavior. Examples of Hookean solids in the food industry include dry pasta, eggshells, and hard candies, which follow Hookean's law when subjected to small strains ($\varepsilon < 0.01$) [6]. However, when subjected to larger strains, these materials may experience brittle fracture or exhibit nonlinear behavior [6]. When energy is applied to a Hookean solid during loading processing and then the processing changed into unloading state before material fractures, it induces deformation stored as potential, fully recoverable energy, which corresponding to elastic strain ε_e; in contrast, for Newtonian fluids and other plastic materials, the applied energy causes the liquid to flow or deformation, dissipating the energy as heat, which means the deformation is not recoverable, which corresponding to plastic strain ε_p, and the total strain is the summation of elastic strain ε_e and plastic strain ε_p [6], as seen in Fig. 3.13. On the stress-strain curve, the strain energy density is represented by the area under the loading curve at a given strain, and resilience is depicted by the area under the unloading curve, indicating the recovered energy per unit volume; a higher resilience signifies that more energy can be recovered from the material [6]. The difference between the strain energy density and resilience is known as hysteresis, and its area represents the energy absorbed during the cyclic loading-unloading stage [6], as illustrated in Fig. 3.13.

Tensile properties are critical for materials during the loading process; however, the unloading path is specific to each material (e.g., semi-crystalline polymers), and the distinctions between loading and unloading paths are not clearly defined. To address this, researchers have developed a nonlinear viscoelastic model—

comprising a Maxwell model connected in parallel with a spring—for deformed semi-crystalline polymers, and this model explains the deformation mechanism through the sliding of amorphous chains at the macromolecular scale [29].

3.3.2 Tensile Measurement Technology

For liquid foods, viscosity is commonly used to evaluate quality, while texture serves as the primary quality indicator for solid foods. Texture assessment is based on the deformation of solid materials, which embodies the study of rheological properties. Depending on the deformation method, various measurement approaches are employed, including compression, tension, and shear. These methods alter the cell structure of solid materials [5], as illustrated in Fig. 3.14. The deformation required to evaluate the quality of solid materials can be measured using a texture analyzer, which is similar to the universal testing machines used in the field of engineering materials.

Texture measurement of solid foods falls under the study of mechanical properties; however, researching the mechanical properties of food materials is more complex than that of engineering materials. This complexity arises because food materials are heterogeneous, and their properties are significantly influenced by external environmental factors. For example, when studying fruits, the rheological properties must be evaluated through compression measurements to prevent mechanical damage during transportation [1]. Nonetheless, several challenges

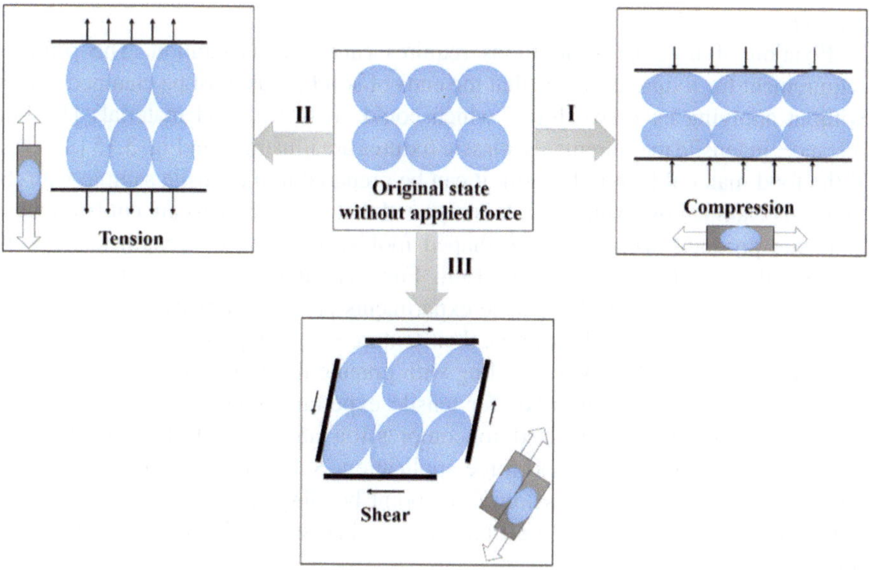

Fig. 3.14 Deformation of cell structure

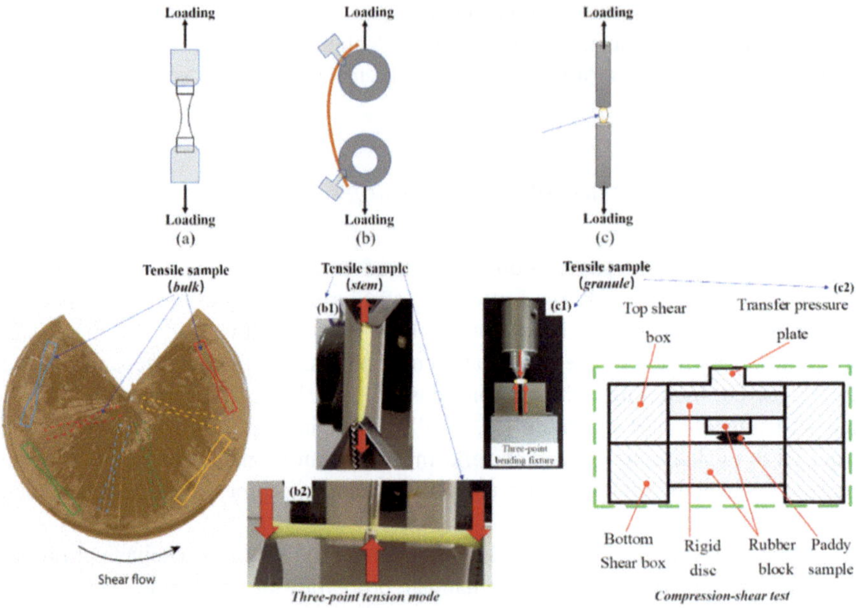

Fig. 3.15 Fixed fixtures of solid foods. (**a**) Tensile and sampling for bulk materials [30], (**b**) Tensile for stem materials [31], (**c**) Tensile for rice [32]

must be addressed first, such as determining whether to use overall sampling or local sampling, accounting for external environmental factors (e.g., moisture content, temperature, variety, storage period), and selecting appropriate fixation methods for solid foods [1].

Fixation of irregular solid foods requires careful consideration. The primary requirement for fixture design is that the ends of solid foods must be clamped firmly without applying excessive force, which could crush the food materials [1]. For instance, in tensile measurements, classic fixtures are illustrated in Fig. 3.15 [30–32]. If the food material is in bulk form, it can be prepared into a tensile sample with its two ends clamped by fixtures [1]. It is reported that the specimens are cut from meat analogue products with dog-bone-shaped molds, and tensile strength analysis is performed by a Texture Analyzer [30]. For stem materials, the sample can be wound onto a ring fixture for tensile experiments [1]. Additionally, by fixing both ends of stem materials with fixtures, their texture can be represented using a three-point tension mode [31]. When dealing with granule materials, the two ends can be glued together with hollow tubes for tensile experiments [1]. In most cases, rice granules are generally subjected to compression and shear forces, resulting in differences in fixture design and force application. Solid mechanics experiments—including quasi-static compression, three-point bending, and compression-shear—are conducted to investigate the breakage mechanism of rice during rubber roll hulling [32].

3.4 Application of Rheological Properties in Food Engineering

3.4.1 *Viscosity and Viscoelasticity of Polysaccharose*

3.4.1.1 Pectin

Rheological properties are important indices during the production and application of polysaccharides, with viscosity serving as an important indicator for the stabilization of pectin. Researchers utilize rheometers equipped with parallel steel plates to measure the viscosity and viscoelasticity of okra pectin. The rheological behavior of okra pectin exhibits shear thinning, meaning that viscosity decreases as the shear strain rate decreases, demonstrating non-Newtonian pseudoplastic behavior [33, 34]. However, different rheological properties of okra pectin arise due to physical treatments applied to the polysaccharide chains [33, 34]. Additionally, factors such as ester groups, methoxyl groups, galacturonic acid content (GalA), molecular weight, and polydispersity significantly influence viscosity. These factors affect the mobility of pectin molecules, enhance the anti-shear ability of pectin solutions, and increase hydrophobic interactions [33, 34]. Okra pectin behaves as a viscoelastic material, exhibiting both solid-like and fluid-like properties. The storage modulus (G') represents the elastic (solid-like) component, while the loss modulus (G'') represents the viscous (fluid-like) component, both of which are related to the stabilization of okra pectin [33, 34]. When the storage modulus (G') and loss modulus (G'') intersect, viscoelastic behavior emerges. The angular frequency at which this intersection occurs indicates the balance between elastic and viscous contributions; a smaller angular frequency at the intersection point signifies a reduced elastic contribution [33, 34].

3.4.1.2 Hydrocolloids

To understand the critical rheological properties of 3D-printed κ-carrageenan-based gummy candies modified with fenugreek gum, several tests are conducted, including temperature ramping (to determine gelation temperature from the crossover point of G' and G''), dynamic viscosity measurements, temperature recovery (to assess the time dependence of G^*), and dynamic viscoelastic property assessments [35]. The studies reveal that the gelation temperature increases with higher concentrations of fenugreek gum, indicating an enhanced sol-gel transition. Additionally, shear thinning behavior is observed, and the addition of fenugreek gum strengthens the gel network. The gelation time decreases as the temperature shifts from printing temperatures (35–55 °C) to room temperature (25 °C) with increasing fenugreek gum concentration. Concurrently, the values of G', G'', and G^* increase, resulting in strengthened mechanical properties. This enhancement is attributed to the correlation between hydrocolloid rheological properties and 3D printing performance [35].

3.4.1.3 Starch

Temperature plays a significant role in affecting viscosity, with viscosity generally decreasing as temperature increases due to changes in molecular structure, in accordance with theories such as Eyring theory and cohesive force theory [6]. The relationship between the viscosity of starch paste and temperature has been analyzed, revealing that as temperature rises from 50 °C to 95 °C, water penetrates the crystalline structure of starch granules alongside cationic cellulose nanocrystals and immobilized α-amylase hydrolysis. This process breaks down intermolecular binding forces, leading to gelatinization. Consequently, viscosity significantly increases due to starch digestion into rich amylose and amylopectin content [36]. When the temperature reaches 120 °C, complete starch gelatinization occurs; however, colloidal viscosity decreases due to coalescence phenomena and the formation of a hydrated starch structure [36].

3.4.2 *3D Printing of Chocolate*

Three-dimensional (3D) printing imposes stringent requirements on the rheological properties of printing inks to ensure they can be extruded and retain their shape effectively. Polysaccharide additives play a crucial role in enhancing these properties by helping the printing ink form a tighter network structure and improving its dynamic modulus. Inks with higher storage modulus (G′) exhibit better mechanical strength, which is essential for maintaining shape fidelity post-printing. However, the thickening effect of polysaccharides often results in high apparent viscosity, which can hinder the extrusion process from the printing nozzle. Therefore, controlling the apparent viscosity is vital during the food processing stage [37].

Chocolate, as a primary material for 3D printing, has garnered significant attention and research focus. The viscosity of chocolate, which is closely related to its fluidity and molding capabilities, is temperature dependent. As temperature increases, chocolate viscosity decreases, enhancing its flowability. To achieve an optimal balance between viscosity and fluidity, various printing parameters must be optimized, including nozzle height, aperture size, moving speed, and extrusion rate [38].

When chocolate is used as a substrate for the 3D printing of drugs, the rheological properties of the drug-loaded chocolate inks are analyzed using a rheometer with parallel plates. The rheological behavior of these chocolate inks mirrors that of traditional inks, exhibiting shear-thinning behavior characterized by decreasing viscosity with increasing shear rate (flow behavior index $n < 1$). Additionally, viscosity decreases with rising temperature, facilitating the extrusion of chocolate inks from the nozzle [39].

3.4.3 Mechanical Properties of Candies

When studying the texture of candies, Texture Profile Analysis (TPA) is a commonly used method that assesses the rheological properties of solid materials. TPA involves setting parameters such as pre-test speed, test speed, post-test speed, strain level, and interval time. Through this method, various TPA indices—including hardness, adhesiveness, springiness, cohesiveness, gumminess, and chewiness—are quantified. These indices are used to evaluate sugar alternatives such as polydextrose, isomalto-oligosaccharides, fructo-oligosaccharides, and xylitol using a texture analyzer equipped with a compression platen probe set to a 75% strain. The application of polydextrose, isomalto-oligosaccharides, and fructo-oligosaccharides has been shown to increase the hardness, adhesiveness, gumminess, and chewiness of gummy candies, while reducing their springiness [40].

In addition to TPA, the fracture behavior of candies provides valuable insights into their mechanical properties. Some candies exhibit brittle fracture when stretched quickly and ductile fracture when stretched slowly. Large strains can induce brittle fracture and nonlinear behavior [6]. The distinction between brittle and ductile fracture lies in the plastic deformation that occurs during ductile processing. Tensile properties of candies are often assessed using dog bone-shaped molds prepared from the candies. Studies have found that the viscosity of the continuous phase and the crystal content may inversely affect the tensile strength of candies [41]. The hardness of jelly candies is another critical mechanical property measured using a texture analyzer with a cylindrical aluminum probe and a penetration distance of 1 mm. Comparisons between Alginate/Pectin jellies and pectin jelly candies reveal that Alginate/Pectin jellies exhibit higher hardness values (16.2 N) compared to pectin jelly candies (6.2 N). This difference is attributed to the denser and more homogeneous network structure of Alginate/Pectin jellies [42].

Stress-Strain Analysis is also employed to evaluate the mechanical properties of candies. Using a texture analyzer with a 2 cm acrylic fixture, candies are uniaxially compressed at a constant displacement rate of 0.2 mm/s until a strain of 50% is achieved. From the resulting stress-strain curves, the compressive strength (peak stress) and elastic modulus (the slope of the stress-strain curve in the linear regime before yielding) can be calculated. These parameters are strongly related to the amount of glycerol in the liquid binder. An increased amount of glycerol reduces both the compressive strength and the elastic modulus of the candies [43].

References

1. Jiang S, Zhao JW. Food physics. Beijing: Chemical Industry Press; 2015.
2. Wang HX, Lin XY, Zhu JC, Yang YX, Qiao SH, Jiao B, Ma L, Zhang YH. Encapsulation of lutein in gelatin type A/B-chitosan systems via tunable chains and bonds from tweens: thermal stability, rheologic property and food 2D/3D printability. Food Res Int. 2023;173:113392.

3. Dahl JF, Gregersen SB, Andersen U, Schulz H-J, Corredig M. Small and large deformation rheology on pizza cheese as an example of application to study anisotropic properties of food soft materials. Food Hydrocoll. 2024;148:109456.
4. Bonilla JC, Sørensen JL, Warming AS, Clausen MP. Quantitative image analysis of protein foam microstructure and its correlation with rheological properties: egg white foam. Food Hydrocoll. 2022;133:108010.
5. Tu K, Zhu WX, Jiang S. Physical properties of foods. Nanjing: Southeast University Press; 2006.
6. Sahin S, Sumnu SG. Physical properties of foods. New York: Springer; 2006.
7. Bean SR, Akin PA, Aramouni FM. Zein functionality in viscoelastic dough for baked food products. J Cereal Sci. 2021;100:103270.
8. Iacovino S, Trivisonno MC, Messia MC, Cuomo F, Lopez F, Marconi E. Combination of empirical and fundamental rheology for the characterization of dough from wheat flours with different extraction rate. Food Hydrocoll. 2024;148:109446.
9. Figura LO, Teixeira AA. Food physics: physical properties-measurement and applications. Springer; 2007.
10. Kießling R, Landgraf R, Scherzer R, Ihlemann J. Introducing the concept of directly connected rheological elements by reviewing rheological models at large strains. Int J Solids Struct. 2016;97-98:650–67.
11. Van Bockstaele F, De Leyn I, Eeckhout M, Dewettinck K. Non-linear creep-recovery measurements as a tool for evaluating the viscoelastic properties of wheat flour dough. J Food Eng. 2011;107:50–9.
12. Zhang L, Yu W, Ma P, Zheng L, Zhang Y. Study on long-term rheological characteristics of polymer gel and prediction of its creep fracture time. J Pet Sci Eng. 2021;201:108445.
13. Noyel J-P, Hajjar A, Debastiani R, Antouly K, Atli A. Impact of viscoelasticity on the stiffness of polymer nanocomposites: insights from experimental and micromechanical model approaches. Polymer. 2024;127443
14. Bird RB, Stewart WE, Lightfoot EN. Transport phenomena. New York: John Wiley & Sons; 1960.
15. Munson BR, Young DF, Okiishi TH. Fundamentals of fluid mechanics. 4th ed. New York: John Wiley & Sons; 2002.
16. Isikli ND, Karababa EA. Rheological characterization of fenugreek paste (cemen). J Food Eng. 2005;69:185–90.
17. White JP, Davidson VJ, Otten L. A capillary viscometer for characterization of fluid foods. Food Res Int. 1993;26:109–13.
18. Thompson JW, Kaiser TJ, Jorgenson JW. Viscosity measurements of methanol–water and acetonitrile–water mixtures at pressures up to 3500bar using a novel capillary time-of-flight viscometer. J Chromatogr A. 2006;1134:201–9.
19. Bramanti E, Ferrari C, Angeli V, Onor M, Synovec RE. Characterization of BSA unfolding and aggregation using a single-capillary viscometer and dynamic surface tension detector. Talanta. 2011;85:2553–61.
20. Kornaeva EP, Stebakov IN, Kornaev AV, Dremin VV, Popov SG, Vinokurov AY. A method to measure non-Newtonian fluids viscosity using inertial viscometer with a computer vision system. Int J Mech Sci. 2023;242:107967.
21. Seifried B, Temelli F. Viscosity and rheological behavior of carbon dioxide-expanded fish oil fatty acid ethyl esters: measurement using a rotational viscometer and modeling. J Supercrit Fluids. 2014;95:519–24.
22. Seifried B, Temelli F. Viscosity and rheological behaviour of carbon dioxide-expanded fish oil triglycerides: measurement and modeling. J Supercrit Fluids. 2011;59:27–35.
23. Solnař S, Štípek J, Alzarzouri F, Žitný R. A simple mechanical spring-driven rheometer for measurement of Newtonian and non-Newtonian fluids—design, validation, and input correction. J Food Eng. 2024;382:112198.

24. Romero A, Felix M, Perez-Puyana V, Choplin L, Guerrero A. Use of a mixer-type rheometer for predicting the stability of O/W protein-based emulsions. LWT-Food Sci Technol. 2017;85:75–81.
25. Liu WH, Zhang H, Sun P, Zeng YP, Gao YY, Wang HF, Zeng RJ. Yield stress measurement of municipal sludge: a comprehensive evaluation of testing methods and concentration effects using a rotational rheometer. Environ Res. 2024;250:118554.
26. Lee CP, Ng MJY, Chian NMY, Hashimoto M. Multi-material direct ink writing 3D food printing using multi-channel nozzle. Future Foods. 2024;10:100376.
27. Bisht A, Goh KKT, Sims IM, Edwards PJB, Matia-Merino L. Does harvesting age matter? Changes in structure and rheology of a shear-thickening polysaccharide from *Cyathea medullaris* as a function of age. Carbohydr Polym. 2024;329:121757.
28. Patra N, Ramesh P, Ramanakumar N. Oleic acid-induced changes in the dynamic viscoelastic behavior of poly(methyl methacrylate) polymer film. Colloids Surf A Physicochem Eng Asp. 2024;701:134889.
29. Lai D, Yakimets I, Guigon M. A non-linear viscoelastic model developed for semi-crystalline polymer deformed at small strains with loading and unloading paths. Mater Sci Eng A. 2005;405:266–71.
30. Schlangen M, Schlangen E, van der Goot AJ. Advanced tensile testing as a new tool to quantify properties of food. Curr Res Food Sci. 2023;7:100577.
31. Tang Z, Zhang B, Wang B, Wang ML, Chen H, Li YM. Breaking paths of rice stalks during threshing. Biosyst Eng. 2021;204:346–57.
32. Chen PY, Jia FG, Zhang JC, Han YL, Li AQ, Wang YL, Fei JM, Shen SH, Feng WY, Hao XZ. Breakage mechanism of brown rice grain during rubber roll hulling. Biosyst Eng. 2023;225:41–53.
33. Zhang L, Hu Y, Wang X, Zhang A, Fakayode OA, Ma HL, Zhou CS. Hybrid techniques of pre and assisted processing modify structural, physicochemical and functional characteristics of okra pectin: controlled-temperature ultrasonic-assisted extraction from preparative dry powders and its field monitoring. Ultrason Sonochem. 2022;88:106080.
34. Xu X, Zhang L, Yagoub AEA, Yu XJ, Ma HL, Zhou CS. Effects of ultrasound, freeze-thaw pretreatments and drying methods on structure and functional properties of pectin during the processing of okra. Food Hydrocoll. 2021;120:106965.
35. Qiu LQ, Zhang M, Ghazal AF, Chu ZY, Luo ZJ. Development of 3D printed k-carrageenan-based gummy candies modified by fenugreek gum: correlating 3D printing performance with sol-gel transition. Int J Biol Macromol. 2024;265:130865.
36. Liu LQ, An XY, Zhang XH, Cheng ZB, Guo JH, Nie SX, Cao HB, Liu HB, Ni YH. Understanding the versatile roles of cellulose nanocrystals in α-amylase immobilization, starch hydrolysis, colloidal stability and printability towards advanced starch based bio-latex performance. Chem Eng J. 2023;460:141808.
37. Zhou L, Meng FB, Li YC, Shi XD, Yang YW, Wang M. Effect of peach gum polysaccharide on the rheological and 3D printing properties of gelatin-based functional gummy candy. Int J Biol Macromol. 2023;253:127186.
38. Mantihal S, Prakash S, Godoi FC, Bhandari B. Optimization of chocolate 3D printing by correlating thermal and flow properties with 3D structure modeling. Innovative Food Sci Emerg Technol. 2017;44:21–9.
39. Chachlioutaki K, Karavasili C, Mavrokefalou E-E, Gioumouxouzis CI, Ritzoulis C, Fatouros DG. Quality control evaluation of paediatric chocolate-based dosage forms: 3D printing vs mold-casting method. Int J Pharm. 2022;624:121991.
40. Maringka CT, Putra ABN, Lo D. Development of gummy candy with polydextrose, isomalto-oligosaccharides, fructo-oligosaccharides, and xylitol as sugar replacers. Int J Gastron Food Sci. 2024;35:100881.
41. Miller E, Hartel RW. Sucrose crystallization in caramel. J Food Eng. 2015;153:28–38.

42. de Avelar MHM, Efraim P. Alginate/pectin cold-set gelation as a potential sustainable method for jelly candy production. LWT-Food Sci Technol. 2020;123:109119.
43. Chadwick E, Barrett AH, Hobson-Rhoades W, Okamoto M, Suleiman Y, Oleksyk LE, Xu H, Shahbazmohamadi S, Shetty A, Baker R, Ma AWK. 3D printing confectionaries with tunable mechanical properties. J Food Eng. 2024;361:111736.

Chapter 4
Texture of Food

Hao Lin and Solomon Yao-Say Selorm Adade

4.1 Introduction

The main qualities of food include texture, appearance, flavor, and nutritional value, which is also known as sensory and intrinsic qualities. Sensory quality mainly refers to the color, aroma, taste, shape, and texture of the food while intrinsic quality also implies the nutritional value, including carbohydrates, fats, proteins, vitamins, minerals and other nutritional quality and quantity. Among them, texture is considered to be the most important food quality. Therefore, in this section, food texture is introduced in detail, including the concept of food texture, the research methods of food texture evaluation, and the practical application of food texture in the food industry.

4.1.1 Basic Concept of Food Textures

Food is not just a source of nutrients but also a means to deliver happiness and satisfaction, making it essential for manufacturers to continuously enhance palatability in product development. Palatability is fundamentally determined by human sensory perception, which involves the five senses—texture, flavor, appearance, sound, and temperature. Texture and flavor are widely recognized as the most influential factors [1]. Texture is vital not only for enhancing enjoyment but also

H. Lin (✉)
School of Food and Biological Engineering, Jiangsu University, Zhenjiang, Jiangsu, China
e-mail: linhao@ujs.edu.cn

S. Y.-S. S. Adade
Food Science and Biological Engineering, Jimei University, Xiamen, Fujian, China

H. Lin et al., *Advanced Technologies for Physical Properties Measurement of Food and its Processing*, https://doi.org/10.1007/978-981-96-2347-1_4

for ensuring safety in consumption, especially in the context of an aging population where swallowing difficulties, or dysphagia, are increasingly prevalent.

Texture encompasses the mechanical, geometrical, and thermal attributes of food that individuals perceive during chewing and swallowing. It involves sensory feedback from the mouth and throat, including interactions with teeth, the tongue, and the hard and soft palate [2]. Despite being a human sensory experience rather than something that can be directly measured with instruments, food science often focuses on correlating human texture perception with structural, rheological, and mechanical properties assessed through instrumental methods. This necessitates using quantitative and reproducible instrumental measurements of "texture-related properties" to approximate human sensory feedback. Texture is not defined by a single characteristic but rather a blend of multiple attributes. Researchers across various fields have aimed to establish consistent terminology and methodologies for objective, reproducible assessments of texture that can be applied broadly to different foods. As part of these efforts, textural properties are often categorized into mechanical, geometric, and surface-related (especially fat) characteristics [3].

The role of texture in influencing the overall acceptance of food varies significantly based on food type. Key aspects of food texture include (1) its derivation from the physical structure of the food; (2) its classification under mechanical or rheological properties; (3) its composite nature, involving a range of attributes; (4) its perception mainly through tactile sensation in the mouth, though other body parts can also contribute; (5) its distinction from chemical senses like taste and smell; and (6) its capacity for objective measurement through factors such as mass, distance, and time [4].

4.1.2 Mechanical Properties for Food Texture Measurement

According to the national standard GB/T10221 "Sensory Analysis Terminology" (which is equivalent to ISO5492 of the International Organization for Standardization), texture is defined as the mechanical, geometric, and surface properties of a product that are perceived through mechanical and tactile methods, or through the visual and auditory senses under specific conditions. The mechanical properties of texture are related to how the product responds when subjected to pressure and can be categorized into five basic properties: hardness, cohesiveness, viscosity, springiness, and adhesiveness.

Hardness: Hardness refers to the amount of force needed to produce deformation or penetration in a product. In the mouth, it is perceived through the compression of the product between the teeth for solids or between the tongue and palate for semi-solids. Adjectives used to describe varying degrees of hardness include soft (low), as in cream or cheese; firm (medium), as in olives; and hard (high), as in a hard candy.

Cohesiveness: Cohesiveness is a mechanical texture characteristic related to the degree of deformation before material fracture, including fragmentation, chewiness, and stickiness.

Viscosity: Viscosity represents a mechanical texture attribute related to a fluid's resistance to flow. It is linked to the effort required to draw a liquid from a spoon onto the tongue or to spread it. Adjectives used to describe different degrees of viscosity include flowing fluid (low), such as water; thin (medium), such as soy sauce; oily unctuous (moderate), such as secondary separated cream; and sticky viscous (high), such as sweet, condensed milk and honey.

Springiness: Springiness refers to the rapid recovery of deformation in a product and is associated with the degree of restitution of deformed materials after the removal of deformation pressure. Adjectives used to describe different degrees of springiness include plastic (inelastic), such as margarine; malleable (moderate), such as marshmallow; and elastic, springy, or rubbery (high), such as squid or mutton.

Adhesiveness: Adhesiveness is a mechanical texture characteristic related to the force required to move a material attached to the mouth or another substance. Adjectives used to describe different degrees of adhesiveness include sticky (low), such as marshmallow food decoration; sticky (medium), such as cream toffee; and sticky, colloidal, gooey, or gluey (high), such as food garnish for caramel fruit ice cream, cooked glutinous rice, and cassava starch pudding.

4.2 Research Methods for Food Texture Evaluation

Texture is crucial in shaping our overall perception of food quality. The evaluation of texture can be generally divided into two main approaches. The first approach is sensory evaluation, also known as the subjective method, which depends on human senses to assess food texture. The second is the objective method, which uses instruments to quantitatively evaluate food texture. Initially, food texture was primarily perceived through sensory experiences. However, the industry has increasingly turned to instrumental measurements for a more precise understanding and description of texture. While sensory evaluations are important, they rely heavily on skilled judges and can be time-consuming and influenced by numerous variables, leading to inconsistency. On the other hand, analytical instrumental testing provides a more accurate representation of food texture and is becoming more widely adopted in the industry.

Sensory evaluation can be further divided into two approaches: analytical and preference. Analytical sensory evaluation uses human senses as tools for classification and scoring during evaluations, closely aligning with the food's physical and chemical properties. To ensure consistent test results, it is important to standardize criteria, maintain uniform test conditions, and select skilled evaluators.

Preference sensory evaluation, on the other hand, assesses food attributes based on consumer likes and dislikes. This method uses samples to measure human sensory responses and preferences. Since this approach relies entirely on human senses and instruments, it can be challenging to achieve an objective assessment of varied sensory perceptions across evaluators. The assessments do not have rigid criteria for food taste or quality, and results are largely based on spontaneous participant reactions. Factors such as individual eating patterns, personal choices, surroundings, and physiological aspects can also influence these outcomes, often reflecting personal preferences. Despite being subjective, these sensory evaluations provide valuable insights when developing new products or conducting market research, guiding the direction of food development and production.

When examining food texture through instruments, there are two main methods: direct detection evaluation and indirect detection evaluation. Direct detection evaluation measures texture-related parameters directly, providing immediate insights. Indirect detection evaluation, on the other hand, measures physical quantities linked to specific texture attributes.

4.2.1 Sensory Evaluation of Food Texture

The basis of food sensory inspection is rooted in human perceptions. This type of assessment involves studying different food attributes through sensory experiences, which produce objective detection results when analyzed using probability and statistical methods. It is important to recognize that this evaluation process is influenced not only by objective factors such as environmental conditions and sample preparations, but also by subjective factors including the basic conditions and quality of the evaluators. The three scientific foundations of sensory testing are statistics, physiology, and psychology.

4.2.1.1 Sensation Introduction

Sensation refers to how the various characteristics and attributes of objects stimulate our sensory organs, which then transmit signals to the nerve center in the cerebral cortex through nerve pathways. Each attribute produces a distinct sensation, and the combined effect of these sensations leads to an individual's perception and evaluation of an object. Human sensory experience comprises five fundamental senses: vision, hearing, touch, smell, and taste. Additionally, humans can also detect sensations such as temperature, pain, and fatigue. Table 4.1 provides further details on the primary sensations related to food.

Food texture evaluation terms are diverse and vary across different countries and regions due to national, cultural, historical, and regional influences. To ensure precise description and effective communication, it is necessary to standardize sensory analysis terms and establish quantitative scales for each term. Since 1988,

Table 4.1 Main classifications of food sensations

Sensory		Sensory organs	Sensory contents
Vision		Eye	Color, shape, size, shine, motion
Aural		Ear	Sound size, height, bite (brittleness)
Olfactory		Nose	Fragrance
Taste		Tongue	Sour, sweet, bitter, salty
Haptic	Thixotropy	Mouth, teeth, tongue, skin	Smoothness, coarseness, hardness
	Thermoception		Ice, cold, hot, scalding

Table 4.2 Food texture evaluation terms

Sensory analysis			
General	Sensory	Sensory characteristics	Analysis method
Sensory analysis; sensory; organoleptic; sensation; assessor; expert; panel; appetite; olfactometer; taster; attribute; acceptability; preference; aversion; palatability; quality factor	Receptor; stimulus; perception; taste; gustatory; olfactory; touch; smell; vision; sensitivity; intensity; kinaesthesis; sensory adaptation; sensory fatigue; ageusia; anosmia; antagonism	Acid; sour; bitter; salty; sweet; alkaline; astringent; flavor; off-flavor; taint; sapid; tasteless; insipid; bland; neutral; flat; mouthfeel; persistence; aroma; odor; note; hue; texture	Test portion; reference point; control; reference; preference test; difference test; triangular test; paired comparison test; duo-trio test; grading; scoring; scale; ratio scale; ordinal scale; random error; standard illuminants
Texture analysis			
General	Mechanical property	Structure	Taste
Structure; texture	Hardness; cohesiveness; brittleness; chewiness; gumminess; viscosity; springiness; adhesiveness; firm; thick; plastic; sticky; glutinous; crunchy	Smooth; fine; powdery; gritty; coarse; lumpy; granularity; flaky; fibrous; strings; pulpy; cellular; puffed; crystalline; glassy; gelatinous; foamed; spongy	Mouthfeel; body; dry; moist; wet; watery; juicy; oily; greasy; waxy; juicy; oily; greasy; waxy; mealy; slimy; creamy; astringent; hot; cold; cooling; moisture; fatness

China has been actively promulgating and revising a range of national standards for sensory analysis that generally align with the relevant standards of the International Organization for Standardization (ISO). These standards include GB/T10221 (Terms of Sensory Analysis), GB/T16860 (Texture Profile Inspection Methods for Sensory Analysis), and GB/T12313 (Flavor Profile Inspection Methods for Sensory Analysis), which correspond to ISO 5492, ISO 11036, and ISO 6564, respectively. By adopting these standardized terms and methods (refer to Table 4.2), food science professionals and related fields can accurately describe and communicate the sensory attributes of food products, promoting greater consistency and understanding across various contexts and regions.

4.2.1.2 Steps for Evaluating Texture Sensation

Food sensory evaluation involves several key steps: evaluating texture samples, establishing an evaluation scale, selecting and training evaluation personnel, choosing an analysis method, conducting the evaluation experiment, processing data, and interpreting the results.

Sample Selection and Preparation: Depending on the goal of the food quality evaluation, whether it's for product sampling or new product development, determine the number of samples needed, the varieties, and the method for sampling and preparing the samples.

Establishing Evaluation Criteria: Based on the texture and characteristics of the samples to be evaluated, establish a scoring system or descriptive vocabulary as the evaluation criteria.

Evaluator Selection and Training: Evaluators are the main participants in sensory evaluation. Conduct physiological and psychological tests on potential evaluators and screen them according to the purpose of the food texture evaluation. Qualified evaluators should receive training and practice in evaluation techniques.

Selection of Evaluation Methods: Choose appropriate evaluation and analysis methods, such as difference testing, descriptive analysis, or emotional testing. Experiment Preparation: Prepare the necessary site and instruments for texture evaluation, organize the sample numbering, and conduct the evaluation experiment.

Data Processing and Result Analysis: After the evaluation, process the data using techniques such as correlation analysis and variance analysis. Analyze the results qualitatively or quantitatively.

4.3 Evaluation of Food Texture Measurement

Instrumental measurement of food texture involves the use of various instruments and equipment to obtain the physical properties of foodstuffs. These physical signals are then correlated with texture parameters using specific analytical evaluation methods to assess the texture of the food. Texture instruments are capable of testing a broad range of physical characteristics of food products, which can be compared to sensory evaluation parameters. Tests can be conducted in various modes such as compression, tensile, shear, bending, and puncture (see Fig. 4.1).

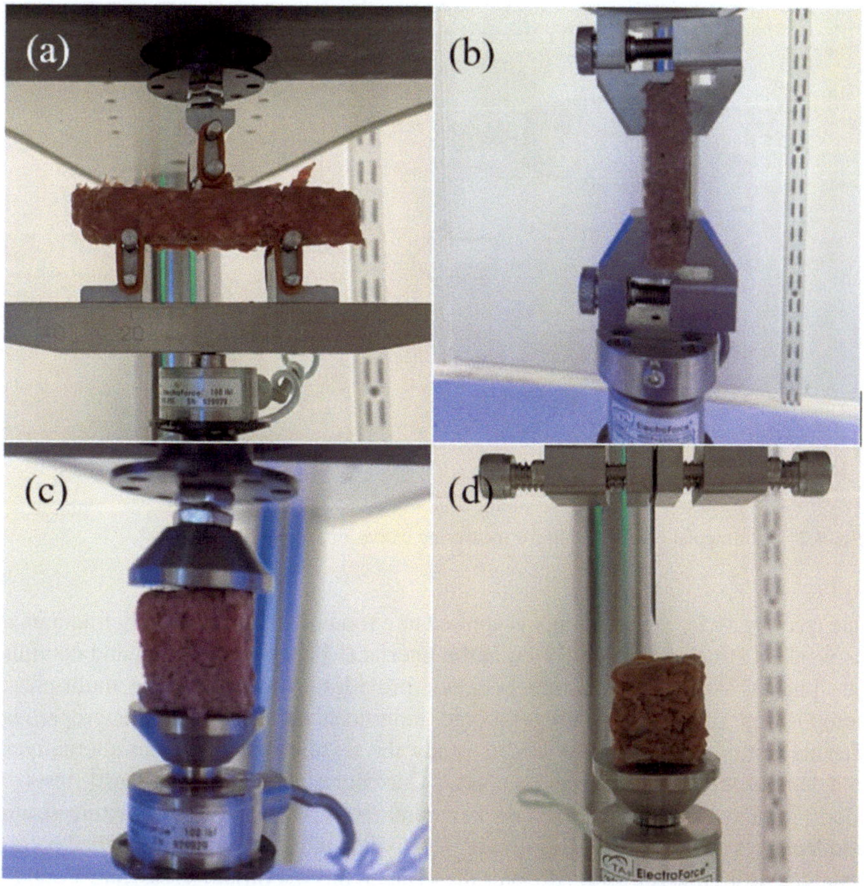

Fig. 4.1 The ElectroForce 3200 Series III experimental setup includes configurations for (**a**) flexion, (**b**) tension, (**c**) compression, and (**d**) cutting [5]

4.3.1 Tensile

Research indicates that extensional flow properties are just as significant as shear flow properties for assessing the texture of semi-liquid or liquid foods during oral processing. Viscoelastic fluids exhibit distinct behaviors under both shear and extensional deformations. For instance, food ingredients containing macromolecules with relatively rigid structures, like xanthan gum, show different flow characteristics based on the type of deformation applied. Extensional flow is a critical factor in both food processing (such as extrusion) and oral processing (like swallowing), suggesting that extensional viscosity has a direct impact on food texture. A tensile test is a mechanical method used to evaluate a material's resistance to tearing. This test involves stretching the material uniaxially between two grips at a constant strain

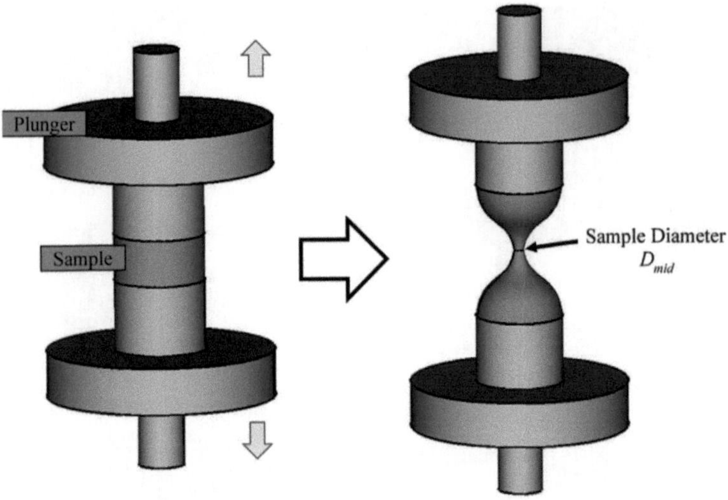

Fig. 4.2 Instrumental configuration for measuring extensional viscosity [3]

rate (see Fig. 4.2). When a force is applied to a food product, it typically fractures at its weakest structural points, often at the interfaces between dispersed and continuous phases. Analyzing fracture behavior provides insights into the multi-phase composition of food, enabling tailored modifications to its fracture properties. Tensile testing is commonly used to study the texture of meat, meat alternatives, and dairy products like mozzarella cheese. Current research involving tensile tests on food products primarily focuses on key parameters: fracture stress, fracture strain, and Young's modulus. Fracture stress refers to the highest stress level reached before a sudden drop, fracture strain denotes the strain at this maximum stress, and Young's modulus represents the slope of the stress-strain curve during the elastic phase.

4.3.2 *Compress*

For soft-textured solid foods, bolus formation involves compressing the food between the tongue and hard palate without the need for chewing. Humans adjust their bolus formation technique based on the food's texture, with hardness being a key factor. Traditional instrumental methods often use rigid geometries, which fail to accurately replicate the breakdown of food via squeezing, leading to notable differences in the fracture properties of foods processed by squeezing versus chewing. This discrepancy may be attributed to the tongue's deformation during eating, which is not accounted for in conventional rigid testing systems.

To address this limitation, an in vitro texture evaluation system has been developed using a simulated tongue made of soft material on a uniaxial compression

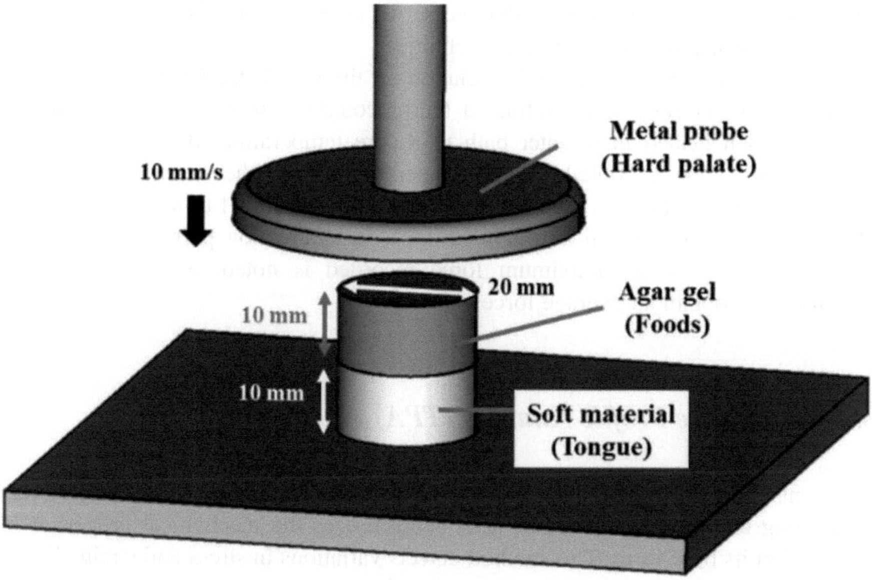

Fig. 4.3 Instrumental simulation of the squeezing action between the human tongue and palate [3]

machine to better mimic the oral squeezing conditions. In this approach, a load is applied to a food sample (e.g., agar gel) by compressing it between a soft tongue-like material and a metal probe that simulates the hard palate (see Fig. 4.3) [6]. If the food sample deforms beyond the level of the simulated tongue, it fractures; however, if its deformation is comparable to or less than that of the simulated tongue, it remains intact. When the apparent elastic modulus of the simulated tongue is around 5.5×10^4 Pa, the fracture or non-fracture of the food sample aligns with the oral strategy humans use for size reduction—either squeezing or chewing. Additionally, research suggests that humans can perceive differences in deformation between food and the tongue within a relatively small strain range (approximately 10%), guiding their oral strategy for breaking down food. This in vitro system is especially beneficial for designing foods tailored to individuals with chewing or swallowing difficulties.

4.3.3 Shear

Tenderness is defined as the force needed to shear through a material. It is primarily assessed using a texture analyzer equipped with a sensor and data collection system, which captures the force applied when a tool slices through a sample. The highest recorded shear force, known as the peak force, is used to quantify the sample's tenderness. The Warner-Bratzler shearometer is the most commonly employed instrument for measuring shear force, as it determines the force necessary to cut through a meat sample of a set thickness. An alternative method for shear force assessment is the slicing shear force technique.

The American Meat Science Association set the initial standards for meat shear measurement. During shear testing, a thermocouple is inserted into the sample, which is then heated in a water bath to a core temperature of 70–72 °C. After heating, the sample is cooled to room temperature using either running water or air cooling. A meat sample is then extracted and aligned parallel to the muscle fibers. The shear force tester cuts through the meat in a direction perpendicular to the muscle fibers, and the maximum force recorded is noted. Multiple trials are conducted to obtain an average force value.

4.3.4 Texture Profile Analysis (TPA)

The uniaxial compression test is one of the simplest methods for objectively measuring texture in solid foods, as it characterizes the mechanical properties of the food in its bulk form. This method detects variations in stress and strain when a sample is compressed vertically without applying lateral pressure. Texture Profile Analysis (TPA) is a widely adopted technique in texture studies that aligns with this approach. Initially developed by a research team at General Foods, the original TPA method involved compressing a food sample, shaped to specific dimensions, twice in a reciprocating motion using a plunger at a constant speed. The plunger's semicircular motion mimics the action of the human jaw during chewing, making TPA suitable for large deformation tests. Bourne later refined this method by incorporating a uniaxial compression device, extending its applications. For accurate measurements, the plunger's surface should match or exceed the sample's surface area, with a recommended deformation of at least 90%. The resulting load (stress) vs. time (strain) graph, known as the TPA curve, allows for the evaluation of key texture parameters such as hardness, brittleness (fracturability), springiness, cohesiveness, and adhesiveness, along with secondary attributes like gumminess and chewiness derived from these primary factors. An example of a TPA curve and the definitions for calculating each parameter can be found in Fig. 4.4.

TPA is extensively used in assessing food texture. In a study by Huidobro et al., TPA was found to be better at predicting hardness in sensory properties compared to WBSF. As two recognized and validated modalities for mechanical assessment of food, TPA can simulate the changes in food as it is being chewed in the mouth, while WBSF simulates the optimal cutting effect. More studies are focusing on the correlation analysis between instrumental measurements (TPA and WBSF) and sensory measurements, as TPA provides more comprehensive quantitative parameters for tenderness prediction and is less costly and time-consuming than sensory evaluation.

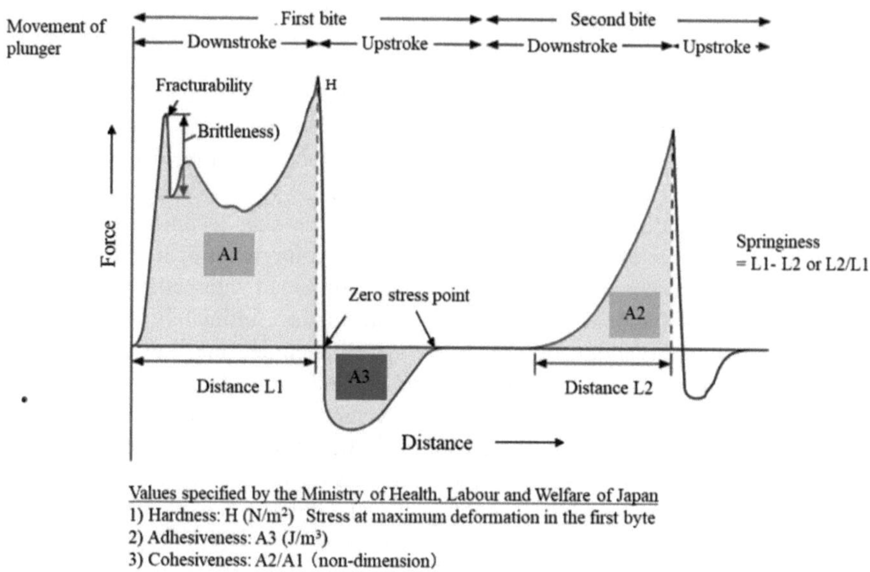

Fig. 4.4 Example of a TPA curve with definitions and calculations for each TPA parameter [3]

4.4 Bionic Sensors Technologies for Food Texture Measurement

Intelligent perception technology is an advanced bionic system that employs sensors to mimic human sensory organs, capturing various sensory attributes of samples. By integrating these sensors with intelligent recognition algorithms for signal processing, this technology enables anthropomorphic, comprehensive evaluations of sample qualities, including food. As a modern extension and enhancement of traditional sensory evaluation methods, this technology offers significant advantages, such as rapid detection, no sample pre-treatment, ease of operation, and real-time on-line analysis.

The main intelligent sensing technologies used in the food sector include E-tongue systems that simulate the sense of taste, the E-nose technology that simulate the sense of smell, the texture analyzer that mimic the sense of touch, and the E-eye as part of bionic vision system. With the continuous advancements in fields like biology, materials science, and computer science, intelligent perception technology is expected to evolve. This evolution will enable it to not only differentiate and identify sensory properties of food but also detect consumer preferences and emotional responses, offering new momentum for innovation and growth in the food industry.

4.4.1 E-Eye in Foods

E-eye technology, initially proposed by Marr in 1982, combines mechanical, optical instruments, electromagnetic sensing, and digital video and image processing technologies to simulate the human eye's function in food detection using computer vision technology. This technology uses optical sensing to collect digital images of the inspected material and then applies image processing techniques to extract feature information related to the food's shape, size, color, texture, and other quality characteristics. The extracted information is then combined with pattern recognition methods to classify and identify the image information, ultimately leading to an overall product quality judgment. The E-eye system includes three key components: an image capture module, a transmission module, and an image processing module. The image capture module utilizes electronic eyepieces and complementary lighting equipment to simulate human eyes, capturing the appearance of the object and obtaining electronic image signals. The transmission module mimics the human optic nerve, responsible for transmitting visual signals. The image processing module corresponds to the human brain's visual center, performing pattern recognition through image recognition algorithms.

E-eye technology offers the advantage of contactless measurement and evaluation, preserving the original condition of the food and minimizing contamination risks associated with traditional methods. With a high-resolution camera and advanced image processing software, it can efficiently handle large volumes of image data with precision and speed. The technology provides objective, repeatable measurement results, eliminating the subjective biases often present in human assessments. This ensures consistent and accurate evaluations of food products. Additionally, the system is flexible, capable of assessing entire products or specific sections, making it suitable for various types and sizes of food production processes. Its ability to store large amounts of image data allows for product traceability by comparison with a database, offering valuable data support for quality management. Figure 4.5 illustrates the use of bionic sensors in food measurements.

4.4.2 E-Nose in Foods

An E-Nose, also referred to as an "odor scanner," is an intelligent bionic perception technology that mimics the human olfactory system. It captures odor characteristics through an array of gas-sensitive sensors and analyzes and processes the collected signals through pattern recognition algorithms to distinguish and classify different odors. The essential components of an E-Nose system include a set of air-sensitive sensors, a system for signal excitation and acquisition, and a pattern recognition algorithm. The gas-sensitive sensor array operates similarly to olfactory receptor cells, detecting chemical odor stimuli. The signal excitation and acquisition system

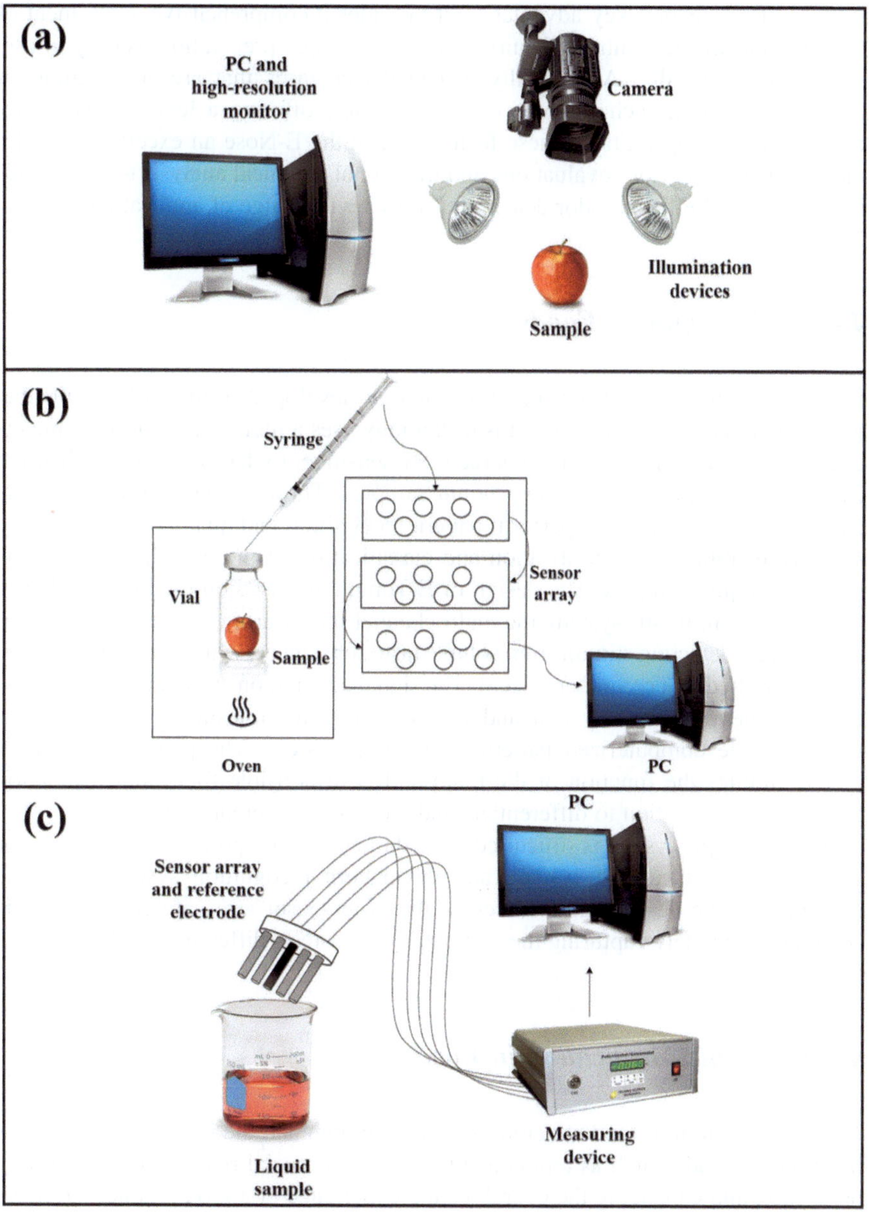

Fig. 4.5 Bionic sensors measurements in foods: (**a**) E-eye; (**b**) E-Nose; (**c**) E-tongue [7]

is analogous to the olfactory bulb, processing the electrical signals measured by the sensors and performing feature extraction.

Pattern recognition algorithms then serve as the "brain," recognizing these features to enable the differentiation and identification of distinct odors. The

E-Nose offers several key advantages: it provides a comprehensive assessment of overall odor profiles without requiring complex sample preparation, is easy to use, and responds rapidly. Additionally, it can detect gases that are challenging for human perception, such as toxic or irritant gases, offering a level of sensitivity beyond human capabilities. These features make the E-Nose an excellent complement to manual sensory evaluations and traditional chemical analysis tools, providing reliable and efficient odor detection across a wide range of applications.

4.4.3 E-Tongue in Foods

E-tongue is an intelligent detection technology developed in the early 1990s that mimics human taste sensations. This technology uses a multi-sensor array, which is low-selective, non-specific, and interactively sensitive, to detect the comprehensive characteristic signals of unknown liquid samples. Through multivariate statistical analysis, the E-Tongue can perform pattern recognition and qualitative/quantitative analysis, enabling the identification and classification of the overall quality information in liquid samples. The design of E-Tongue is based on the biological taste sensory system. In this system, the multi-channel sensor array acts as taste cells and taste nerves, detecting stimuli from soluble taste substances, which are then collected and transmitted to a computer system. The signal acquisition system is similar to the gustatory neurosensory system and is responsible for transmitting these sensed signals to the computerized pattern recognition system. The pattern recognition system mimics the function of the brain and is responsible for feature extraction and pattern recognition to differentiate and recognize different samples.

The E-Tongue system is structured around three key components: the taste sensor array, the signal capture system, and the pattern recognition system. Of these components, the design and research of the taste sensor array are crucial, as it is central to accurately capturing the complexity of taste in different samples.

4.4.4 Bionic Mastication in Foods

The disintegration of food structure within the mouth and the subsequent release of flavor compounds (such as aroma and taste) play a crucial role in shaping sensory perception and enhancing the overall eating experience of a food product. Over the past decade, extensive research has been conducted on food oral processing to uncover the principles that govern eating and sensory perception. However, this research is often hindered by ethical restrictions and practical challenges associated with studying the mouth. he oral cavity's complex, enclosed space complicates the direct observation of food behavior during consumption. Furthermore, the continuous secretion of saliva, combined with the variability in individual oral physiology, makes it challenging to establish direct correlations between food properties and

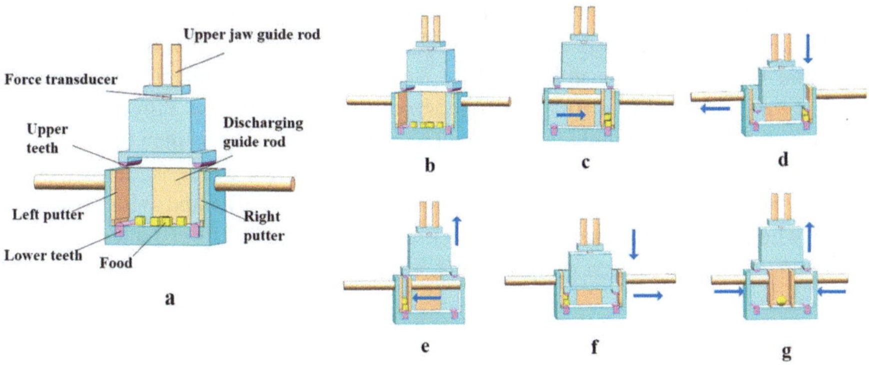

Fig. 4.6 Schematic representation of the simulated mastication process: (**a**) depicts the structure of the mastication unit; (**b**) food is introduced into the oral chamber; (**c**) the left pushing plate directs the food toward the right teeth; (**d**) the left plate returns to its initial position, while the upper jaw compresses the food vertically on the right side; (**e**) the right pushing plate moves food toward the left teeth; (**f**) the right plate returns to its original position, with the upper jaw pressing down on the food located on the left side; (**g**) the combined action of the left and right pushing plates drives the food bolus out of the oral chamber [4]

sensory attributes [8]. To overcome these obstacles, various masticatory devices have been developed to replicate the oral chewing process. These devices enable in vitro analyses of eating and sensory responses, providing valuable insights while avoiding the limitations inherent in in vivo studies (Fig. 4.6).

4.5 Application of Texture Research in Food Engineering

4.5.1 Tenacity of Wet Dough

The process of preparing and consuming food often involves significant deformations, which can make analyzing its mechanical and fracture behavior challenging. However, understanding these properties is essential for product innovation. A variety of mechanical tests, including compression tests, Warner-Bratzler shear tests, and tensile tests, are used to study the structural integrity of food products. Among these, tensile tests are particularly effective for examining the anisotropic structures found in items such as dairy products and plant-based meat alternatives.

A tensile test involves evaluating a material's resistance to tearing. In this test, the sample is secured between two grips and extended along a single axis at a controlled strain rate. When stress is applied, the food sample typically breaks at its weakest sections, which are often located at the interfaces between dispersed and continuous phases. By studying how foods fracture, we can gain insights into their complex multi-phase structures, allowing for the customization of fracture characteristics to meet specific product requirements.

In this section, we use wheat dough as an example of food tensile testing. The dough's moisture level was adjusted to 55% by incorporating distilled water. The flour and water mixture was blended for 4 minutes and subsequently kneaded for 1 minute at 37 °C using a BL15 Dough Mixer (manufactured by Yinying Cooking Machinery Co., Ltd., Shandong, China). The prepared dough was then rested at 37 °C for 30 minutes before testing. The extensional properties were evaluated using a Texture Analyzer (TA.XT plus, Stable Micro Systems, UK) fitted with an A/KIE probe. The testing parameters included a pre-test speed of 2.0 mm/s, a test speed of 3.3 mm/s, a post-test speed of 10.0 mm/s, and a maximum probe travel distance of 60.0 mm.

4.5.2 Anti-bending Capacity of Dried Noodles

Bending strength is an important measure for evaluating the quality of various food items, including cookies, dried noodles, dried rice noodles, vermicelli, and chocolate. For evaluating the bending capacity of elastic and elongated straight food products, the method of flexion after compression bar is more suitable, particularly for assessing the bending capacity of hanging noodles, straight rice noodles, and straight vermicelli. During the test, the specimen is slowly loaded until it fractures, and the fracture load P is determined. The maximum stress at bending fracture σ can be calculated using the following formulas.

$$\text{circular section}: \sigma = \frac{8PL}{\pi D^3} \quad (4.1)$$

$$\text{rectangular section}: \sigma = \frac{3PL}{2ab^2} \quad (4.2)$$

$$\text{rectangular circular section}: \sigma = \frac{8PLD_2}{\pi\left(D_2^4 - D_1^4\right)} \quad (4.3)$$

Where L is the distance between the supports, D is the diameter of the circular specimen, a and b are the width and thickness of the rectangular specimen, and D_1 and D_2 are the inner and outer diameters of the hollow circular cross-section specimen.

In bending tests of food materials, factors such as the loading rate, the effective length L of the specimen, and the shape and size of the support can affect the destructive stress test. The standard guidelines indicate that the ratio between the support seat diameter (d) and the specimen's effective length (L) should be approximately 1%, while the ratio of deflection (Y) to effective length (L) should fall within the range of 5–10%. Table 4.3 presents the bending test conditions for several types of noodles.

In this section, the bending resistance of Vermicelli was evaluated using the post-compression bar flexion method. Given the stable buckling deformation behavior

Table 4.3 Bending test conditions of different kinds of noodles

Noodle	L/(mm)	D/(mm)	S/(mm/s)
Macaroni	130	7.5	8
Soba	60	3	8
Naengmyeon	130	5	8
Vermicelli	40	3	8

Table 4.4 Numerical generalization of large deflection theory

$\theta_0/(°)$	a	$F(a, \pi/2)$	$E(a, \pi/2)$	w_{max}/l	P/P_{cr}	$\Delta l/l$
0	0.0000	1.5708	1.5708	0.00000	1.00000	0.0000
5	0.0436	1.5716	1.5700	0.02775	1.00102	0.0020
10	0.0872	1.5738	1.5678	0.05538	1.00383	0.0076
15	0.1305	1.5776	1.5641	0.08274	1.00868	0.0172
20	0.1736	1.5828	1.5589	0.10971	1.01534	0.0302
25	0.2164	1.5898	1.5522	0.13615	1.02428	0.0472
30	0.2588	1.5981	1.5442	0.16195	1.03507	0.0675
35	0.3007	1.6083	1.5347	0.18697	1.04832	0.0916
40	0.3420	1.6200	1.5238	0.21112	1.06363	0.1188

after compression, a one-to-one correspondence exists between the deformation parameters. In engineering mechanics, the bending fracture of Vermicelli is considered akin to a slender compression rod hinged at both ends. By applying the theory of post-buckling deflection of the compression rod, the relationship between the post-buckling deformation parameters of the hanging surface was established. The key parameters for the post-buckling deformation of Vermicelli are as follows. The relationships among the end turn angle θ_0, end axial displacement Δl, midpoint deflection w_{max}, and end axial pressure P are given by:

$$\Delta l: \frac{\Delta l}{l} = 2\left[1 - \frac{E\left(a, \frac{\pi}{2}\right)}{F\left(a, \frac{\pi}{2}\right)}\right] \tag{4.4}$$

$$w_{max}: \frac{\Delta l}{l} = 2\left[1 - \frac{E\left(a, \frac{\pi}{2}\right)}{F\left(a, \frac{\pi}{2}\right)}\right] \tag{4.5}$$

$$P: \frac{P}{P_{cr}} = \frac{4}{\pi^2}\left[F\left(a, \frac{2}{\pi}\right)\right]^2 \tag{4.6}$$

Where, Δl is the end axial displacement, mm; l is the length of straight food, mm; w_{max} is the midpoint deflection, mm; P is the end axial pressure, N; P_{cr} is the end axial critical pressure; $F\ (a, \pi/2)$ represents the complete elliptic integral of the first kind; $E\ (a, \pi/2)$ is the complete elliptic integrals of the second category; $a = \sin\theta_0/2$, θ_0 is the end angle.

According to Table 4.4, there is a one-to-one correspondence between the end turning angle θ_0 and the end axial displacement Δl. Therefore, when assessing the bending resistance of straight food products, it can be directly evaluated by either the end axial displacement Δl or the end turning angle θ_0. For Vermicelli specifically, the standard LS/T3212-1992 stipulates that if the thickness is less than 0.9 mm and the length is 180 mm, and the end turning angle θ_0 reaches 30° without breakage, the Vermicelli are deemed qualified. In terms of end axial displacement, a Δl corresponding to an end turning angle of 30° is 12.14 mm. Thus, if the end axial displacement reaches 12.14 mm, the Vermicelli are considered qualified. Table 4.4 shows the numerical generalization of large deflection theory.

4.5.3 TPA of Gel

Gel is an essential form of matter in foods, comparable to juices, soy sauce, buttermilk, oils, other liquid items, and products like cookies, shortbread, hard candy, and various solid foods. Most food items are consumed in the gel state, making the texture of gel foods a critical factor in determining food quality. Additionally, food manufacturing frequently utilizes colloidal additives such as gum, agar, gelatin, gum arabic, seaweed gum, starch, soy protein, among others. The gel properties of these additives greatly influence the overall product quality. One of the key indices for characterizing gel performance is gel strength. This is measured by inserting a 12.7 mm diameter cylindrical probe 4 mm below the surface of a jelly containing 6.67% gelatin, and the force applied during this process is recorded as the gel strength. Figure 4.7 depicts the gel strength test device used for this measurement.

Fig. 4.7 Gel strength test device

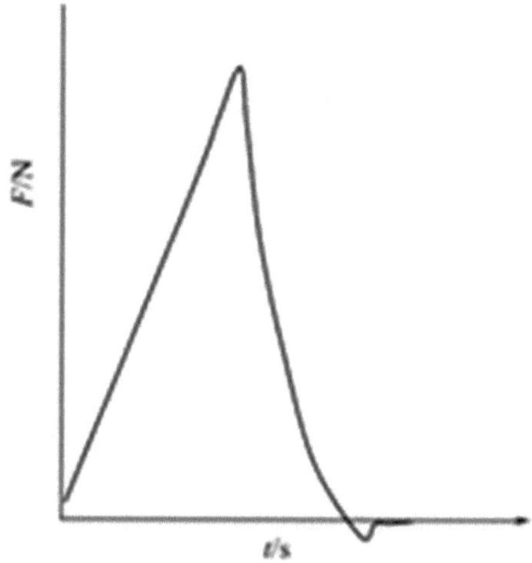

Fig. 4.8 Schematic diagram of gel strength test curve

The specific measurement method is as follows: Take a precise amount of gelatin and add the specified amount of water at approximately 20 °C. Let it sit for 2 hours to allow the gelatin to absorb water and expand. Afterward, place the mixture in a water bath set at *(65 ± 1)°C* for 15 minutes to dissolve it into a uniform solution. Adjust the resulting solution to a concentration of 6.67% to prepare 150 mL of gel (using a triangular flask). Pour 120 mL of the prepared solution into a standard test jar (150 mL capacity), cover it, and cool it in a constant temperature bath at (10 ± 0.1) °C for 16–18 hours. When the sample is ready, position the test jar under the probe. Configure the texture analyzer with the following settings: a pre-test speed of 1.5 mm/s, a test speed of 1.0 mm/s, and a return speed of 1.0 mm/s. Set a test penetration distance of 4 mm, apply a trigger force of 5 g, and adjust the data acquisition rate to 200 pps. Utilize a column probe (P/0.5R, with a diameter of 12.7 mm). Once initiated, the probe is inserted into the gel at a speed of 1.0 mm/s until it reaches a depth of 4 mm. The force-depth curve is recorded during this insertion process, as illustrated in Fig. 4.8. The value of the force at 4 mm (or 4 seconds) is recorded as the gel strength.

4.5.4 Bionic Sensors in Foods Analysis

Ensuring food quality is essential for promoting social development, yet the complexity of the food supply chain makes it a challenging task. Traditional monitoring methods for food quality are no longer sufficient to meet the growing demands of the industry. However, with the rapid advancement of bionic sensing technology, food quality monitoring has become more efficient and reliable. In the field of food, several intelligent bionic sensing technologies have emerged, including E-tongue that simulate taste perception, E-Nose that mimic olfactory perception, texture

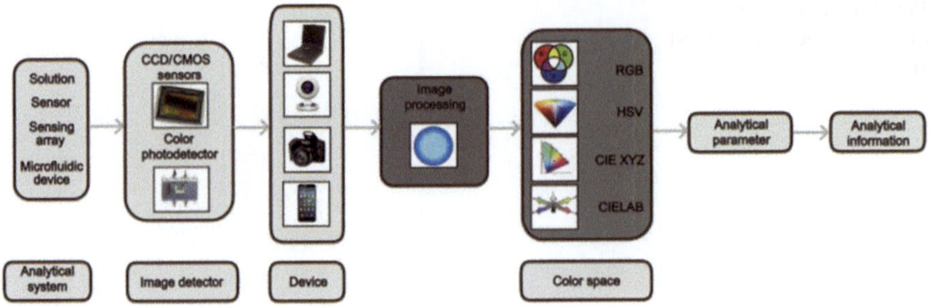

Fig. 4.9 Structure of E-eye system [6]

analyzers that replicate texture perception, and E-eye that simulate visual perception. By emulating human senses such as taste, smell, touch, and vision, these instruments can comprehensively capture various characteristics of food and provide intelligent sensory evaluations of food quality.

E-Eye in Foods Analysis

Recently, image recognition technology using machine vision has been increasingly applied to food classification, defect identification, and quality assessment. For instance, Meenu et al. examined how image processing technology can be used to classify and predict the safety and quality of a range of food products, such as grains, vegetables, fruits, and oils.

In addition, Su et al. applied image recognition methods to detect and classify defects in the appearance of food, achieving an accuracy of 90% in predicting potato quality and 88% in detecting appearance defects [9]. Azarmdel et al. additionally utilized image recognition techniques to extract geometric, color, and texture characteristics from mulberry images. Using feature selection, mulberries were categorized into three groups—mature, immature, and overripe—employing ANN and SVM models, with the ANN model outperforming the others [10]. Figure 4.9 illustrates the structure of the E-eye system.

E-Nose in Foods Analysis

The E-Nose, a method first presented by Persaud and Dodd in 1982, has experienced considerable development over time. It comprises a set of gas sensors linked to automated detection systems aimed at stimulating the human olfactory process (see Fig. 4.10). It uses a combination of sensor arrays that react with chemical substances on the surfaces of sensitive materials, producing distinct responses that are then transformed into digital signals for the identification of "odor" [12]. In contrast to traditional techniques like GC coupled with MS and flame ionization detectors, the E-nose provides several benefits, including lower cost, portability, adaptability, and faster analysis, making it ideal for the quick and precise identification of VOC profiles.

Several studies have utilized the E-nose for food quality assessment. For instance, Cervellieri et al. used multivariate statistical techniques and SVM to analyze gas data from the E-nose system, enabling swift assessment of flour quality [13]. Avian et al.

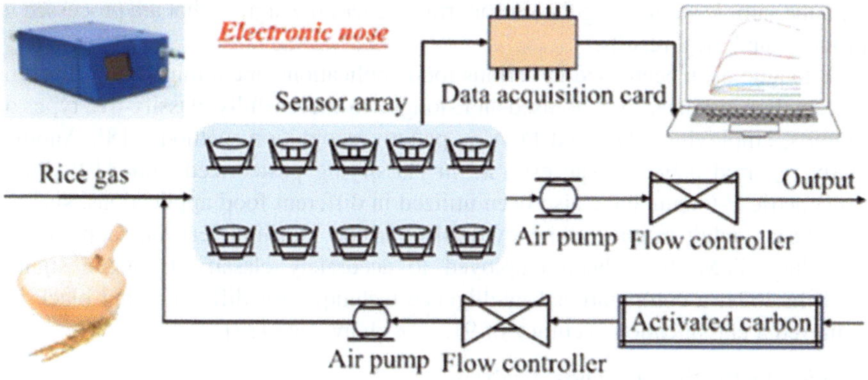

Fig. 4.10 Structure of E-nose system [11]

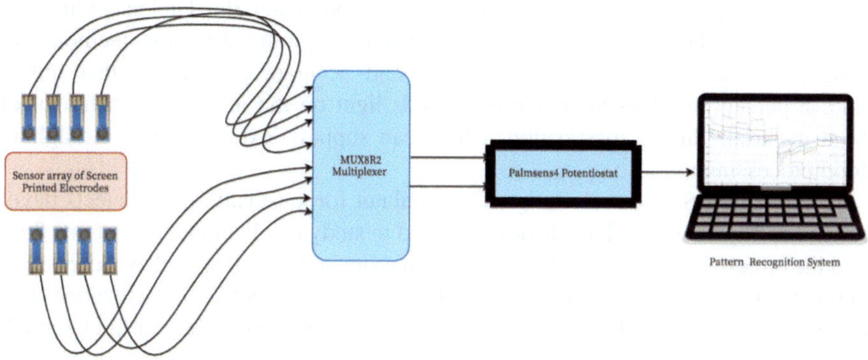

Fig. 4.11 The detection process of E-tongue [17]

applied principal component analysis along with a deep extreme learning machine to identify pork adulteration in beef [14]. Tian et al. used an E-nose system to detect the adulteration of raw milk with vegetable oil, utilizing a RF algorithm for classification [15]. In a separate study, Kang et al. developed a combined filter-wrapper technique with SVM to classify Chinese dry-cured ham into distinct quality levels [16].

E-Tongue in Foods Analysis

The design of the E-tongue mimics the biological mechanism of human taste perception. In humans, the tongue is the organ responsible for detecting taste. Similarly, in an E-tongue system, a sensor array functions analogously to the human tongue, detecting various chemicals and transmitting this information to a computer. This computer performs a function similar to that of the brain in biological systems, analyzing and processing the data via software to differentiate and identify the overall characteristics of diverse substances, ultimately providing sensory information for each. As illustrated in Fig. 4.11, the E-tongue detection process begins with the sample stimulating the sensor array, generating signals that are processed to produce analytical results.

E-tongue have been used in various food applications, including yogurt discrimination. For instance, Li et al. used an E-tongue to successfully classify five types of yogurts, employing LDA and PCA as pattern recognition methods [18]. Another study reported a 98.7% success rate in classifying pasteurized milk [19]. Pulse voltametric E-tongue have also been utilized in different food applications, such as monitoring adulterations in honey [20]. Various machine learning approaches, including SVM, have been employed to accurately classify E-tongue signals [21]. In 2021, a computational intelligence technique for differentiating black tea achieved a classification accuracy of 91.34% on average [22].

Bionic Mastication in Foods Analysis

Chewing is a complex process involving the breakdown and sensory assessment of food, providing rapid and detailed insights into the hardness, elasticity, viscosity, chewability, and other physical attributes of food materials. Understanding the mechanism of human mastication and developing bionic detection methods for food properties enables quick, objective, and accurate evaluations that mimic human perception. This approach also sheds light on the mechanical behavior of food materials during mastication, which can support texture control throughout food processing.

For instance, Salles et al. designed a simulator for mastication that detects flavor release during chewing. This device was used to study the disintegration of pea flour and the effects of different oral conditions on aroma release. It was also successfully applied to investigate how fat influences aroma release in beef [23]. Alemzadeh et al. developed a device that simulates human chewing in a controlled, airtight setting, closely replicating both chewing action and tooth structure. During a chewing gum experiment, the gum and artificial saliva were gathered afterward to measure xylitol content, providing valuable information about flavor release during the chewing process [24]. Recently, Zhou et al. created a six-axis parallel mechanism that mimics jaw motion and recreates human chewing efficiency by modifying oral parameters [25]. Figure 4.12 illustrates the design of the mastication simulator.

Texture is essential in food product development, and having an in-depth understanding of texture is fundamental to leading innovation in this field. In the industry, instrumental quantification of textural attributes is critical as it provides an objective framework for texture design, enabling a standardized approach to product quality. It is the authors' hope that this section offers food manufacturers valuable insights to drive novel product development, ultimately enhancing the QOL and ADL for consumers.

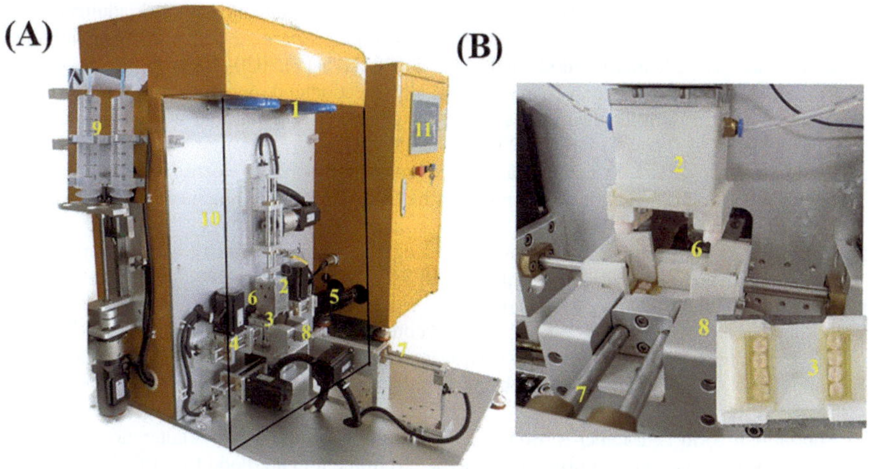

Fig. 4.12 The set-up diagram and enlarged local view of the mastication simulator: (1) heat lamp; (2) palate; (3) oral cavity chamber; (4) left putter; (5) right putter; (6) discharging guide rod; (7) feeding guide rod; (8) food platform; (9) syringe pump; (10) acrylic insulation cover; (11) touch screen operation panel [4]

References

1. Szczesniak AS. Classification of textural characteristicsa. J Food Sci. 2006;28(4):385–9.
2. Kohyama K. Oral sensing of food properties. J Texture Stud. 2014;46(3):138–51.
3. Funami T, Nakauma M. Instrumental food texture evaluation in relation to human perception. Food Hydrocoll. 2022;124
4. Bourne MC. Texture, viscosity, and food; 1982.
5. Souppez J-BRG, Dages BAS, Pavar GS, Fabian J, et al. Mechanical properties and texture profile analysis of beef burgers and plant-based analogues. J Food Eng. 2025;385
6. Bourne MC. Chapter 3 - Physics and texture. In: Bourne MC, editor. Food texture and viscosity (second edition). London: Academic Press; 2002. p. 59–106.
7. Di Rosa AR, Leone F, Cheli F, Chiofalo V. Fusion of electronic nose, electronic tongue and computer vision for animal source food authentication and quality assessment—a review. J Food Eng. 2017;210:62–75.
8. Gal JA, Gallo LM, Palla S, Murray G, et al. Analysis of human mandibular mechanics based on screw theory and in vivo data. J Biomech. 2004;37(9):1405–12.
9. Su Q, Kondo N, Li M, Sun H, et al. Potato quality grading based on machine vision and 3D shape analysis. Comput Electron Agric. 2018;152:261–8.
10. Al-Mallahi A, Kataoka T, Okamoto H, Shibata Y. An image processing algorithm for detecting in-line potato tubers without singulation. Comput Electron Agric. 2010;70(1):239–44.
11. Tang X, Wang N. Electronic nose combines an effective deep learning method to identify the rice quality under different storage conditions and storage periods. Sens Actuators A Phys. 2024:379.
12. Li Z, Song J, Ma Y, Yu Y, et al. Identification of aged-rice adulteration based on near-infrared spectroscopy combined with partial least squares regression and characteristic wavelength variables. Food Chem X. 2023;17:100539.
13. Cervellieri S, Lippolis V, Mancini E, Pascale M, et al. Mass spectrometry-based electronic nose to authenticate 100% Italian durum wheat pasta and characterization of volatile compounds. Food Chem. 2022;383:132548.

14. Avian C, Leu J-S, Prakosa SW, Faisal M. An improved classification of pork adulteration in beef based on electronic nose using modified deep extreme learning with principal component analysis as feature learning. Food Anal Methods. 2022;15(11):3020–31.
15. Tian H, Wu D, Chen B, Yuan H, et al. Rapid identification and quantification of vegetable oil adulteration in raw milk using a flash gas chromatography electronic nose combined with machine learning. Food Control. 2023;150:109758.
16. Qian K, Bao Y, Zhu J, Wang J, et al. Development of a portable electronic nose based on a hybrid filter-wrapper method for identifying the Chinese dry-cured ham of different grades. J Food Eng. 2021;290:110250.
17. Leon-Medina JX, Anaya M, Tibaduiza DA. Yogurt classification using an electronic tongue system and machine learning techniques. Intell Syst Appl. 2022;16
18. Li M, Jin Y, Wang Y, Meng L, et al. Preparation of Bifidobacterium breve encapsulated in low methoxy pectin beads and its effects on yogurt quality. J Dairy Sci. 2019;102(6):4832–43.
19. Carrillo-Gómez JK, Durán Acevedo CM, García-Rico RO. Detection of the bacteria concentration level in pasteurized milk by using two different artificial multisensory methods. Sens Bio-Sens Res. 2021;33:100428.
20. Sobrino-Gregorio L, Bataller R, Soto J, Escriche I. Monitoring honey adulteration with sugar syrups using an automatic pulse voltammetric electronic tongue. Food Control. 2018;91:254–60.
21. Zhang J, Xue Y, Zhang T, Chen Y, et al. Detection of hazardous gas mixtures in the smart kitchen using an electronic nose with support vector machine. J Electrochem Soc. 2020;167(14)
22. Acharya S, Das D, Chatterjee TN, Mukherjee S, et al. Voltametric electrode array optimization for black tea discrimination using computational intelligence approach. IEEE Sens J. 2021;21:20589–95.
23. Zhang X, Chen J, Panda S, Benjamin O. Feasibility analysis of a newly developed multifunctional mastication simulator. Innovative Food Sci Emerg Technol. 2023;87
24. Alemzadeh K, Jones SB, Davies M, West N. Development of a chewing robot with built-in humanoid jaws to simulate mastication to quantify robotic agents release from chewing gums compared to human participants. IEEE Trans Biomed Eng. 2021;68(2):492–504.
25. Zhou X, Yu J. Development and validation of a chewing robot for mimicking human food oral processing and producing food bolus. J Texture Stud. 2022;53

Chapter 5
Optical Characteristics and Color in Foods

Hao Lin

In foods, the appearance is a primary criterion in making purchasing decisions [1]. Appearance is utilized throughout the production-storage-marketing-utilization chain as the primary means of judging the quality of individual units of product [2]. The appearance of unities of products is evaluated by considering their size, shape, form, color, freshness condition, and finally the absence of visual defects [3]. Especially, color is an important sensorial attribute to provide the basic quality information for human perception and has close association with quality factors such as freshness, maturity, variety and desirability, and food safety, and therefore is an important grading factor for most food products [4]. Color is used as an indicator of quality in many applications.

The main raw materials of food come from agricultural products and livestock products. The vast majority of agricultural products grow under natural conditions, and their leaves, stems, stalks, fruits, etc., form their inherent colors under the nurturing of sunlight. These colors are influenced by factors such as irradiation, nutrition, moisture, growth environment, pests and diseases, damage, and maturity level, which may deviate from or change their inherent colors. In other words, people can identify and evaluate the quality characteristics of plants or fruits and their quality characteristics (including internal component content, such as sugar, acidity, starch, protein, etc.) through changes in the color of agricultural products. For example, well-growing rice and wheat are green, and when lacking nutrition or experiencing drought, the plants start to turn yellow from the leaf tips. The larger the yellow area, the more severe the situation. Therefore, distinguishing and calculating the ratio of yellow area to green area becomes an indicator for quantitatively evaluating field management of rice and wheat, which is more scientific than empirical estimation; the color change from unripe to ripe fruits, the formation of

H. Lin (✉)
School of Food and Biological Engineering, Jiangsu University, Zhenjiang, Jiangsu, China
e-mail: linhao@ujs.edu.cn

H. Lin et al., *Advanced Technologies for Physical Properties Measurement of Food and its Processing*, https://doi.org/10.1007/978-981-96-2347-1_5

brown spots on rotten fruits, thus using the color of ripe fruits to judge their maturity, finding ripe fruits from unripe ones or from the branches and leaves of fruit trees, selecting rotten fruits from good ones; the color difference between milled rice and brown rice, the color difference between high-quality rice and rice with white belly, chalky white, immature rice, insect spots, moldy rice, etc., are all quantitative indicators for the degree of rice milling and selecting high-quality rice; distinguishing broken peanut kernels, corn kernels; distinguishing the stems and fruits of green and red peppers, tomatoes, etc., based on the color of food and agricultural products.

The color of the food surface undergoes changes in physical and chemical properties before, during, and after processing. Standardized process control, selected processing raw materials, and specified use of additives all affect the surface color of food. During the storage of food, the surface color also changes, and this change is often closely related to the quality of the food. In other words, the surface color of food is an important indicator reflecting its quality. In Chinese food culture, emphasis is placed on "color, aroma, taste, and shape," with "color" being the first, showing how important the color of food is in people's minds. Color is the first quality attribute of food evaluated by consumers and is therefore an important component of food quality relevant to market acceptance. Rapid and objective measurement of food color is required in quality control for the commercial grading of products.

5.1 Basic Properties of Light

5.1.1 The Nature of Light

Long ago, people studied optical phenomena and recognized the characteristic of light traveling in straight lines. By understanding the properties of light reflection and refraction on the surface of media, they mastered the basic laws of imaging. These all fall within the scope of geometric optics.

Issues regarding the nature and propagation of light have long attracted people's attention. Ancient Greek philosophers proposed the following view: the sun and all other luminous and heat-emitting objects emit tiny particles, which can cause people to perceive light and heat. In the seventeenth century, there were two different theories about the nature of light. One was the particle theory advocated by Newton, which posited that light is a particle emitted from a luminous body and propagates through space at a certain speed, using this viewpoint to explain the laws of reflection and refraction. The other was the wave theory of light advocated by Huygens, which suggested that light is a wave propagating in a medium, although the exact nature of this wave was still unclear. This theory could explain phenomena such as interference, diffraction, and polarization of light. It was not until Maxwell's electromagnetic theory was established in 1860 that light was recognized as an electromagnetic wave, thereby essentially proving the unity of light and

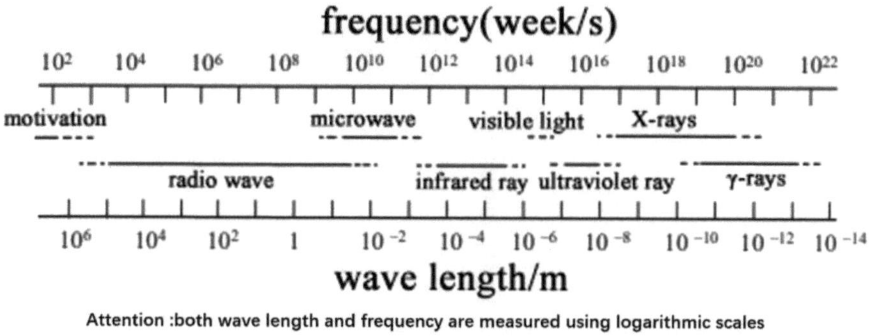

Fig. 5.1 Classification of electromagnetic waves by wavelength and names of each wavelength region

Fig. 5.2 The relationship between color and spectrum

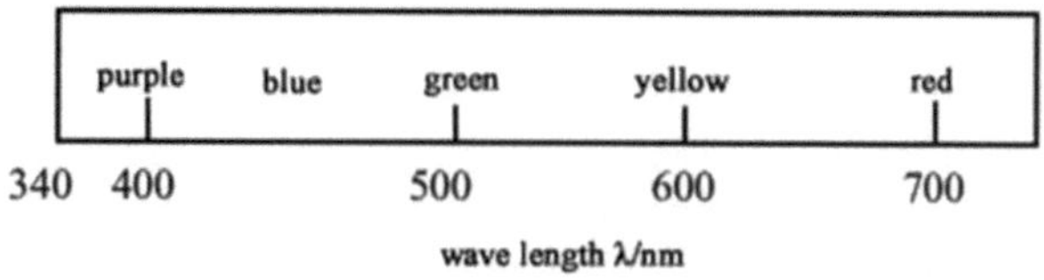

electromagnetic waves and effectively explaining the phenomena of emission, polarization, and propagation of light in anisotropic media.

In the electromagnetic spectrum, electromagnetic waves with wavelengths between 250 and 340 nm are generally referred to as ultraviolet light, those with wavelengths between 380 and 780 nm as visible light, and those with wavelengths between 0.78 and 1000 μm as infrared light. The range from 0.78 to 2.50 μm is called the near-infrared region; 2.5 to 40 μm is the mid-infrared region; and 40 to 1000 μm is the far-infrared region, as shown in the specific division of the electromagnetic spectrum in Fig. 5.1.

5.1.2 *Monochromaticity of Light*

Light with a single wavelength is called monochromatic light. In the visible light region, monochromatic light of different wavelengths is perceived as different colors by the human visual organs. The human eye's sensitivity to visible light of different wavelengths varies, with the sequence from longer wavelengths being red, orange, yellow, green, cyan, blue, and violet. The specific relationship between wavelength and color is shown in Fig. 5.2. Ultraviolet and infrared light, being outside the range detectable by the human eye, are therefore colorless.

Thc human cyc's scnsitivity to visiblc light of the samc powcr radiation varies, and under special circumstances, the visual range can extend to parts of the near-infrared and ultraviolet spectrum. When the eye is exposed to high-energy radiation, the visual range can extend to the ultraviolet end at 312.5 nm and the infrared end at

Fig. 5.3 Spectral luminous efficiency curves of bright and dark vision

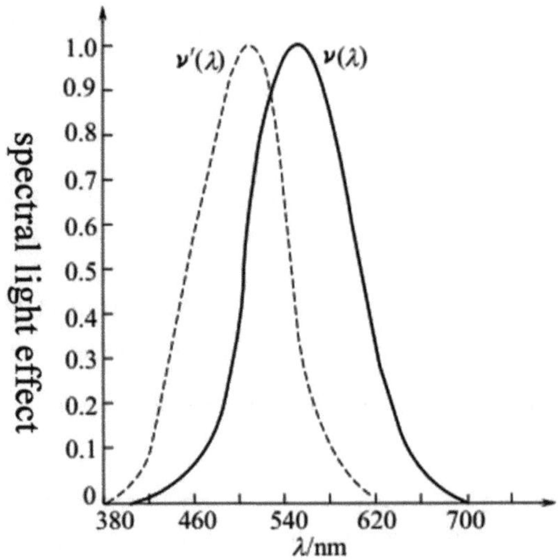

1150 nm. Tests on many people have shown that the visual perception of different wavelengths changes when moving from a bright environment to a dark one [5]. In 1924, the International Commission on Illumination (International Commission on Illumination, abbreviated as CIE) established the photopic vision curve, abbreviated as the photopic vision curve [ν(λ)]. In 1951, the CIE also established the scotopic vision relative brightness curve, abbreviated as the scotopic vision curve [ν'(λ)]. The officially recommended photopic spectral luminous efficiency curve [ν(λ)] and scotopic spectral luminous efficiency curve [ν'(λ)] are two approximately symmetrical smooth bell-shaped curves (Fig. 5.3).

In optical experiments and the application of optical principles, it is often necessary to use monochromatic light of a certain wavelength, which can be separated from composite light spectra. Common separation methods include prisms, plane gratings, interference filters, Michelson interferometers, and acousto-optic tunable methods.

5.1.3 Polarization and Coherence of Light

Light is a transverse wave, with the light vector perpendicular to the direction of light propagation. If the light vector vibrates in a fixed direction within a fixed plane, this light is called linearly polarized light or plane-polarized light (abbreviated as polarized light). The plane formed by the vibration direction and the propagation direction is called the vibration plane, and the plane perpendicular to the vibration direction that contains the propagation direction is called the polarization plane. Light emitted by a molecule at a certain moment is originally polarized, but the light emitted by a large number of molecules or atoms in a light source is intermittent, with one "extinguishing" and another "igniting." During this transition, the light vector

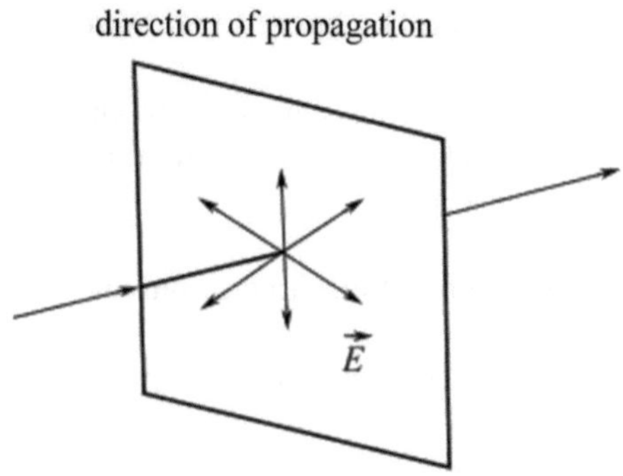

Fig. 5.4 The relationship between light vector and propagation direction

cannot remain in the same direction but instead takes all possible directions in a very rapid and irregular sequence, with no direction being more dominant than others. Therefore, natural light is unpolarized, with equal amplitude in all possible directions, as shown in Fig. 5.4.

Light in space exhibits superposition. Suppose two beams of monochromatic natural light meet in space; their vibrations will be the vector sum of their respective vibrations. However, for overlapping light beams to produce interference phenomena, that is, to form stable alternating bright and dark interference fringes in space, it is much more difficult. This is because, as discussed in physics regarding mechanical waves, coherent waves must meet conditions such as having the same frequency, the same vibration direction, the same phase, or a constant phase difference. However, as mentioned earlier, the state changes of individual molecules or atoms in a light source are not the same, and the emission of light by atoms or molecules is intermittent. Therefore, light waves from two light sources do not meet the coherence conditions, and even using two different parts of the same light source cannot produce coherent light waves. Only light emitted from the same part, after passing through certain devices, can obtain two beams of light that meet the coherence conditions [6].

5.1.4 Light Sources

Devices that emit visible light are called light sources, but in practice, devices that emit ultraviolet and infrared rays are also commonly referred to as light sources. There are many types and forms of light sources. In photoelectric detection technology, the following four types of light sources are frequently seen:

5.1.4.1 Thermal Radiation Light Sources

Light sources that emit light due to an increase in temperature, such as the sun, black bodies, incandescent lamps, and halogen lamps.

5.1.4.2 Gas Discharge Light Sources

Light sources that emit light when current passes through gas (including metal vapor), such as fluorescent lamps, mercury lamps, and xenon lamps.

5.1.4.3 Solid-State Light Sources

Such as electroluminescent lamps and semiconductor light-emitting diodes (LEDs).

5.1.4.4 Lasers

Among which semiconductor lasers (LD) are rapidly developing and are a new type of solid-state quasi-monochromatic light source that will be widely used.

5.1.5 Reflection, Refraction, Scattering, and Absorption Laws of Light

5.1.5.1 Reflection of Light, Law of Reflection

As shown in Fig. 5.5, a beam of parallel light is incident along the direction BA on the vertical surface DE. Part of the light is reflected from the surface along the direction AF; part of the light passes through the surface and changes direction along

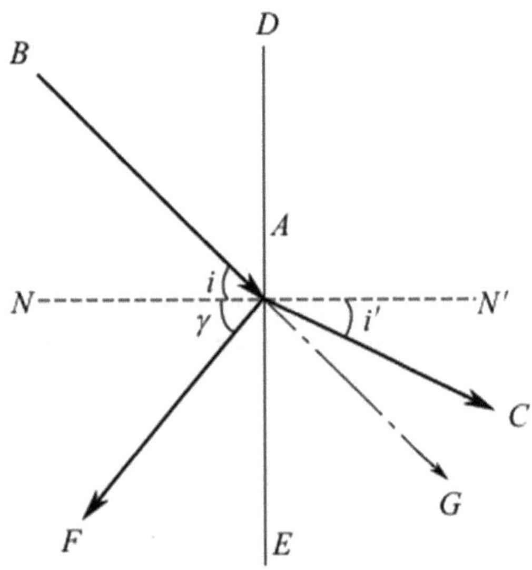

Fig. 5.5 The interaction between light and medium

AC, exhibiting the phenomenon of refraction. The line NN' perpendicular to the surface AD is called the normal [7].

The law of reflection states: The incident ray, the reflected ray, and the normal all lie in the same plane, and the angle of incidence i is equal to the angle of reflection. The ratio of the energy of the reflected light to the energy of the incident light is called the reflectance R:

$$R = E_f / E_0 \tag{5.1}$$

In the formula: R is the reflectance; E_f is the energy of the reflected light; E_0 is the energy of the incident light.

5.1.5.2 Refraction of Light, Law of Refraction

When a beam of parallel light enters another medium from one medium, it changes its direction of propagation, a phenomenon known as refraction. The reason for this phenomenon is that the speed of light varies in different media.

If the speed of light in a vacuum is taken as the standard, the relative speed of light in other media is defined as speed of light in vacuum/speed of light in the medium. This ratio is called the absolute refractive index, usually denoted by n. Since the speed of light in any transparent material is less than in a vacuum, the absolute refractive index of all media is greater than 1, ranging between the refractive index of water (excluding air) at 1.33 and that of diamond at 2.42.

When light enters from one medium into another, such as from air into glass or from glass into water, it is more convenient to express the change in speed as the ratio of the speed of light in the two media. This is the relative refractive index, which equals the ratio of the absolute refractive index of the second medium to that of the first medium. Typically, a medium with a high refractive index is called a "dense" medium, while one with a low refractive index is called a "rare" medium.

The law of refraction states that the incident ray, the refracted ray, and the normal all lie in the same plane, with the incident ray and the refracted ray on opposite sides of the normal. The ratio of the sine of the angle of incidence to the sine of the angle of refraction is given by the following mathematical expression [8]:

$$\frac{\sin i}{\sin i'} = \frac{n'}{n} \tag{5.2}$$

In this formula, n'/n represents the relative refractive index from the first medium to the second medium, and for two media and a specific wavelength of light, this value is constant.

5.1.5.3 Scattering of Light

When light waves strike the surface of a general object (non-optical surface), due to the object's dimensions being much larger than the wavelength of the light waves, diffuse reflection (also known as diffuse scattering) occurs, which is a common phenomenon. When light waves hit small particles, according to Huygens' principle, secondary wavelets are excited from each point on the particle's surface, forming light waves of the same wavelength that spread out in all directions, as shown in Fig. 5.6a. This phenomenon is known as light scattering. The scattering substance does not undergo resonant absorption of the incident light, so this phenomenon is not resonant radiation but rather "reflection" directly from the surface of the illuminated particles. However, it does not comply with the law of reflection, so it is not entirely reflected light. In fact, light scattering is closely related to reflection and diffraction. For example, when light waves enter a turbid medium (a transparent substance containing many suspended particles), because many particles in the medium have dimensions larger than the wavelength and are distributed irregularly, some light waves are scattered. The scattered light waves will bend around the particles and diverge in all directions, similar to single-slit diffraction. However, in a turbid medium, the presence of suspended particles disrupts the optical homogeneity of the medium, presenting as a scattering phenomenon. This type of light scattering is called Tyndall scattering. As shown in Fig. 5.6b, when a few drops of soy milk are added to a glass of clear water, forming a turbid transparent medium, light passing along the X-axis can be seen emitting brightness in the Y-axis direction, which is an example of Tyndall scattering [9].

For instance, certain media that appear uniform and pure on the surface can also exhibit scattering phenomena when light waves pass through them, although the intensity of scattered light is not as strong as that of a turbid medium. This scattering phenomenon is caused by medium molecules whose size is smaller than the wavelength of light, known as molecular scattering or Rayleigh scattering. For example, air molecules in the atmosphere scatter blue light waves from sunlight particularly prominently (as shown in Fig. 5.6c), resulting in a blue sky. It is evident that the intensity of scattered light is inversely proportional to the fourth power of the

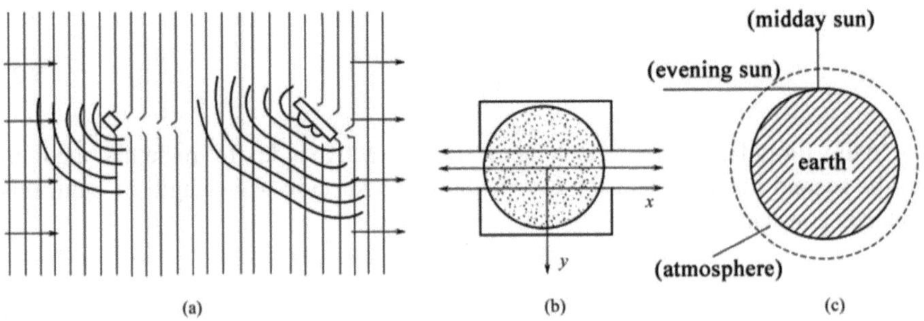

Fig. 5.6 Tyndall scattering phenomenon

wavelength, known as Rayleigh's law. Thus, it can be understood that the scattering effect of short-wave components in white light is more significant, and the larger the wavelength, the less significant the scattering. Therefore, the scattering of blue light by air molecules is particularly prominent, presenting a blue color in the scattering of sunlight.

5.1.5.4 Light Absorption and Absorption Laws

When a beam of light interacts with a medium, in addition to reflection and refraction phenomena, light absorption will also occur. For most media, after light travels a certain distance within the medium, part of the light energy is converted into forms such as thermal energy, electrical energy, or chemical energy. In other words, the light has been absorbed by the material. The mechanism of light absorption involves theories from other fields such as physical optics and biological optics, some of which are not yet fully understood. When explaining the photoelectric effect produced by light absorption, the theory of wave optics cannot provide an explanation, as light exhibits more particle-like properties in this context.

Lambert's energy law provides that when light passes through a distance in a uniform substance, the same proportion of energy will be absorbed by that layer of substance (as shown in Fig. 5.7), mathematically expressed as [9]:

$$dI/I = -\alpha_a \cdot dx \tag{5.3a}$$

$$\text{Assume: } I = I_0 e^{-\alpha_a l} \tag{5.3b}$$

Where: α_a—The light absorption coefficient of the substance
I_0—Incident light energy, J
I—Light energy reaching a distance, J
l—The distance light travels through a substance, mm

Considering the inhomogeneity of the substance and the phenomenon of light reflection, Lambert's law can be rewritten as:

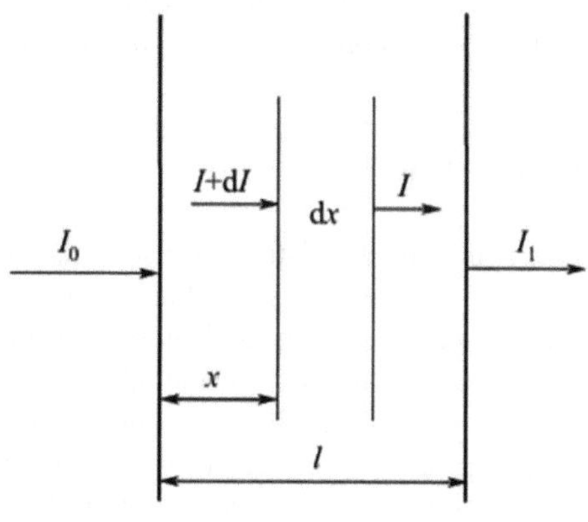

Fig. 5.7 The propagation of light in a uniform medium

$$I = I_0 e^{-(\alpha_a+\alpha_s)\cdot l} = I_0 e^{-\alpha\cdot l} \tag{5.4}$$

α_s is called the scattering coefficient, which is a function of the composition of the object and the wavelength; $\alpha$is called the attenuation coefficient of the object.

5.2 Basic Representation System of Color

5.2.1 Classification and Characteristics of Color

Color can be divided into two major categories: achromatic and chromatic. Color is a general term for achromatic and chromatic. Achromatic colors are white, black, and various shades of gray. They can be arranged into a series, gradually transitioning from white to light gray, to medium gray, then to dark gray, and finally to black. This series is called the white-black series. The change from white to black in the white-black series can be represented by a vertical line, with one end being pure white and the other end being pure black, with various transitional grays in between. Pure white is an ideally completely reflective object, with a light reflectance equal to 1; pure black is an ideally non-reflective object, with a light reflectance equal to zero. In real life, there are no pure white and pure black objects. Magnesium oxide can only approach pure white, and the color of black velvet approaches pure black. The white-black series represents the change in light reflectance of objects, which visually corresponds to the change in brightness. The closer to white, the higher the brightness, and conversely, the closer to black, the lower the brightness.

Chromatic colors refer to all colors outside the white-black series. Chromatic colors have three characteristics: brightness, hue, and saturation.

5.2.1.1 Brightness

The higher the brightness of colored light, the brighter it appears to the human eye, or in other words, it has higher brightness. The higher the light reflectance of the surface of a colored object, the higher its brightness.

5.2.1.2 Hue

Hue is the characteristic that distinguishes one color from another. The radiation of different wavelengths of visible light manifests visually as various hues, such as red, orange, yellow, green, and blue. The hue of a light source is determined by the spectral composition of the radiation and the sensation it produces in the human eye. The hue of an object is determined by the spectral composition of the light source and the proportion of various wavelength radiations reflected (or transmitted) by the object's surface, and the sensation it produces in the human eye.

5.2.1.3 Saturation

Saturation refers to the purity of a color. Various monochromatic lights in the visible spectrum are the most saturated colors. The more white light components mixed into the spectral color, the less saturated it becomes. When a large proportion of white light components are mixed into the spectral color, it no longer appears as a colored light to the eye but becomes white light.

The saturation of an object's color is determined by the degree of selectivity in the object's surface reflection spectrum. If an object has a high reflectance in a narrow band of the spectrum and low or no reflectance at other wavelengths, it indicates a high spectral selectivity, resulting in high color saturation.

Non-colored objects only have differences in brightness, lacking the characteristics of hue and saturation.

5.2.2 Color Equations

In 1854, Grassmann (H. Grassmann) summarized the phenomenon of color mixing into the laws of color mixing.

Human vision can only distinguish three variations of color: brightness, hue, and saturation. In a mixed color composed of two components, if one component changes continuously, the appearance of the mixed color also changes continuously. From these two principles, two laws are derived:

5.2.2.1 Complementary Color Law

Every color has a corresponding complementary color. If a color is mixed with its complementary color in an appropriate ratio, it produces white or gray; if mixed in other ratios, it produces a non-saturated color similar to the predominant color component.

5.2.2.2 Intermediate Color Law

Any two non-complementary colors mixed together produce an intermediate color, with the hue determined by the relative quantities of the two colors, and the saturation determined by their proximity in the hue sequence. Light with the same color appearance, regardless of its spectral composition (i.e., metamerism), has the same effect in color mixing. That is, colors that are visually identical are equivalent. From this law, the substitution law of color is derived.

Substitution Law: Similar colors mixed remain similar. If color A = color B; color C = color D, then:

$$\text{Color A} + \text{color C} = \text{color B} + \text{color D} \tag{5.5}$$

The substitution law indicates that as long as colors are similar in perception, they can replace each other, resulting in the same visual effect. Suppose A + B = C, if B is absent, and X + Y = B, then A + (X + Y) = C. This mixed color produced by substitution has the same visual effect as the original mixed color.

Representing the mixed color as (C), and the primary colors producing the mixed color as (R), (G), (B) for red, green, and blue respectively, and using R, G, B to represent the quantities of the three primary colors (the quantities of the primary colors are called tristimulus values), the color equation can be written as:

$$(\mathrm{C}) = \mathrm{R}(\mathrm{R}) + \mathrm{G}(\mathrm{G}) + \mathrm{B}(\mathrm{B}) \tag{5.6}$$

In the equation, the "=" sign indicates a match, meaning visually equal.

5.2.3 *The Law of Three Primary Colors*

Experiments have shown that various other colors can be mixed using the three basic colors: red, green, and blue. These three basic colors that compose various colors are called the three primary colors. Experiments also prove that these primary colors do not necessarily have to be red, green, and blue; they can be any other three basic colors, as long as any one of the three basic colors cannot be mixed from the other two. However, red, green, and blue are the most convenient for producing other colors, making them the optimal choice for the three primary colors.

5.2.4 *Color Representation Systems*

5.2.4.1 Color Coordinates and Chromaticity Diagram

Quantitative analysis of color falls within the scope of colorimetry. In colorimetry, colors are not directly represented by the quantities of the three primary colors (R, G, B specific values); instead, they are represented by the relative proportions of each primary color in the total amount of R + G + B. The relative proportions of each primary color in the total amount of R + G + B are called chromaticity coordinates. The chromaticity coordinates r, g, b of a specific color are:

$$r = \frac{\mathrm{R}}{\mathrm{R}+\mathrm{G}+\mathrm{B}}, \quad g = \frac{\mathrm{G}}{\mathrm{R}+\mathrm{G}+\mathrm{B}}, \quad b = \frac{\mathrm{B}}{\mathrm{R}+\mathrm{G}+\mathrm{B}} \tag{5.7}$$

Since $r + g + b = 1$, a color can be represented using only r and g. The equation for a specific color (C) can be written as:

$$(C) = r\,(R) + g\,(G) + b\,(B)$$

In color matching experiments, specific white light (such as daylight white light) is used as a standard, and three specific wavelengths of red, green, and blue primary colors are mixed until the primary colors match the standard white light in appropriate proportions. At this point, the luminance values of the primary colors may not be equal, but each primary color's luminance value can be considered as a unit, and the ratio is set to an equal relationship of 1:1:1. In other words, to match the standard white light, the quantities of the three primary colors R, G, B are equal, i.e., $r + g + b = 1$. Substituting the tristimulus values of the standard white light (W) into Eq. (5.7), its chromaticity coordinates are [10]:

$$r = \frac{1}{1+1+1} = 0.333,\quad g = \frac{1}{1+1+1} = 0.333,\quad b = \frac{1}{1+1+1} = 0.333 \quad (5.8)$$

Therefore: $(W) = 0.333$ (R) + 0.333 (G) + 0.333 (B).

To specify a particular color, it can also be located on a chromaticity diagram using chromaticity coordinates. Maxwell (J.C. Maxwell) first proposed using a triangular chromaticity diagram to represent colors, so this chromaticity diagram is called the Maxwell color triangle. This chromaticity diagram is a plane coordinate diagram of a right triangle (Fig. 5.8). The three vertices of the triangle respectively represent the three primary colors (R), (G), (B) (1 unit of red primary color, 1 unit of green primary color, 1 unit of blue primary color), and the chromaticity coordinates r and g respectively represent the relative proportions of R and G in the total amount of R + G + B. There is no b coordinate on the triangular chromaticity diagram because $r + g + b = 1$, so the chromaticity coordinate $b = 1 - (r + g)$. Therefore, it is sufficient to provide only the r and g coordinates. The chromaticity coordinates of the standard white light (W) produced by the equal addition of the three primary colors are $r = 0.333$, $g = 0.333$. The Maxwell color right triangle is now officially adopted internationally as the standard chromaticity diagram [11].

Fig. 5.8 Color triangle

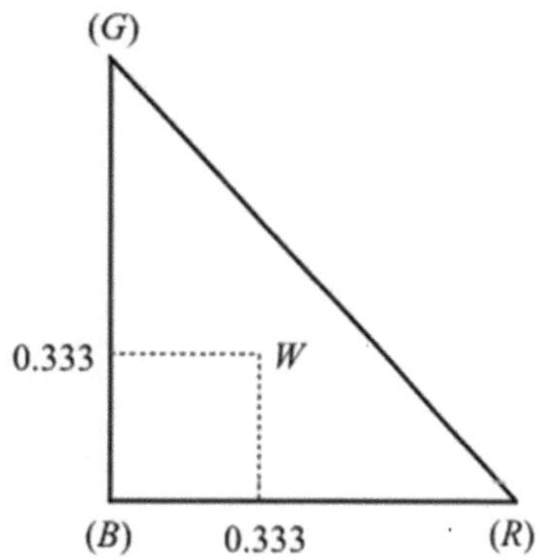

5.2.4.2 CIE Standard Colorimetric System

CIE1931-RGB System

The optical radiation from the external environment acts on the human eye, producing a color sensation. Therefore, the color of an object is determined both by external physical stimuli and by the visual characteristics of the human eye. The measurement and specification of color should conform to the observation results of the human eye. To specify color, it is first necessary to study the color vision characteristics of the human eye. However, the color characteristics of different observers vary to some extent, which requires determining a set of primary color data needed to match an equal-energy spectrum color based on the color vision experiments of many observers, namely the "standard colorimetric observer spectral tristimulus values," to represent the average color vision characteristics of the human eye, used for colorimetric calculations to specify color. In 1931, the International Commission on Illumination (CIE) established the CIE1931 standard colorimetric observer spectral tristimulus values. In 1931, CIE adopted the average results of two people to determine the matching equal-energy spectrum color of $\overline{r}$, $\overline{g}$, $\overline{b}$. Spectral tristimulus values, the function of tristimulus values is called the "CIE1931-RGB System Standard Colorimetric Observer Spectral Tristimulus Values," abbreviated as "CIE1931-RGB System Standard Observer." Figure 5.9 is a chromaticity diagram drawn based on the spectral tristimulus values of the CIE1931-RGB system standard observer.

In the chromaticity diagram, the horseshoe-shaped curve is the spectral locus. It should be noted that a large portion of the *r* coordinates of the spectral locus are

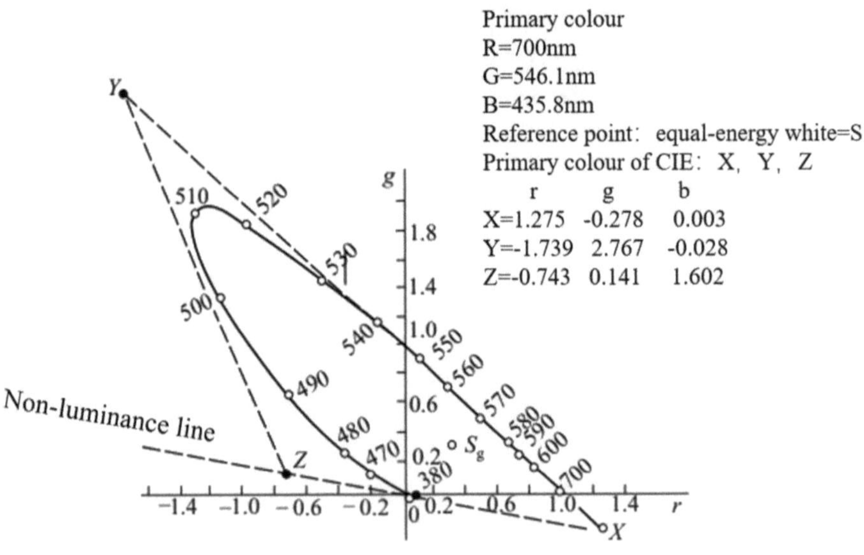

Fig. 5.9 Chromaticity diagram of spectral tristimulus values in CIE1931-RGB system [12]

negative values. The CIE1931-RGB system uses 700 nm, 546.1 nm, and 435.8 nm as the (R), (G), (B) primary colors. The B system stipulates using equal amounts of (R), (G), (B) to match equal-energy white light, with the unit luminance ratio of the (R), (G), (B) primary colors for matching equal-energy white light being 1.000: 4.591:0.060; their radiance ratio is 72.096:1.379:1.000. The CIE1931-RGB system's $\bar{r}$, $\bar{g}$, $\bar{b}$ spectral tristimulus values are derived from experiments and can originally be used for colorimetric calculations and color calibration. However, since the primary colors used to calibrate spectral colors appear as negative values, calculations become extremely inconvenient and difficult to understand. Therefore, in 1931, the CIE discussed and recommended a new internationally accepted colorimetric system—the CIE1931-XYZ system.

CIE1931-XYZ System

In 1931, based on the RGB system, the CIE established a new chromaticity diagram—the CIE1931-XYZ chromaticity diagram—by using three hypothetical primary colors (X), (Y), (Z). The establishment of the CIE1931-XYZ system was mainly based on the following three considerations.

(a) To avoid the negative values in the $\bar{r}$, $\bar{g}$, $\bar{b}$ spectral tristimulus values and chromaticity coordinates in the CIE1931-RGB system, it is necessary to select three primary colors other than (R), (G), (B). The triangle formed by these three primary colors in the chromaticity diagram should encompass the entire spectral locus; that is, these three primary colors must fall outside the spectral locus on the chromaticity diagram, rather than within the range of the spectral locus. This determined the selection of three hypothetical primary colors (X), (Y), (Z) [(X) approximates the primary red, (Y) approximates the primary green, (Z) approximates the primary blue]. The positions of these three new primary colors on the RGB chromaticity diagram are shown in Fig. 5.8. Although they do not exist in reality, the dashed triangle formed by X, Y and Z encompasses the entire spectral locus. Therefore, in this new system, the chromaticity coordinates on and within the spectral locus are all positive values.
(b) The spectral locus from 540 to 700 nm is essentially a straight line on the RGB chromaticity diagram. By mixing two colors on this line, various spectral colors between the two can be obtained. The XY edge of the new XYZ triangle should coincide with this straight line. In this way, the colors on this straight line spectral locus only involve changes in the (X) primary color and (Y) primary color, without involving the (Z) primary color, making calculations convenient.
(c) It is stipulated that the brightness of (X) and (Z) is 0, and the XZ line is called the line of no brightness. Points on the line of no brightness only represent color without brightness, but Y represents both chromaticity and brightness. Thus, when calculating chromaticity using X, Y, and Z, since Y itself also represents brightness, it makes brightness calculation more convenient. For ease of use, the

XYZ triangle is transformed into the Maxwell right triangle, which is the currently internationally used CIE chromaticity diagram. In the CIE1931-XYZ chromaticity diagram, the basic properties and relationships of the RGB system are still maintained. For the spectral stimulus of a certain wavelength λ, the relationship between $r(\lambda)$, $g(\lambda)$, $b(\lambda)$ and $x(\lambda)$, $y(\lambda)$, $z(\lambda)$ chromaticity coordinates is given by:

$$
\begin{aligned}
x(\lambda) &= \frac{0.4900r(\lambda) + 0.3100g(\lambda) + 0.2000b(\lambda)}{0.6669r(\lambda) + 1.1324g(\lambda) + 1.2006b(\lambda)};\\
y(\lambda) &= \frac{0.1769r(\lambda) + 0.8124g(\lambda) + 0.0106b(\lambda)}{0.6669r(\lambda) + 1.1324g(\lambda) + 1.2006b(\lambda)};\\
z(\lambda) &= \frac{0.0000r(\lambda) + 0.0100g(\lambda) + 0.9900b(\lambda)}{0.6669r(\lambda) + 1.1324g(\lambda) + 1.2006B(\lambda)};
\end{aligned}
\tag{5.9}
$$

Using the above formula, calculate the chromaticity points of the same wavelength spectral stimulus in the CIE1931-RGB chromaticity diagram on the CIE1931-XYZ chromaticity diagram, and then connect each point to form the spectral locus of the CIE1931-XYZ chromaticity diagram.

CIE1964-XYZ Supplementary Chromaticity System

Under large field of view observation conditions (>4o), due to the involvement of rod cells and the influence of the central fovea's yellow pigment, color vision undergoes certain changes. This mainly manifests as a decrease in saturation and the appearance of non-uniformity in the color field. Therefore, to suit the chromaticity measurement of a 10o large field of view, in 1964, the CIE specified another set of "CIE1964 Supplementary Standard Chromaticity Observer Spectral Tristimulus Values (Color Matching Functions)" and the corresponding chromaticity diagram. This system is called the "CIE1964 Supplementary Standard Chromaticity System."

The CIE1964 Supplementary Standard Chromaticity Observer Spectral Tristimulus, abbreviated as "CIE1964 Supplementary Standard Observer," has its spectral chromaticity diagram shown in Fig. 5.9.

Comparing the CIE1931 chromaticity diagram with the CIE1964 Supplementary Chromaticity System chromaticity diagram (Fig. 5.10), the spectral loci of both are very similar in shape. A careful comparison reveals that the spectral colors of the same wavelength have considerable differences in their positions on their respective spectral loci. For example, in the range of 490 ~ 500 nm, the approximate coordinate values on the two diagrams differ by more than 5 nm in wavelength. Only at 600 nm do the spectral colors have roughly similar coordinate values. The only overlapping chromaticity point on the 1931 2° field of view and the 1964 10° field of view chromaticity diagrams is the equal energy white point.

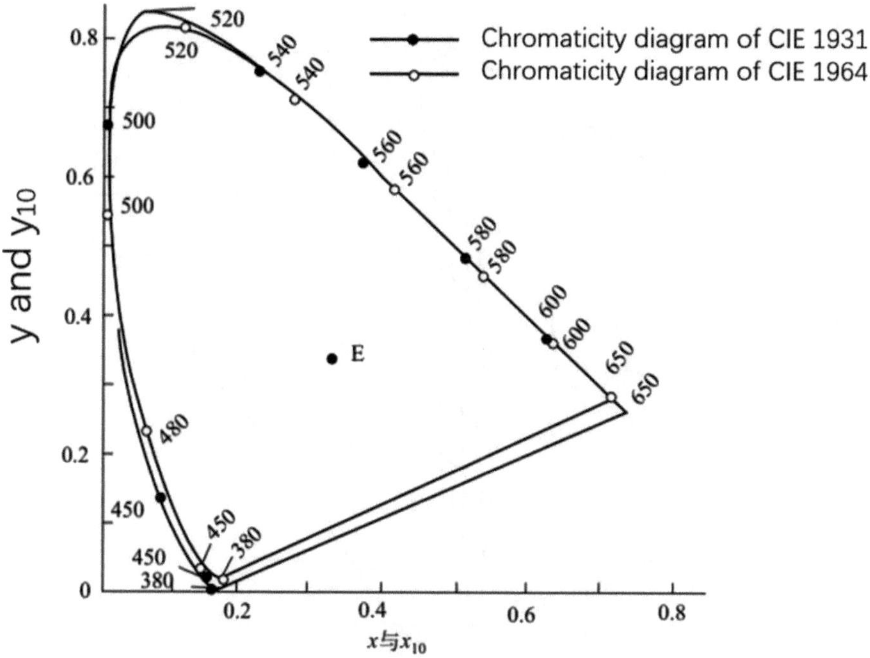

Fig. 5.10 Comparison of CIE1931 and CIE1964 chromaticity diagrams [12]

Munsell Chromaticity System

The Munsell Color System, created by A. H. Munsell, is a method of representing surface colors using a color solid model. It uses a three-dimensional space similar to a spherical model to represent all three basic characteristics of surface colors: hue, value, and chroma. In the solid model, each part represents a specific color and is given a certain label. The Munsell Color System is now widely used internationally as a method for classifying and calibrating surface colors.

The Munsell color solid model resembles a double cone, with its central axis representing achromatic colors, i.e., the value levels of neutral colors. It transitions from black at the bottom to white at the top, divided into 11 perceptually equal gray levels, called Munsell value. The horizontal distance of a specific color from the central axis represents chroma, called Munsell chroma. It indicates the degree to which a color with the same value departs from a neutral color. The chroma of neutral colors on the central axis is 0, and the further away from the central axis, the greater the chroma value. The angle projected horizontally from the central axis represents hue. Figure 5.11 Munsell Color Solid Model Horizontal Section is the horizontal section of the Munsell Color Solid Model, with each central angle representing ten hues. These include five principal hues: Red (R), Yellow (Y), Green (G), Blue (B), and Purple (P), and five intermediate hues: Yellow-Red (YR), Green-Yellow (GY), Blue-Green (BG), Purple-Blue (PB), and Red-Purple

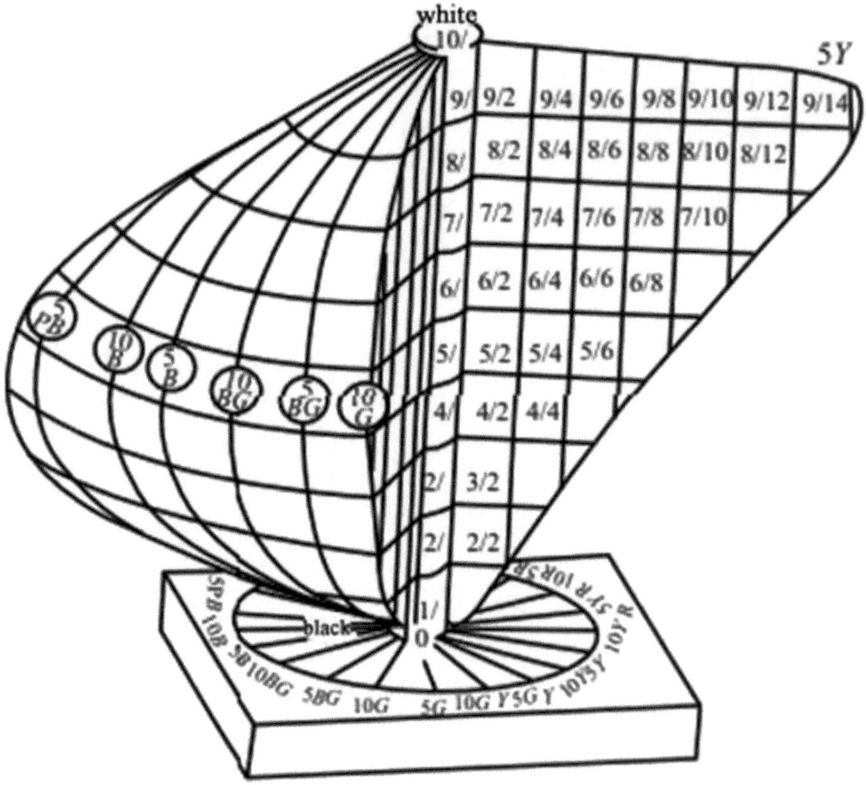

Fig. 5.11 Munsell color system

(RP). Each hue can be divided into 10 levels, with each principal and intermediate hue level set to 5.

The Munsell color solid's horizontal section represents ten basic colors. As shown in Fig. 5.12, it contains five primary colors: Red (R), Yellow (Y), Green (G), Blue (B), and Purple (P) and five secondary colors: Yellow-Red (YR), Green-Yellow (GY), Blue-Green (BG), Purple-Blue (PB), Red-Purple (RP). Based on these 10 main colors, it is further subdivided into 40 colors, and the entire atlas includes 40 color tone samples.

Any color can be specified using the three coordinates of hue, value, and chroma on the color solid, and assigned a code. The method of specification is to first write the hue H, then the value V, and the chroma C after a slash.

$$HV/C = \text{Hue Value/Chroma}$$

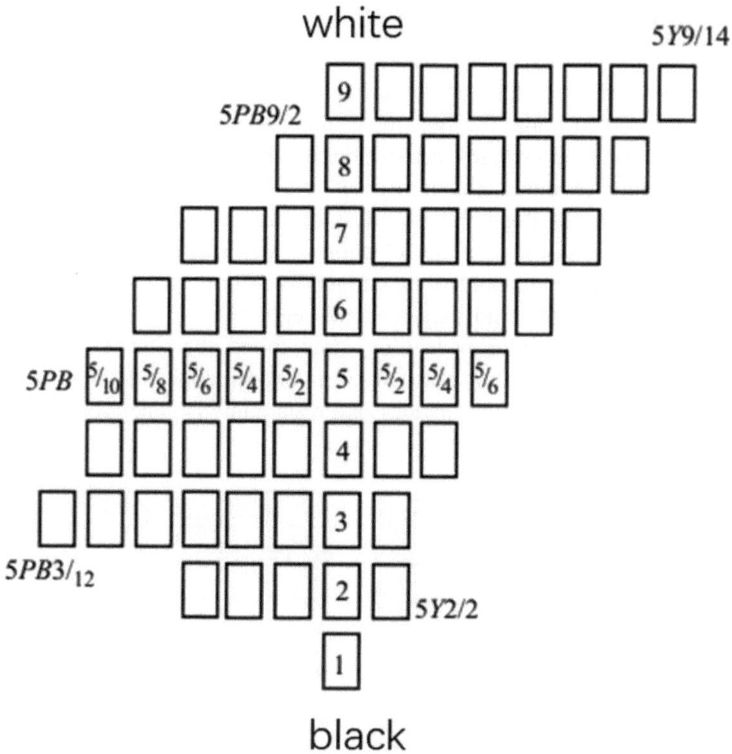

Fig. 5.12 Munsell color stereoscopic Y-PB vertical section

For example, a color coded as 10Y8/12: its hue is the intermediate color between Yellow (Y) and Green-Yellow (GY), the value is 8, and the chroma is 12. This code also indicates that the color is relatively bright and has a high chroma. The code 3YR6/5 indicates: the hue is between Red (R) and Yellow-Red (YR), leaning towards Yellow-Red, with a value of 6 and a chroma of 5.

For the non-colored black and white series (neutral colors), N is used, with the value V indicated after N, and the chroma is not written after the slash.

$$NV/ = \text{Neutral Color Value}/$$

For example, the meaning of the code N5/: a gray color with a value of 5.

The "Munsell Color Atlas" is arranged with the vertical section of the color solid as one page in sequence. The entire solid is divided into 40 vertical sections, with the atlas comprising 40 pages, each page including samples of different values and chromas of the same hue.

5.2.5 Color and Evaluation of Food

The color of food is one of the important indicators in the evaluation of food quality. Color can be used to judge the freshness, maturity, processing precision, variety characteristics, and their changes in food. For example, the freshness of a cake, the processing quality of bread, the purity of edible oil, the oxidation changes in beverages, the grade of flour, the freshness of vegetables, the maturity of fruits, the processing precision of rice, the selection of high-quality rice and chalky rice, immature rice, insect spots, moldy rice, variety distinction, etc. [13–15]. Figure 5.13 shows the changes in color and shape of chili peppers.

Modern research in color psychology has found that human emotions and feelings are highly dependent on color. Red can excite people and increase appetite, so the decoration of some public places like restaurants is generally dominated by

Fig. 5.13 Changes in color and shape of chili peppers

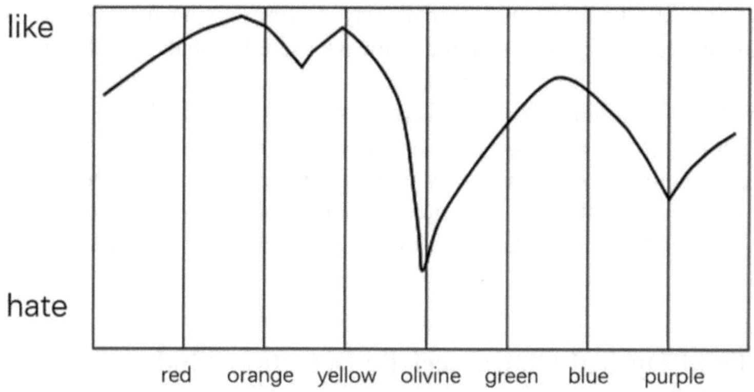

Fig. 5.14 The relationship between food color and appetite

reddish (warm) colors, with the purpose of stimulating diners' emotions, exciting their nerves, and achieving a joyful and harmonious atmosphere, while hospital wards generally use light blue walls, aiming to stabilize the mood of patients and their families, which is conducive to the treatment and recovery of patients.

Although there are certain differences in the understanding of the relationship between color and appetite due to differences in living habits, social traditions, and visual perception between Eastern and Western people, the overall feeling in color psychology is still basically the same. American researcher Birren studied the relationship between food color and appetite, believing that the colors most stimulating to appetite are between red and orange, with a low peak between orange and yellow, and yellow being a high peak again, while yellow-green is a low peak. Yellow-green is not popular in food; cool green and cyan are better; the feeling of purple decreases again, as shown in Fig. 5.14. Colors that can stimulate appetite include red, scarlet, peach, coffee (yellow-brown), milky white, light green, bright green, while blue-green foods are rare. Colors considered unfavorable to appetite are purplish-red, purple, deep purple, yellow-green, yellow-orange, and gray.

When monochromatic light of different wavelengths interacts with a medium, various physical phenomena occur. For example, when the midday sunlight (a composite white light) shines on an object, it imparts a certain color characteristic to the illuminated object. Simultaneously, the sunlight that passes through the object also exhibits a certain color. This color manifestation occurs because the object does not reflect, refract, scatter, and absorb all wavelengths of monochromatic light uniformly. Selective reflection, selective refraction, selective scattering, and selective absorption phenomena often occur. The spectrum of wavelengths that are reflected determines the color of the spectrum displayed. Similarly, in the absorption process of food, the wavelengths of the spectrum that are selectively absorbed result in the absence of those spectral colors. In fact, the color that food exhibits falls under the category of reflection spectrum analysis, while the color displayed after passing through the food falls under the category of transmission spectrum analysis. Therefore, to detect the surface color of food, the reflection detection mode of light is always used, and to detect the color displayed after passing through the food, the transmission detection mode of light is always used. By analyzing the composition of the reflection spectrum or transmission spectrum, and using the method of colorimetry to represent color, the tristimulus values X, Y, and Z can be calculated, along with a series of color representation parameters such as brightness and whiteness, thereby comprehensively evaluating the color characteristics of food. Thus, it can be seen that the detection of food's color characteristics is essentially the detection of the optical characteristics of food within the visible light range [16].

5.3 Food Optical Characteristic Detection Mode and System

5.3.1 Food Optical Characteristics and Detection Mode

5.3.1.1 Optical Characteristics of Food

The optical characteristics of food refer to the properties of food that cause reflection, absorption, transmission, diffusion, or excitation of other wavelengths of light when light is projected onto its surface. Materials are composed of many tiny internal intermediate layers, and different materials have different types and compositions, resulting in varied responses in optical characteristics.

When a beam of light is directed at a material, approximately 4% of the light is directly reflected by the material's surface, while the remaining light enters the surface layer of the material, encountering the internal network structure and scattering in all directions. Most of the scattered light is refracted back to the material's surface and exits near the point of incidence; this reflection is known as "body reflection." A small portion of the scattered light diffuses deeper into the interior, with some being absorbed by the material and some penetrating the fruit. The amount absorbed depends on factors such as the material's properties, the wavelength of the light, and the length of the propagation path. For certain materials, some absorbed light is converted into fluorescence, delayed emission light, etc. Therefore, the light leaving the material's surface consists of the following components: direct reflection light, body reflection light, transmission light, and emission light (Fig. 5.15). When incident light Io acts on the material, it produces [17]:

1. *Body Reflection* $I_{\rho 1}$: This reflection is caused by the inhomogeneity of the material's contents, also known as diffuse reflection, where $I_{\rho 1}$ is the energy of the body reflection light.
2. *Specular Reflection* $I_{\rho 2}$: This reflection is determined by the direction of the incident light, the incident surface, and the law of reflection, where $I_{\rho 2}$ is the energy of the specular reflection light.
3. *Absorption* I_a: This absorption is the selective absorption of light by the internal structure, where I_a is the energy of the absorbed light.

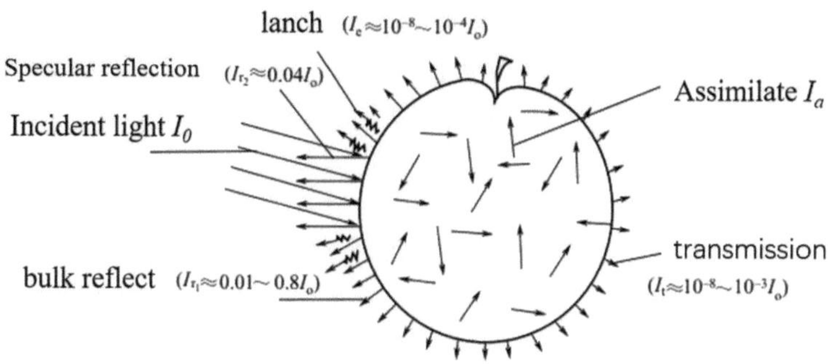

Fig. 5.15 Light-fruit interactions

4. *Transmission* I_τ: This part of the light is the light that passes through the material and exits, where I_τ is the energy of the transmitted light.
5. *Emission* I_e: The light in this section refers to the energy absorbed by the material and then re-emitted as light energy of a specific wavelength, mainly as delayed luminescence (DLE) and fluorescence emission, with I_τ being the emitted light energy.

According to the law of conservation and transformation of energy:

$$I_o = I_{\rho 1} + I_{\rho 2} + I_a + I_\tau + I_e \tag{5.10}$$

The commonly used optical density OD (OD = log (I o /I τ)) is used to represent the ratio of incident light energy to transmitted light energy. The lower the light transmittance of the material, the higher the optical density. For a certain food material, a beam of parallel incident light I_o projected onto the material produces reflected light I_ρ, transmitted light I_τ, absorbed light I_a, and emitted light I_e, whose spectral distribution and intensity are closely related to the properties such as the type and structure of the material, exhibiting strong selectivity. This principle can be used for food quality discrimination.

5.3.1.2 Food Optical Characteristics Detection Modes

The modes for detecting the optical characteristics of food and agricultural products mainly include transmission and reflection detection. Generally, transmission detection places the sample to be tested between the light source and the detector, with the detector measuring the light transmitted through the sample; reflection spectrum detection places the detector and the light source on the same side of the sample to be tested, with the detector measuring the light reflected back by the sample in various ways.

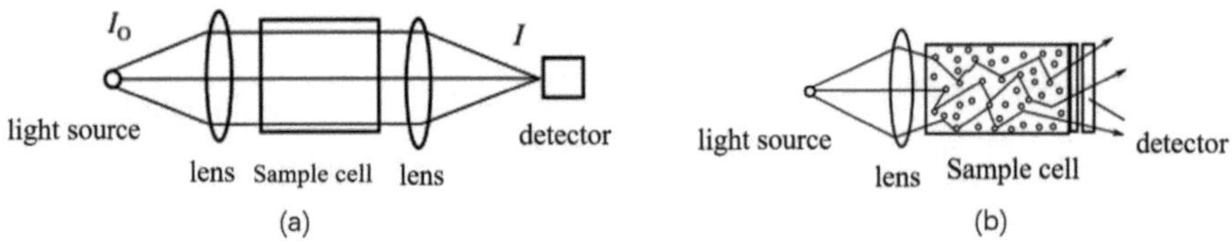

Fig. 5.16 Projection spectrometry: (**a**) transmittance measurement, (**b**) diffuse transmittance measurement

Transmission Mode

The transmission spectrum detection mode, as shown in Fig. 5.16, is mainly suitable for clear and transparent liquid foods, such as wines and beverages [18]. Some relatively homogeneous mixed liquid foods can also use diffuse transmission detection methods to detect their optical characteristics, such as milk and dairy products, pure natural fruit juices, and other liquid foods.

The relationship between the transmitted light energy and the various components (concentration) of liquid foods conforms to the Lambert-Beer law, and the absorption path length needs to be determined based on the basic optical characteristics of the liquid food being tested:

$$A = \log \frac{I_0}{I} = \varepsilon c l \tag{5.11}$$

Where: A—Absorbance;

I_0—incident light energy, J;

I—transmitted light energy, J;

ε—molar absorptivity;

c—concentration of the sample, g/L or mL/L;

l—thickness of the sample cell, mm.

When the optical path length is fixed, the Lambert-Beer law can be written in the following simple form:

$$A = Ec \tag{5.12}$$

In the formula, E is called the extinction coefficient.

Reflection Mode

As shown in Fig. 5.17, depending on the surface state and structure of the food, there are three different situations: specular reflection, diffuse reflection, and transmitted diffuse reflection. In practical analytical work, the diffuse reflection measurement method is most commonly used and can be applied to foods with general surface conditions, powder samples, etc. [19].

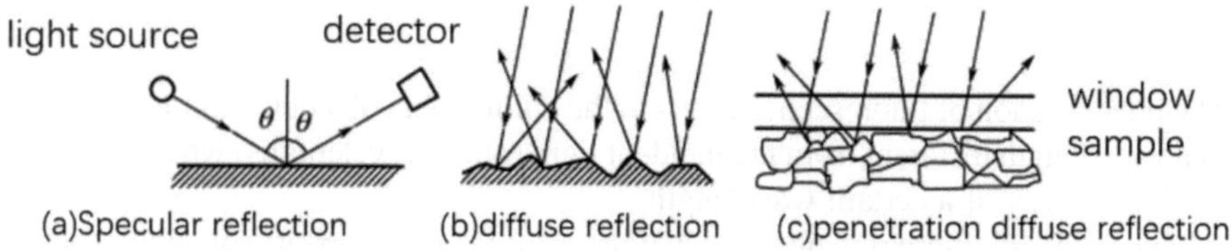

Fig. 5.17 Commonly used food surface reflectance spectroscopy detection methods

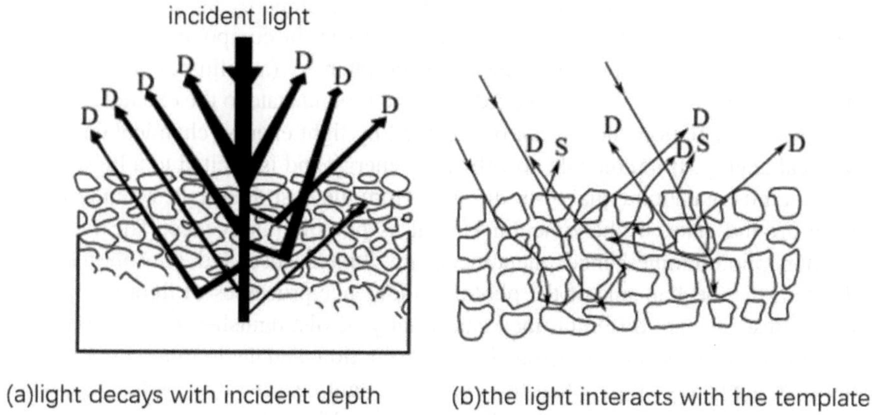

Fig. 5.18 Schematic diagram of diffuse reflection of light from a powder sample: (**a**) light decays with incident depth; (**b**) the light interacts with the template

If a beam of light is directed at a diffuse reflector composed of powder, after interacting with the substance (including reflection, absorption, transmission, diffraction, etc.), part of the light will be reflected back in various ways. As shown in Fig. 5.18 (with S and D representing specular reflection and diffuse reflection, respectively). In Fig. 5.18a, the width of each arrow indicates the change in the degree of attenuation of the reflected light corresponding to different incident depths (optical paths). Figure 5.18b shows the interaction between light and the sample. Diffuse reflected light is the light that has interacted with the molecules inside the sample, thus carrying information about the structure and composition of the sample, which can be used for analysis.

The energy intensity of diffuse reflected light depends on the sample's absorption of light and the scattering determined by the sample's physical state. The diffuse reflected light does not conform to the Lambert-Beer law with respect to the sample's component content, so it is necessary to study the relationship between diffuse reflection spectrum parameters and sample concentration to apply diffuse reflection spectroscopy to quantitative analysis [20].

Fluorescence

The phenomenon of fluorescence refers to the ability of an object to emit light energy of a wavelength different from the incident light (other wavelengths) when irradiated by light energy of a certain wavelength.

Typically, high-energy ultraviolet light is used as the excitation light source, which is directed onto the food material to be tested through an optical system. Due to certain chemical components (such as chlorophyll) and certain microorganisms (or molds) emitting fluorescence under ultraviolet excitation, the wavelength and intensity of the fluorescence (or phosphorescence) emitted by the object are related to the composition of the object, the content of certain components, and external factors such as the frequency (wavelength) and intensity (amplitude) of the exciting ultraviolet light. Molecules are excited from the ground state to the excited state, and the required excitation energy can be supplied by light energy, chemical energy, or electrical energy. If a molecule absorbs light energy and is excited to a high energy state, it emits radiation equal to or different from the absorbed light when returning to the ground state, a phenomenon known as photoluminescence. The two most common photoluminescence phenomena are fluorescence and phosphorescence. Although the mechanisms of the photoluminescence processes of fluorescence and phosphorescence are different, they can usually be distinguished by their phenomena. Fluorescence occurs immediately after excitation, and the luminescence process almost immediately stops when the excitation light ceases ($10^{-9} \sim 10^{-6}$ s), while phosphorescence lasts for a period of time (10^{-4} to 100 s). Fluorescence analysis and phosphorescence analysis are analytical methods established based on these photoluminescence phenomena [21].

Apart from ultraviolet light sources, other light sources can also excite fluorescence. Depending on the different wavelengths emitted by fluorescent substances under the irradiation of excitation light, fluorescence can be divided into X-ray fluorescence, ultraviolet fluorescence, visible light fluorescence, and infrared fluorescence. The fluorescence used for rapid detection of food and agricultural products is mainly excited by ultraviolet light sources.

Fluorescence is categorized into atomic fluorescence and molecular fluorescence based on whether it is emitted by atoms or molecules. Here, we mainly introduce molecular fluorescence.

Molecules in the ground state usually have multiple pairs of spin-paired electrons. According to the Pauli exclusion principle, the spins of these two electrons in one orbital are opposite. The electron spin state can be represented by the spin quantum number s, where $s = \pm 1/2$. As a result of spin pairing, the total electron spin is zero. If all spins in a molecule are paired, the electronic energy state of the molecule is called a singlet state, denoted as $2s + 1 = 1$, represented by s_0. When one electron of a paired electron in a ground state molecule absorbs radiation and is excited, its spin usually remains unchanged, and the excited state is called an excited singlet state, denoted as S. If the spins of the electrons in the excited state are unpaired, i.e., the spins are parallel, $s = 1$, then $2s + 1 = 3$, this state is called an excited triplet state, denoted as T.

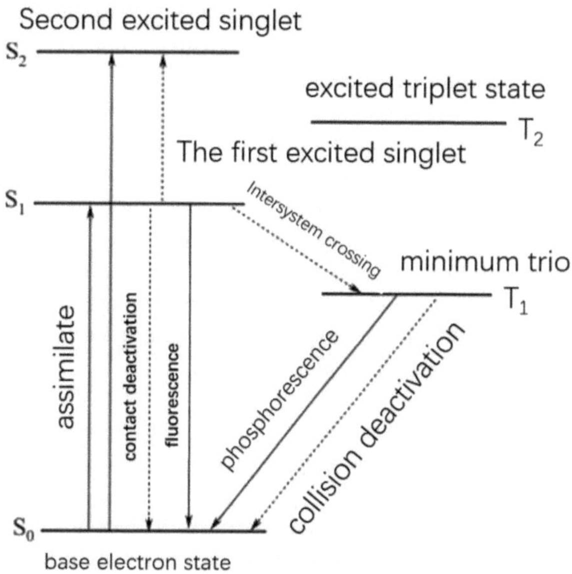

Fig. 5.19 Schematic diagram of fluorescence and phosphorescence generation process

The properties of the excited singlet state and the excited triplet state are significantly different. Singlet state molecules are diamagnetic, while triplet state molecules are paramagnetic; the average lifetime of an excited singlet state is about 10^{-8} s, whereas the average lifetime of an excited triplet state is as long as 10^{-4} to over 100 s; excitation from the ground state singlet to the excited singlet state occurs easily and is an allowed transition, while the probability of excitation from the ground state singlet to the excited triplet state is only about 10^{-6} of the former, essentially a forbidden transition [22].

When a molecule absorbs energy and transitions to a higher energy state s_1 or s_2 (as shown in Fig. 5.19), the excited state molecule is unstable. It first transfers excess energy to other molecules through collisions, rapidly undergoing non-radiative transitions to the lowest vibrational energy level of the same energy state (s_1 or s_2), a process known as vibrational relaxation. If a molecule in the lowest vibrational energy level of the excited singlet state emits a photon and returns to various vibrational energy levels of the ground state within about 10^{-9} to 10^{-6} s, fluorescence is produced, a process known as fluorescence emission [23].

During the process of fluorescence generation, due to various forms of non-radiative transitions, some energy is lost, causing their maximum emission wavelengths to shift towards longer wavelengths, with phosphorescence wavelengths shifting the most, and its intensity is also relatively weaker.

Delayed Luminescence

It has been discovered that all photosynthetic organisms can exhibit a type of "afterglow." For example, when plants are exposed to light for a period and then placed in the dark, they can continue to emit faint red light. This afterglow is called "delayed luminescence," also known as "delayed fluorescence," which is distinct from "prompt fluorescence" that stops once the excitation light is removed. These two types of luminescence have the same spectral distribution. Delayed luminescence is a type of fluorescence phenomenon closely related to various life activities of food and agricultural products, such as oxidative metabolism, cell division, apoptosis, growth regulation, and photosynthesis. Therefore, the chlorophyll content in agricultural products will directly affect the state of delayed luminescence. Although the mechanism of delayed luminescence cannot be fully explained yet, it is certain that it is closely related to the primary reactions of photosynthesis, electron transfer, and photophosphorylation processes, and delayed luminescence is also produced along with photosynthesis.

Light Scattering Phenomenon in Food

Most foods are neither transparent substances nor fully reflective mirror objects, but rather translucent bodies. Therefore, when light hits food, a portion is reflected, a portion is absorbed, and another portion is diffused. The phenomenon of diffusion affects not only the transparency but also the reflective properties. Light undergoes changes not only in energy but also in the complex changes in the direction of light energy propagation. When tomatoes are locally illuminated with monochromatic light, both the intensity and direction of light transmission through the tomatoes change.

5.3.2 Food Optical Characteristics Detection System

A complete optical characteristics detection system mainly consists of a light source (radiation source part) unit, optical path, photoelectric detection unit, electronic circuit, data acquisition unit, digital-to-analog conversion unit, data processing unit, computer control, data output, and storage unit. Whether it is a conventional spectroscopic detection instrument or a specialized optical characteristics detection instrument used in actual production, the light source, spectroscopic components, and detectors are the three most essential units. There are many aspects to consider when selecting them. Below, the main indicators of these three units are discussed for reference when making a selection.

5.3.2.1 Light Source (Radiation Source)

Devices that emit visible light are called light sources, and now devices that emit ultraviolet and infrared rays are also generally referred to as light sources. There are many types and forms of light sources. The following four types are commonly seen:

(1) Thermal radiation light sources. Light sources that emit light due to an increase in the temperature of an object, such as the sun, blackbody, incandescent lamps, and halogen lamps; (2) Gas discharge light sources. Light sources that emit light when current passes through gas (including metal vapor), such as fluorescent lamps, mercury lamps, and xenon lamps; (3) Solid-state light sources such as electroluminescent lamps and semiconductor light-emitting diodes; (4) Lasers.

Characteristic Parameters of Radiation Sources or Light Sources

The characteristic parameters representing radiation sources mainly include:

Radiation Efficiency and Luminous Efficiency

In a given λ_1~λ_2 wavelength range, the ratio of the radiant flux emitted by any radiation source to the electrical power required to produce this radiant flux is the radiation efficiency of the radiation source within the specified spectral range. The ratio of the luminous flux emitted by any light source to the electrical power required to produce this luminous flux is the luminous efficiency of the light source, with the unit being l m/W.

Spectral Distribution of Radiant Flux

To describe the distribution of radiant flux at various wavelengths in the spectrum emitted by a radiation source, the term spectral distribution of radiant flux is introduced. In practical use, the "relative spectral distribution" of radiant flux is often used because the absolute value of radiant flux changes with the distance from the light source, but its relative value remains basically unchanged, so using relative spectral distribution better explains the issue. Similarly, the spectral distribution or relative distribution of luminous flux can be used to describe the spectral distribution in the visible light region of the light source.

Radiation Intensity Distribution

Generally speaking, the radiation intensity distribution of a point radiation source is uniform; it is independent of direction. For other forms of radiation sources, the radiation intensity has a certain relationship with the angle (the angle between the central perpendicular and the direction), and this relationship curve is called the radiation intensity distribution of the radiation source. Sometimes we hope that the radiant flux of the radiation source is concentrated within a limited angle. For example, searchlights used in navigation, guidance, and stage use hope to control the radiant flux of the radiation source within as small an angle as possible.

Distribution Temperature

The relative light source distribution of a radiator within a certain wavelength range is equal to or approximates the relative light source distribution of a blackbody at a certain temperature, and this temperature is called the distribution temperature of the radiator.

Color Temperature

The color temperature of a radiation source refers to the temperature at which a blackbody emits the same color appearance as the radiator. Since a certain color appearance can be produced by different spectra, the color temperature does not indicate the spectral distribution of the radiation source.

Color Rendering Index

When different light colors of light sources illuminate an object, the color displayed by the object (strictly speaking, the color sensation produced by the object in the human eye) is different. This characteristic of the illuminated object determined by the light source is called the color rendering of the light source. To quantitatively compare the color rendering of light sources, the concept of color rendering index is introduced. The color rendering index of sunlight is set to 100, and the color rendering index of other artificial light sources is determined by comparison with it, with their color rendering index being less than 100. Generally speaking, a color rendering index of 100 to 75 is considered an excellent light source; 75 to 50 is considered a general light source, but if it falls below 50, it is considered a poor light source.

Color Appearance

The color seen when directly observing a light source with the eyes is called the color appearance of the light source. For example, the color appearance of an incandescent lamp is reddish-yellow, and the color appearance of a xenon lamp is white, etc.

Thermal Radiation Sources

The main source of thermal radiation is the heating of an object (primarily using electrical energy) to an incandescent state to emit light. As long as the object's temperature is above absolute zero, it can emit thermal radiation. At low temperatures (less than 400 °C), objects emit invisible infrared radiation. When the object's temperature rises to 550 °C, it emits a small amount of long-wavelength dark red light. As the temperature continues to increase, it emits orange-red light and orange-yellow light. When the temperature exceeds 1500 °C, shorter-wavelength green-blue light and a small portion of ultraviolet light begin to be emitted, and the object appears white. Thermal radiation sources all comply with the laws of blackbody radiation.

The Sun

When solar radiation is transmitted from outside the atmosphere to the Earth's surface, it is attenuated by various factors, mainly including ozone absorption, Rayleigh scattering by molecules, and absorption by substances such as oxygen and carbon dioxide. Figure 5.20 describes the spectral irradiance curve of the sun at standard sea level.

Extensive measurements of solar irradiance conducted by Brown on the Earth's surface indicate that when the sky is clear and the sun is at the zenith, the illuminance produced on the sea's horizontal surface is $E_v = 1.24 \times 10^5$ lx.

Incandescent Lamp

An incandescent lamp consists of a filament and a glass bulb that is either vacuum-sealed or filled with inert gas. The light radiation flux is usually emitted by an electrically heated tungsten filament, with an operating temperature distribution in the range of 2000 to 3000 K. The melting point of tungsten is approximately 3600 K. The luminous flux and lifespan of an incandescent lamp largely depend on the

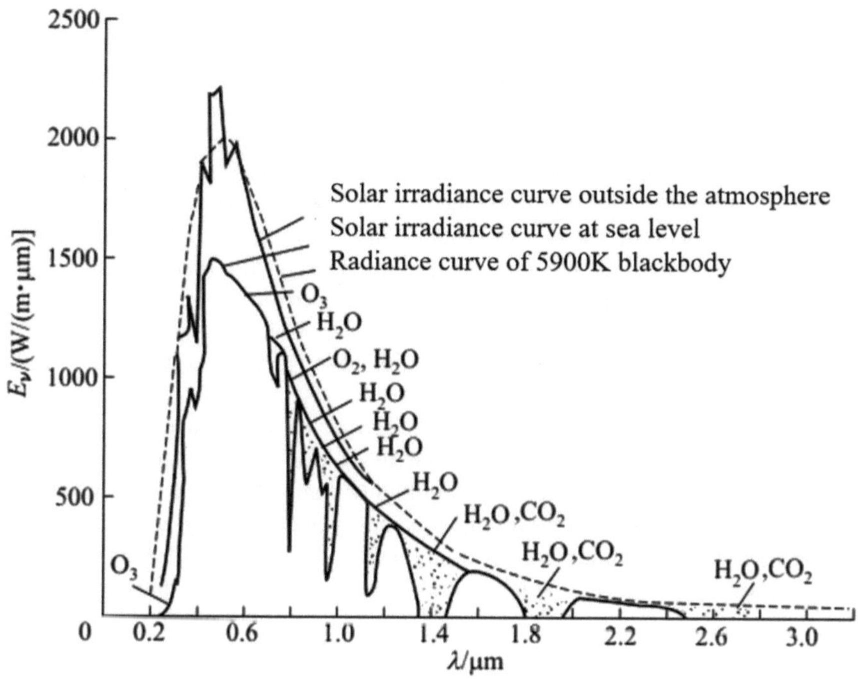

Fig. 5.20 Spectral illuminance curve of the sun

voltage. There are many varieties of incandescent lamps available to meet various needs. The luminous efficiency of incandescent lamps is very low, about 10 to 20 lm/w, with an emission efficiency below 50%. The spectral distribution of the radiant flux of incandescent lamps is continuous.

Halogen Lamp

A halogen lamp is a new type of light source made using the halogen cycle principle discovered in 1959. Its luminous principle is the same as that of an incandescent lamp. Generally, the bulb is filled with a compound of halogen elements, such as hydrogen bromide. When the lamp is in operation, in the low-temperature area near the glass bulb, halogen atoms combine with tungsten atoms to form colorless volatile tungsten halides. With the convection of gas inside the lamp, tungsten halides diffuse to the hot area near the filament, where halogen and tungsten decompose into halogen atoms and tungsten atoms, and tungsten redeposits onto the filament. As a result, the evaporated tungsten is returned to the filament. The halogen atoms are then carried by the airflow to the low-temperature area of the glass bulb and recombine with the evaporated tungsten. This type of reaction between halogen and tungsten is cyclic, i.e., combination-decomposition-combination, hence it is called the halogen cycle principle.

Gas Discharge Light Source

A light source made using the gas discharge principle is called a gas discharge light source. During manufacturing, the lamp is filled with a gas used for illumination, which can also be a liquid or solid that becomes gaseous at high temperatures. Under the influence of an electric field, electrons and ions are excited, turning the gas into a conductor. When ions move towards the cathode and electrons towards the anode, they gain energy from the electric field. When they collide with gas atoms or molecules, new electrons and ions are excited. During this process, some inner electrons transition to higher energy levels, causing atomic excitation. When the excited atoms return to lower energy levels, they emit visible radiation or ultraviolet and infrared radiation. This luminescence mechanism is known as the gas discharge principle.

Gas discharge light sources have the following common characteristics: high luminous efficiency, compact structure, long lifespan, and strong adaptability to light color. Therefore, gas discharge lamps are highly competitive and have developed rapidly. Currently, common gas discharge lamps include mercury lamps (including low-pressure mercury lamps, high-pressure mercury lamps, and ultra-high-pressure mercury lamps), fluorescent lamps, sodium lamps, xenon lamps, and metal halide lamps.

5.3.2.2 Spectral (Dispersion) Devices

In optical property detection, many situations require the use of a monochromatic light source (radiation of a single wavelength), and devices that decompose composite light into monochromatic light are called dispersive devices. Common dispersive devices include filters, prisms, gratings, Michelson interferometers, and acousto-optic tunable filters (AOTF), among others.

Filters

Devices that can filter out a certain wavelength range of the spectrum are called filters. Filters can be classified based on their working principle into absorption, interference, reflection, combination types, etc. Based on the spectral characteristics of filters, they can be divided into bandpass filters and cutoff filters. Figure 5.21 shows the characteristics of bandpass filters. In the figure, $T_{max.}$ represents the maximum transmittance, and λ_0 is called the central wavelength, which corresponds to the wavelength of the maximum transmittance T_{max}. $\Delta\lambda_{0.5}$ is called the half-width, and $\Delta\lambda_{0.1}$ is called the spectral width.

Among filters, the development of interference filters is the most notable, and currently, interference filters can be manufactured for the entire spectral range from ultraviolet to mid-infrared. Modern interference filters have evolved to use multilayer film structures. Materials used for making thin films include dielectrics, metals, and semiconductors. The narrowest spectral width of narrowband interference filters can now reach about 0.1 nm.

Prisms

Spectral instruments using prisms as dispersive elements are no longer produced or used. However, as the earliest method of dispersion, we should understand it. The

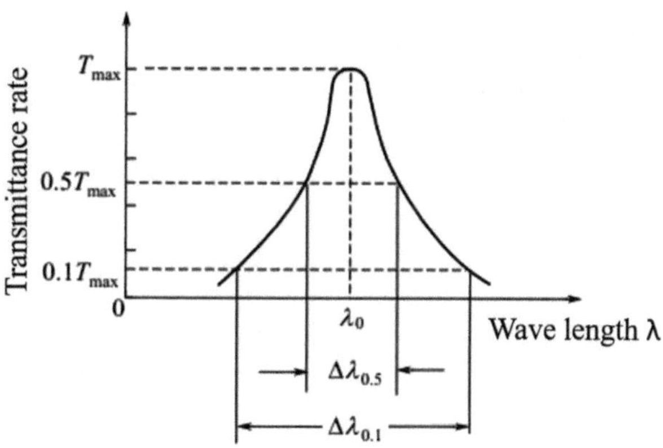

Fig. 5.21 Spectral properties of band filters

principle of prism dispersion is based on the different relative refractive indices of different wavelengths of the spectrum. The wavelength resolution is too low.

Gratings

Gratings come in two forms: transmission and reflection gratings. Currently, transmission gratings are no longer used in spectral instruments. Figure 5.22 shows a schematic diagram of the light path of a reflective plane diffraction grating [24]. When parallel light rays hit the grating, diffraction occurs, and the light scatters in all directions, as shown in Fig. 5.22a. In the figure, d is the spacing of the grooves, called the grating constant, and its reciprocal 1/d indicates the number of grooves per unit length, known as the grating's groove density. Analyzing two adjacent parallel incident light rays 1 and 2 and their two parallel diffracted light rays 1′ and 2′, as shown in Fig. 5.22b. In the figure, θ_i is the angle between the incident light ray and the normal to the grating plane, called the angle of incidence; θ_m is the angle between the diffracted light ray and the normal to the grating plane, called the diffraction angle. The optical path difference Δ between light rays 11′ and 22′ is:

$$\Delta = \mathrm{AB} - \mathrm{CD} = d\,(\sin\,\theta_m - \sin\,\theta_1) \tag{5.13}$$

According to wave optics, the condition for obtaining the maximum value after coherent light beams interfere is:

$$\Delta = \pm k\lambda = d(\sin\theta_m - \sin\theta_i)$$

In the formula, λ is the wavelength of light; k is a positive integer. When this condition is met, a bright line is obtained on the focal plane, and the light beams reinforce each other.

With the improvement and advancement of grating ruling technology, gratings have been widely used in spectroscopic instruments. In addition to plane gratings,

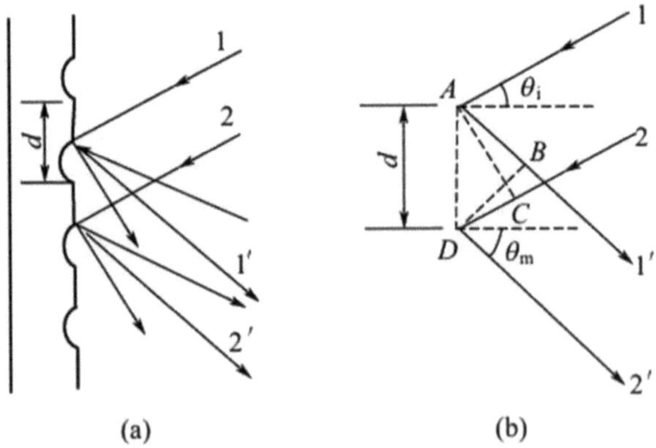

Fig. 5.22 Schematic diagram of the optical path for reflective plane diffraction grating spectroscopy

there are also concave gratings. Reflective gratings can be applied throughout the entire optical spectrum region.

Michelson Interferometer

As early as 1902, Michelson proposed the idea of converting the fundamental wave of an interference pattern into spectral information. However, it was not until 1949 that British astrophysicist Fellgett first converted the interference pattern into a spectrum through mathematical calculation methods. The establishment of the fast Fourier transform calculation method in 1966 and the improvement of computer performance truly realized Michelson's idea, and after 1970, Fourier transform spectrometers truly developed into commercial products [25].

The principle of Fourier transform spectroscopy is shown in Fig. 5.23. The core part of the optical system is a Michelson interferometer. The light beam emitted from the light source O forms parallel light after passing through the collimating lens. The semi-transparent and half-reflective beam splitter divides the light beam into two paths. One path passes through the beam splitter and is projected onto the movable plane mirror M_1 and then reflected back. The other path is reflected on the beam splitter, passes through the compensation plate P_2 and the fixed mirror M_2 and is then reflected back. The two beams are recombined and interfere with each other. When the optical path difference between the two beams is λ/2 times. This n-stage dynode

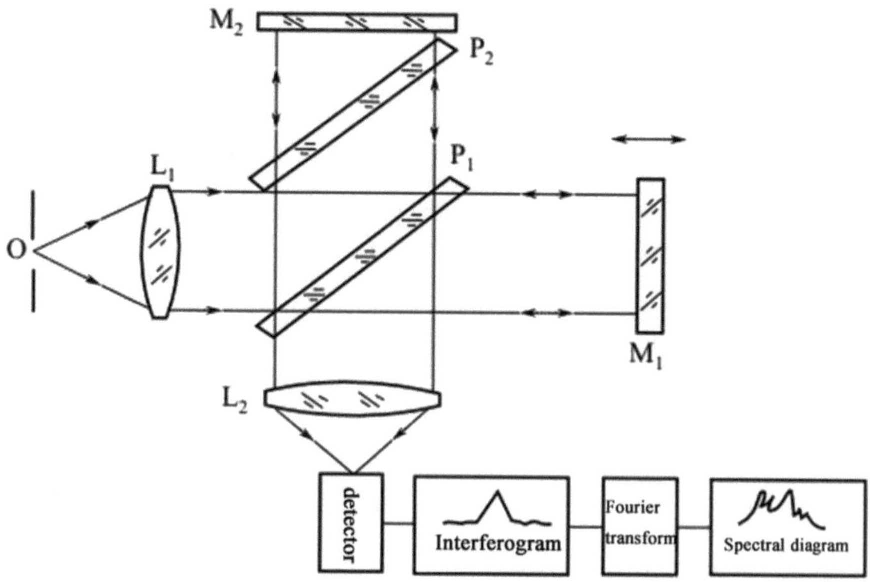

Fig. 5.23 Principle of Michelson interferometer and Fourier transform spectrometer (O—light source; L1, L2—lens; M1—movable mirror; M2—fixed mirror; P1—semi-transparent and half-reflective beam splitter; P2—Compensation plate)

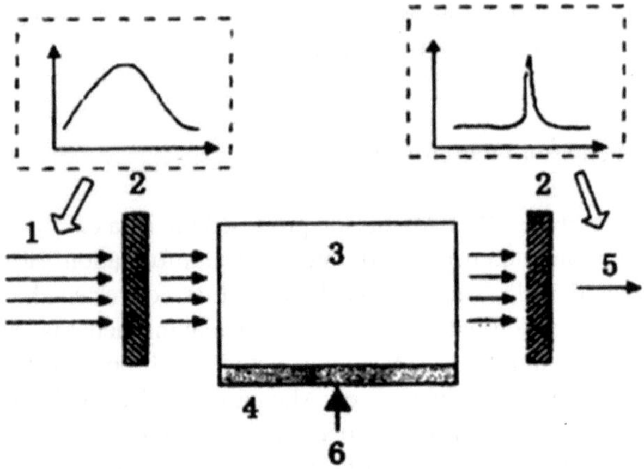

Fig. 5.24 Schematic diagram of AOTF modulated monochromatic light (1-incoming light; 2-polariser; 3-TeO2 crystal; 4-piezoelectric transducer; 5-outgoing light; 6-high frequency source)

assembly forms the electron multiplier, which amplifies the electron beam with low noise, and the anode A collects the amplified electron beam for output [26].

Acoustic -Optical Tunable Filter (AOTF)

The use of acousto-optic tunable filters (AOTF) as spectroscopic devices was the most prominent advance in spectroscopic instrumentation in the 1990s. AOTFs are optoelectronic devices that use ultrasonic waves to interact with specific crystals to produce spectroscopy. Although its working principle was experimentally confirmed as early as in the early 1930s, it was not until the last decade or so that it has been more and more widely used in the field of defense and industry. Compared with several other spectroscopic devices, AOTF uses acousto-optic modulation to produce monochromatic light and achieves spectral scanning through changes in ultrasonic RF, no moving parts in the optical system, fast wavelength switching, good reproducibility, and programmable wavelength control to make the application of this type of instrument has greater flexibility [27].

When the frequency of the RF signal is changed, the center of the optical passband will be changed accordingly, so that a fast scanning of the diffracted light wavelength can be achieved by continuously changing the ultrasonic frequency. Figure 5.24 shows a schematic diagram of the AOTF generating tunable monochromatic light.

5.3.2.3 Craft

In the optical characteristic detection, the detector is the optical signal into other forms (mainly electrical signals) of the device. Various technical parameters of the detector directly affect the sensitivity and stability of the detection system, a reasonable choice of detector is one of the key issues in the composition of the optical characteristics of the detection system. With the continuous progress of science and technology, high-quality, high-precision photoelectric detection components continue to appear, especially the progress of the semiconductor manufacturing process, to people in the formation of optical characteristics of the detection system to create a larger space. The following several commonly used detectors, such as photomultiplier tubes, photoresistors, photodiodes, phototransistors and photocells.

Detector Quality Factors

The quality factor of a detector can be used to characterize the performance of a detector, and at the same time can be used to compare the performance parameters of similar detectors or different types of detectors.

Spectral Responsivity

Spectral responsivity is the ratio of the output voltage (current) of a detector to the irradiance of monochromatic light incident on the detector (luminous flux). Spectral responsivity can also be interpreted as the detector output voltage (current) per unit irradiance incident on the detector. The larger this value is, the more sensitive the detector is.

Quantum Efficiency

The quantum efficiency of monochromatic radiation is the quotient of the number of photoelectrons output from the detector and the number of photons incident on the surface of the detector, i.e., the number of photoelectrons produced by a single photon, which should be less than 1.

Spectral Responsivity Function (Curve)

The spectral responsivity of a detector indicates the responsivity of that detector versus wavelength. Figure 5.25 shows two typical spectral responsivity function curves. One is a typical curve for a silicon photovoltaic cell, where the longest wavelength that can be responded to is called the long-wavelength limit or red limit, and this detector is called a selective detector. If this curve is normalized by the peak wavelength responsivity and characterized by the units on the right side of the figure, then it can also be called the relative spectral responsivity function. The other is the spectral responsivity function of the thermopile, which is essentially wavelength independent, and this detector is called a non-selective detector.

Response Rate (Integral Response Rate)

Response rate is the ratio of the detector's output signal to the incident radiant flux. Since the distribution of spectral fluxes emitted from different radiation sources,

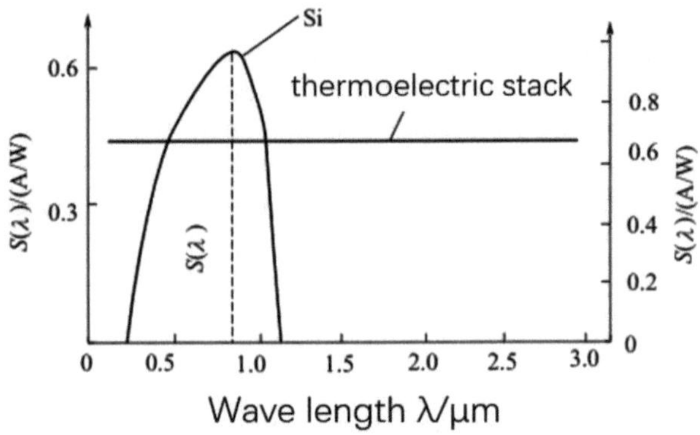

Fig. 5.25 Spectral responsivity functions of two typical photodetectors

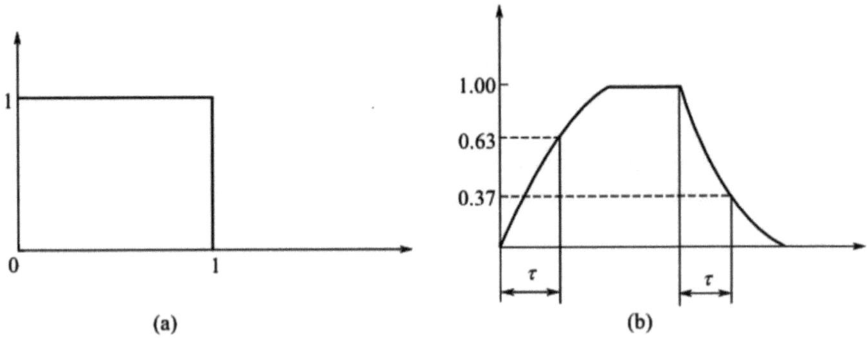

Fig. 5.26 Definition of detector time constant

even the same source with different color temperatures, is different, the source and color temperature should be specified when providing the data.

Linearity

One of the most important requirements in many photometric and radiometric measurements is the linearity of the detector, which is defined as the performance of a detector whose output power is precisely proportional to the input power over a specified range. Linearity can also be interpreted as the fact that the response rate of the detector is constant over a specified range. This specified range is known as the linear region.

Time Constant

When the detector is irradiated with a pulse of radiation, which can be considered as a square wave if both the rise and fall times of this pulse are short (Fig. 5.26a), the electrical output waveform shown in Fig. 5.26b is obtained from the detector, with rising and falling strings occurring due to inertia of the detector. We refer to the time required to rise from zero to 63% of the peak as the detector rise time constant, and the time required to fall from the peak to 37% of the peak as the fall time constant, which tend to be equal and are therefore collectively referred to as the detector time constant.

Photomultiplier Tube

Although semiconductor photoelectric devices have developed rapidly since the 1940s, they have replaced most of the applications of photoelectric tubes and some of the applications of photomultiplier tubes with their low price, stable performance, and easy-to-use circuits. However, the photomultiplier based on the principle of photoelectric emission still maintains its absolute advantage in the following fields, i.e., the detection of weak light signals and the detection of weak signals at high frequencies or pulses, and therefore it remains an important detector device [28].

The main reasons for the restricted use of photomultiplier tubes are the relatively narrow spectral response range of photomultiplier tubes, which is generally 0.5 to 1.1 m; the requirement of a highly stable power supply of 500 to 2000 V; and the need for shielding.

The principle of operation of a photomultiplier tube is shown in Fig. 5.27. It is composed of a photocathode, an electro-optical system, an electron multiplier, and an anode contained within a vacuum. The photocathode receives weak light signals and emits photoelectrons, 1,2,3 ......*n* is the multiplier pole, and after applying a DC power supply and dividing the voltage-by-voltage divider resistors R_1, R_2 and R_3...... the potential is increased step by step from the photoelectric anode, the multiplier pole to the cathode [29]. Multiplier pole 1 utilizes the electro-optical system to direct emitted electrons from the photoelectric cathode, ensuring the highest possible collection efficiency. These electrons gain energy as they are accelerated by the electric field, enabling secondary electron emission. When these electrons strike multiplier pole 2, they trigger the emission of additional electrons, further amplifying the signal. This process repeats across multiple stages (*n-pole* multiplier poles), resulting in a significant amplification of the photoelectrons. The

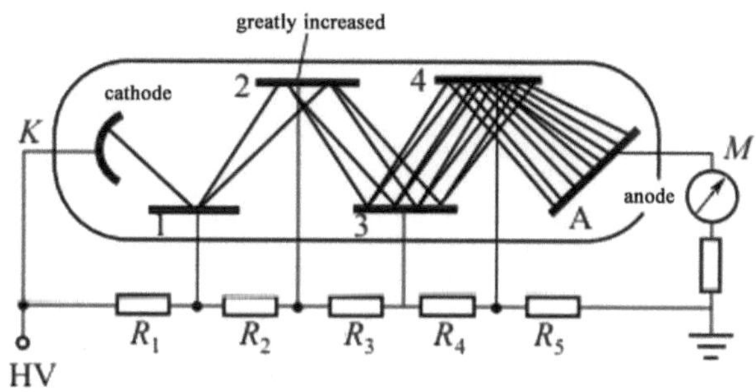

Fig. 5.27 Photomultiplier tube working principle

overall amplification factor typically ranges from 10^5 to 10^8 times. This n level multiplier pole composed of electron multiplier, plays a low noise amplification of the role of the electron beam, the anode A is the amplification of the electron beam collected to the external output.

Photoconductive Resistor

Semiconductor photoconductivity refers to the phenomenon where the conductivity of a semiconductor changes after irradiation. Objects with photoconductive effects are called photoconductors. When electrodes are plated on both ends of a photoconductor, it becomes a photoconductive resistor, also known as a photoconductive cell.

Photoconductive resistors are characterized by their small size, low bias voltage, and wide spectral response range. Their time constant is generally 10^{-2}–10^{-7} s, specially made can reach 10^{-10} s.

In the ultraviolet and visible light regions, the advantages of photoresistors are not prominent. In the infrared region, they have a higher response rate and good mechanical properties, thus they are widely used and are an important type of infrared detector.

Figure 5.28 shows the photoelectric characteristic curve of a typical CdS photoresistor in rectangular coordinates. From the figure, it can be seen that as the illumination increases, the resistance value rapidly decreases and then gradually approaches saturation. However, within a certain range of illuminance in logarithmic coordinates, the characteristic curve of resistance versus illuminance is basically a straight line.

Under certain illumination, the relationship curve between the photocurrent of a photoresistor and the applied voltage is called the volt-ampere characteristic. Generally speaking, the volt-ampere characteristic of a photoresistor is a linear relationship. The time constant of a photoresistor has a significant dependence on

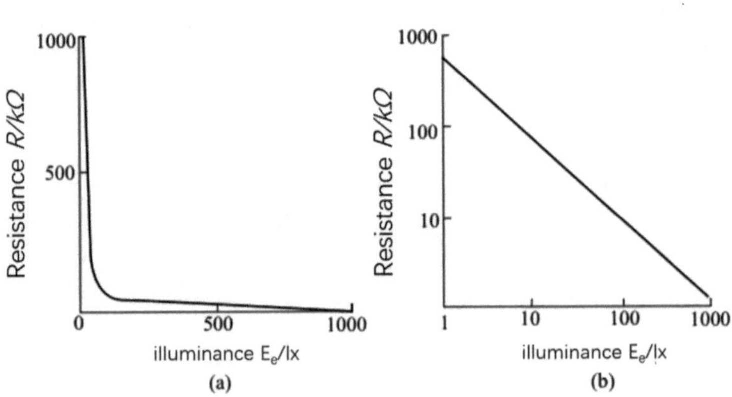

Fig. 5.28 Photoelectric characteristics of CdS photoresistors

illuminance and operating temperature, and the rise time is often different from the fall time. The characteristics of a photoresistor are greatly affected by the operating temperature; even a slight change in temperature will cause changes in its response rate, spectral response function, peak wavelength, long-wave limit, etc., and these changes lack a certain regularity, which is a major drawback of photoresistors. Therefore, to improve its performance stability, reduce noise, and increase detection rate, the use of specialized cooling devices is very necessary.

5.4 Spectral Detection Technology

5.4.1 Near-Infrared Spectroscopy Technology

The near-infrared region refers to electromagnetic waves with wavelengths in the range of 780 to 2526 nm, which is the first non-visible light region discovered by humans, with a history of over 200 years. In the early twentieth century, people first obtained the near-infrared spectrum of organic compounds using spectrographic methods and explained the spectral characteristics of related groups, indicating that near-infrared spectroscopy (NIR) might be applied as an analytical technique. Due to the lack of instrumental foundation, research on near-infrared spectroscopy was limited to a few laboratories before the 1950s and did not find practical application [30]. After the 1950s, with the emergence of simple near-infrared spectrometers and the extensive work done by Norris and others on near-infrared diffuse reflection technology, a small wave of near-infrared spectroscopy applications began. Research on the quality determination of agricultural products (including grains, feed, fruits, vegetables, meat, eggs, milk, etc.) using near-infrared spectroscopy received widespread attention. Since these applications were based on traditional spectroscopic quantitative methods, when the background, granularity, matrix, etc., of the sample changed, the measurement results often produced significant errors. After the 1980s, with the rapid development of computer technology, which drove the digitization of analytical instruments and the development of the chemometrics discipline, chemometric methods achieved good results in solving the extraction of spectral information and reducing background interference [31]. Coupled with the unique characteristics of near-infrared spectroscopy in detection technology, people have re-recognized the value of near-infrared spectroscopy, and its application research in various fields has been gradually carried out.

Near-infrared spectroscopy analysis technology has the advantages of directly measuring granular, powdery, pasty, opaque samples, flowing samples, and even complete samples without strict pre-treatment of the samples. This technology has low analysis cost, fast speed, high precision, reliable data, and no chemical pollution, making it convenient for real-time monitoring and online analysis. When traditional quality detection technology is difficult to meet production needs, this technology, as a new quality detection method, has gradually attracted more and more attention.

5.4.1.1 Principle of Near-Infrared Spectroscopy

In the near-infrared spectral region (0.78 to 2.0 μm), besides exhibiting general characteristics when interacting with electromagnetic waves and objects (transmission, diffuse reflection, absorption, etc.), the most significant feature is that this spectral region is the overtone and combination frequency absorption area for hydrogen-containing groups (OH, SH, CH, NH, etc.). Since the components of plant and animal foods and feeds are mostly composed of these groups, the absorption spectrum of the groups characterizes the chemical structure of these components. For example, the absorption spectrum of common components in food and agricultural products such as water, fat, starch, and protein reflects the characteristic peaks of the CH group. Based on the position of these characteristic peaks, absorption intensity, etc., the chemical structure of a certain component can be qualitatively and quantitatively described. On the basis of known chemical structures, a certain component can be detected, and its content determined, thus this spectrum has special significance for food and agricultural products.

The absorption of infrared by substances, with very few exceptions, is caused by the harmonic or combination vibration absorption of the normal stretching vibrations between two atoms connected by a bond, most of which are linearly related to the normal stretching vibrations of hydrogen atoms in the substance. That is, when the frequency of the light wave is the same as or a multiple of the vibrational frequency of the atomic bonds in the molecular structure, the wave of that wavelength is absorbed.

The near-infrared absorption spectrum is mainly due to the overtone (0.78 to 2.0 μm) absorption of fundamental frequency vibrations. Fundamental frequency vibrations have a series of overtone absorption bands, and as the overtone order increases, the intensity of the spectral bands gradually decreases. Combination frequency absorption is caused by the simultaneous transition of multiple vibrational energy levels, and the more energy levels that transition simultaneously, the lower the intensity of the absorption band. Most near-infrared spectral absorption bands are caused by the resonance and overtone absorption of the stretching and bending vibrations of X-H (X = C, D, N, S, P) bonds. Other important near-infrared absorption bands are caused by hydrogen bonds, hydroxyl groups, carbon-nitrogen bonds, carbon-carbon bonds, and metal halides.

The most common bands in the near-infrared spectrum include the combination frequency and second and third overtone absorption bands of the fundamental frequencies of O-H, N-H, and C-H bonds. The form of hydrogen bonds will change the force constant of the X-H bond, thereby causing a shift in the position of the absorption band and broadening the spectral band. Given that combination frequencies are due to the addition of multiple fundamental frequencies and overtones are caused by vibrational energy level transitions of more than one energy level difference, the formation of hydrogen bonds will have a greater impact on combination frequencies and overtones than on fundamental frequencies.

The information contained in the near-infrared spectrum is related to bond strength, chemical species, dielectric properties, and hydrogen bonds. For solid samples, the near-infrared spectrum also contains information on scattering, diffuse reflection, specular reflection, refraction, and polarization of reflected light. The spectrum will also contain information related to hydrogen bonds and hydrated hydrogen ions.

Methods for Measuring Near-Infrared Spectra

The main methods for measuring near-infrared spectra are transmission spectroscopy and reflection spectroscopy. Transmission spectroscopy (mostly referring to the short-wave near-infrared region, generally in the wavelength range of 700 to 1100 nm) involves placing the sample to be measured between the light source and the detector, where the light detected by the detector is the light that has interacted with the sample molecules (carrying information about the structure and composition of the sample, see Fig. 5.29) [32].

When the substance being measured is a transparent substance (solution), and only light absorption occurs within the substance (solution) without other phenomena such as reflection, scattering, or fluorescence, its absorbance follows the Lambert-Beer law:

$$T(\lambda) = I_t(\lambda)/I_s(\lambda) \tag{5.14}$$

Modelled on Eq. (5.14), this can be calculated as:

$$A(\lambda) = lg[I_s(\lambda)/I_t(\lambda)] = lg[1/T(\lambda)] \tag{5.15}$$

Where $A(\lambda)$—Relative absorbance, commonly referred to as absorbance in application.

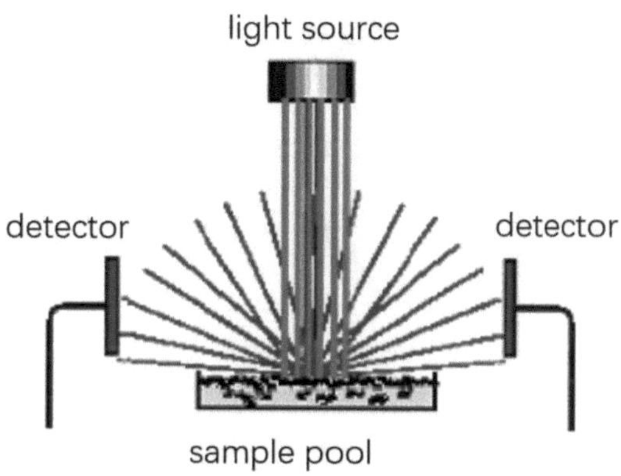

Fig. 5.29 Schematic diagram of the reflection method

Reflectance spectroscopy (mostly referring to the long-wave near-infrared region, with wavelengths generally in the range of 1100 ~ 2500 nm) involves placing the detector and light source on the same side of the sample, where the detector measures the light reflected back from the sample in various ways (Fig. 5.29). The reflection of light by an object is divided into specular reflection (mirror reflection) and diffuse reflection. Specular reflection refers to the reflection that occurs on the surface of an object according to the law of reflection, where the angle of incidence equals the angle of reflection; diffuse reflection occurs when light is projected onto an object (often a powder or other particulate matter), resulting in reflection in various directions on the surface or inside the object. Analysis using diffuse reflected light is called diffuse reflectance spectroscopy.

Steps for Near-Infrared Spectroscopy Detection

Similar to all methods that utilize the physical properties of agricultural products to detect the quality of food and agricultural products, these methods are indirect detection methods. All indirect detection methods are based on standard legal detection data. Theoretically, although various detection methods utilizing physical properties are fast and convenient, their accuracy will not exceed that of standard legal detection methods. The main reason is that detection methods using physical properties require establishing a mathematical relationship—a calibration model—between the physical quantities and the indicators to be measured. The process of establishing a detection method is shown in Fig. 5.30. The specific operational steps are as follows:

(a) Select a reasonable detection system and set the parameters of the detection system.

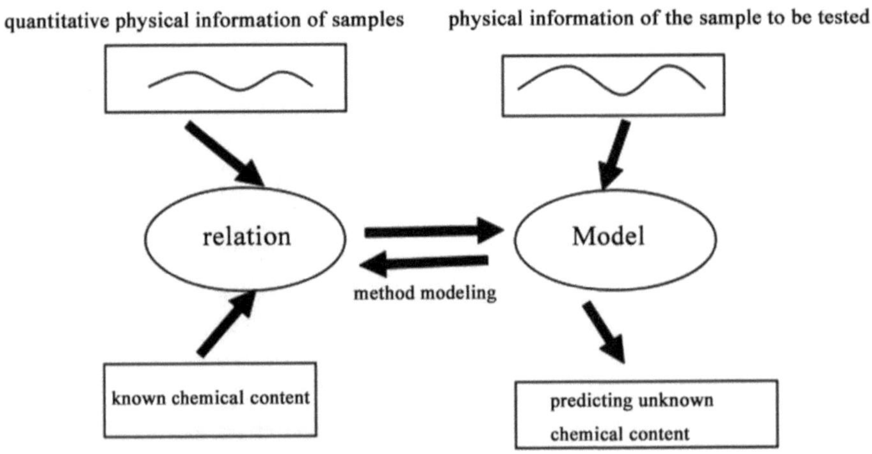

Fig. 5.30 Steps in building a predictive model for NDT

(b) Collect samples that include all possible states, generally requiring about 50 ~ 80 samples, some of which are used to establish the calibration equation, while others are used to test the calibration equation.
(c) Measure the physical information of all samples using a unified detection method.
(d) Measure the content of the components to be tested in all samples using legal chemical methods according to national or industry standards.
(e) Use chemometric methods to establish a calibration equation between the chemical measurement values of the calibration samples and the physical detection data.
(f) Substitute the physical detection data of the prediction samples into the calibration equation to calculate the theoretical values of the components to be tested.
(g) Calculate and analyze the error between the predicted calculation values and the chemical measurement values of the prediction samples. If the error does not meet the predetermined requirements, it is necessary to change the method of establishing the calibration equation and return to step (5) to re-establish the calibration equation.

5.4.1.2 Applications in Quality Analysis of Food and Agricultural Products

Near-infrared spectroscopy has been widely applied in the quality analysis of food and agricultural products. The common applications are shown in Table 5.1.

Due to the high price and large profit margin of premium teas, to a certain extent, this has led to the phenomenon of inferior products being passed off as superior ones and counterfeit products being passed off as genuine ones in China's tea market, especially in the premium tea market. By utilizing the differences in near-infrared spectral absorption peaks between genuine and counterfeit teas, they can be distinguished.

In the experiment, the production date of the genuine Biluochun samples used was all in May 2004, and the production location was Jiangsu. The counterfeit Biluochun samples were made by imitating the processing techniques of Biluochun, with all counterfeit samples produced between May and July 2004, originating from Jiangsu, Anhui, Hunan, and Jiangxi. To ensure uniform sampling, before the experiment, 5 ± 0.5 g was randomly weighed as a sample according to the quartering method. A total of 228 tea samples were selected, with 138 samples as the training set, including 30 genuine Biluochun samples, and the remaining 90 samples as the prediction set, including 25 genuine Biluochun samples.

The near-infrared detection system used in the experiment mainly consisted of a near-infrared spectrometer (Nexus 670 FT-IR, Nicolet, USA). Scanning range: 11000 ~ 3800 cm^{-1}; number of scans: 64 times; resolution: 4 cm^{-1}. During the experiment, the indoor temperature and humidity were kept basically consistent, and the samples were poured into sample cups and fully compacted. Each sample was

Table 5.1 Common applications in quality analysis of food and agricultural products

Food items	Mode	Type of study	References
Bayberry	Transmission	Titratable acidity, malic and citric acid	[33]
Coffee	Reflectance	Detection of addition of barley to coffee	[34]
Tomato	Absorption	Lycopene, total acid, sugar, phenols, and antioxidant properties	[35]
Radix	Reflectance	Free amino acid determination	[36]
Milk powder	Reflectance	Quantification of adulterants	[37]
Tea	Reflectance	Total polyphenol quantification	[38]
Honey	Reflectance	Antioxidant properties	[39]
Maize	Reflectance	Kernel rot and mycotoxins determination	[40]
Vinegar	Transmittance	Soluble solid contents and pH measurement	[41]
Tea	Reflectance	Quantification of caffeine and total polyphenol	[42]
Summer squash	Reflectance	Ascorbic acid, polyphenols, and chlorophyll	[43]
Olives oil	Reflectance	Total polyphenols	[44]
Sorghum grain	Reflectance	Polyphenols, condensed tannin, and3-deoxyanthocyanidins	[45]
Karkade	Reflectance	Antioxidant properties	[46]
Goji berry	Reflectance	Chemical composition measurement	[47]
Goji berry	Reflectance	Antioxidant activity measurement	[48]
Black tea	Reflectance	Amino acid, water content, caffeine, and theaflavin	[49]
Cocoa beans	Transmittance	Quantification of total fungal content	[50]
Black tea	Reflectance	Prediction of fermentation quality	[51]
Coffee	Reflectance	Caffeine, caffeic acid, chlorogenic acid	[52]
Cocoa beans	Reflectance	Novel total fat content	[53]
Apple	Reflectance	Glucose, fructose, and sucrose	[54]
Bayberry	Transmittance	Glucose, fructose, and sucrose	[55]
Citrus	Transmittance	Quantification of sugars	[56]
Date fruit	Transmittance	Determination of sucrose	[57]
Fuji apple	Reflectance	Soluble solid content	[58]

collected four times, and the average of the four collections was taken as the original spectral data of the sample.

Although the appearance of genuine and counterfeit Biluochun tea can be completely similar, the content and proportion of most organic substances (such as polyphenols, plant alkaloids, amino acids, proteins, and cellulose) inside them are affected by factors such as geographical location, climate, soil conditions, and picking time, always resulting in certain differences from genuine Biluochun tea. These organic substances' hydrogen-containing groups (such as C-H, O-H, S-H, and N-H) can produce overtone and combination frequency absorption in the near-infrared region. Their first overtone near-infrared spectral band is located at

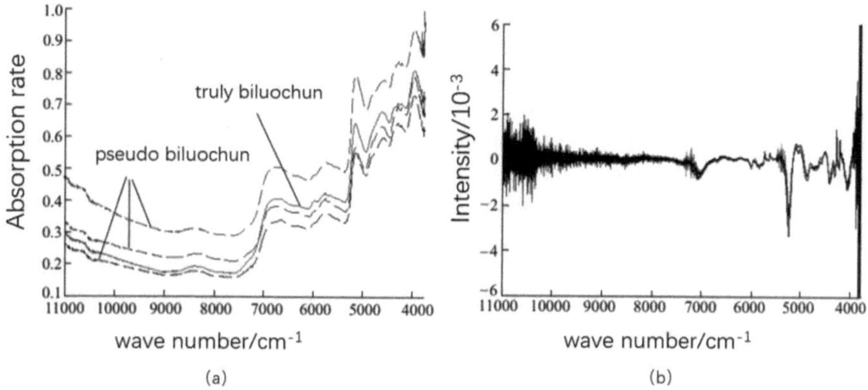

Fig. 5.31 Raw spectra (**a**) plot and first-order derivative spectra (**b**) plot of tea leaves

7200–5500 cm^{-1}; the second, third, and fourth overtones are located at 12800–8300 cm^{-1}; and the combination frequency is located at 5000–4000 cm^{-1}. If the content and proportion of organic substances inside the tea differ, they will exhibit different absorption signals in the near-infrared spectrum. Based on this principle, the SVM pattern recognition method can be used to identify genuine and counterfeit Biluochun tea.

Figure 5.31 shows the raw spectra (a) and first-order derivative spectra (b) of genuine and several counterfeit Biluochun samples. From the figure, it can be seen that the original near-infrared spectrum of tea leaves has two obvious absorption peaks near 5155 cm^{-1} and 6944 cm^{-1}, and the first-order derivative spectrum also shows significant fluctuations near these points. Since the first overtone of the O-H stretching vibration in pure water is located near 6944 cm^{-1}, and one of its combination frequency regions is near 5155 cm^{-1}, these two wavelengths are sensitive areas for water absorption. From the figure, it can be seen that in these two regions, the moisture in dry tea has a significant impact on the absorption peaks of the near-infrared spectrum. In the experiment, the moisture content of dry tea was about 5%. To reduce the influence of moisture, the spectral wavelength range was chosen to avoid the characteristic wavelength regions of water absorption peaks as much as possible during analysis. By analyzing the selected wavelength segments, it was found that using spectral data in the range of 6500–5300 cm^{-1} in the first overtone region avoided the influence of moisture and achieved better experimental results.

In the experiment, due to the inconsistency in the particle size, uniformity, and compactness of the tea samples, the corresponding near-infrared spectrum would be affected to some extent. Therefore, it is necessary to preprocess the original spectral data of the samples. There are many spectral preprocessing methods, including Multiplicative Scatter Correction (MSC), Standard Normal Variate (SNV), first-order derivative, and second-order derivative, among others. Through analysis and comparison, it was found that the preprocessing effect of SNV was significantly better than that of the first-order and second-order derivatives and slightly better than MSC. Ultimately, the SNV preprocessing method was adopted.

Many hydrogen-containing groups of organic compounds in tea leaves can produce overtone and combination band absorption in the near-infrared region. As a result, there is a significant correlation among the near-infrared spectral data of tea leaf samples, leading to a large amount of information redundancy. In model building, the inclusion of this redundant information can reduce the predictive performance of the model. Principal Component Analysis (PCA) is a statistical method that transforms multiple indicators into a few comprehensive indicators. It projects from a multi-dimensional spectral data space to a lower-dimensional data space along the direction of maximum covariance, with each principal component vector being orthogonal to each other. By selecting reasonable principal components, one can avoid information redundancy in modeling without excessively losing spectral information, while also achieving simplification in data analysis. During the training process of the model, the number of principal components has a certain impact on the model's predictive performance.

The experiment uses the number of misclassifications during training and prediction as an indicator to measure the model's quality. The impact of the number of principal components on the model is shown in Fig. 5.32. From Fig. 5.32, it can be seen that when the number of principal components is less than 11, the number of misclassifications in both the training set and the prediction set decreases as the number of principal components increases. However, when the number of principal components is greater than or equal to 11, the number of misclassifications during training remains basically unchanged as the number of principal components increases, while the number of misclassifications during prediction shows an upward trend. This is because when the number of principal components reaches 11, the cumulative variance contribution rate is 99.86%, and these 11 principal component factors can almost fully reflect the overall spectral information. Therefore, the model established with the first 11 principal component factors is optimal. Based on the above analysis, spectral data in the range of 6500–5300 cm^{-1} is selected, standardized (SNV) preprocessed, and 11 principal component factors are extracted. The 11 principal component information of 138 samples forms a matrix $X_{138 \times 11}$ as the input factor for SVM. In the design of the training algorithm, based on the above

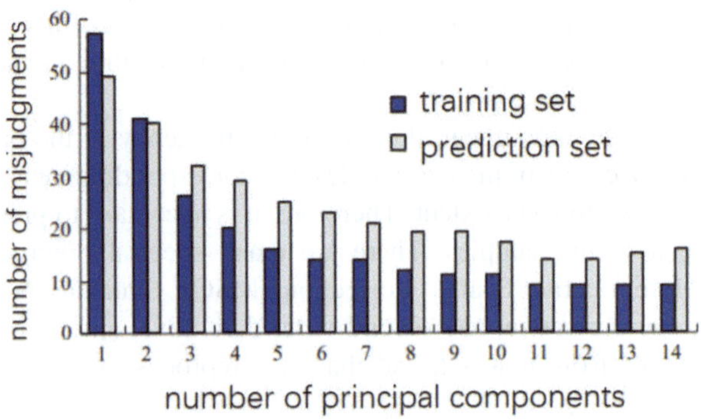

Fig. 5.32 Relationship between the number of model misclassifications and the number of principal components during training and prediction

analysis, the radial basis function (RBF) is selected as the kernel function for SVM, with a penalty constant =100, and a kernel function width parameter $\delta = 0.5$, to train and establish the discrimination model. This model achieves a reclassification discrimination rate of 93.48% for the training set samples and a prediction discrimination rate of 84.4% for the prediction set samples.

5.4.2 *Ultraviolet-Visible Spectroscopy Technology*

5.4.2.1 Principle of Ultraviolet-Visible Spectroscopy

Ultraviolet absorption spectra and visible absorption spectra both belong to molecular spectra, and they are both generated by the transition of valence electrons. The ultraviolet-visible spectrum and the degree of absorption produced by the absorption of ultraviolet and visible light by molecules or ions of a substance can be used to analyze, measure, and infer the composition, content, and structure of the substance.

When a substance M (atom or molecule) absorbs ultraviolet-visible light, it is excited to an excited state M, and returns to the ground state through radiative or non-radiative relaxation processes. Relaxation can also be achieved by the decomposition of M into new components, a process known as a photochemical reaction. It is noteworthy that the lifetime of M* is generally very short, so its concentration can be neglected at any moment. Moreover, the released heat is often immeasurable. Therefore, except for photochemical decomposition, the measurement of absorbance has the advantage of causing minimal disturbance to the studied system.

Since the ultraviolet-visible absorption spectrum of a substance is determined by the transition of valence electrons in the molecule, different molecular compositions, especially different valence electron properties, will result in different absorption spectra. Therefore, the wavelength of the absorption peak can be correlated with the type of bond present in the studied substance, achieving the purpose of identifying functional groups in the molecule. More importantly, ultraviolet-visible absorption spectroscopy can be used for the quantitative determination of compounds containing absorbing functional groups [59].

5.4.2.2 UV-Visible Spectrophotometer Testing Method

The UV-Visible spectrophotometer can be designed into four types: single-beam, double-beam, dual-wavelength spectrophotometer, and multi-channel spectrophotometer.

Single-Beam Spectrophotometer

This type of spectrophotometer uses the same single beam to sequentially pass through the reference cell and the sample cell, setting the absorbance of the reference

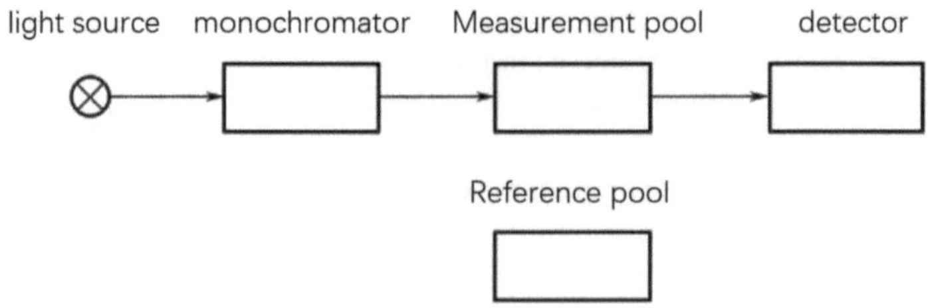

Fig. 5.33 Schematic diagram of the optical path of a single-beam spectrophotometer

cell to zero to measure the absorbance of the sample. Taking the 752 models instrument as an example, its wavelength range is 200 ~ 1000 nm, with a hydrogen lamp as the UV light source and a tungsten lamp as the visible light source. The grating serves as the dispersive element, and the detectors include a UV-sensitive phototube (suitable for the wavelength range of 200 ~ 625 nm) and a red-sensitive phototube (suitable for the wavelength range of 625 ~ 1000 nm). The absorption cells are made of glass and quartz, suitable for the visible and UV regions, respectively. It is a relatively precise and reliable instrument suitable for quantitative analysis and can be used for the conversion measurement of absorption coefficients.

Figure 5.33 is a schematic diagram of the optical path of a single-beam spectrophotometer. Its structure is relatively simple and easy to operate, but its drawback is that the light source and detection system must have high stability, and it cannot perform automatic scanning. Each wavelength change requires blank correction. Other spectrophotometers of this type include the 751 and 753 models produced by Shanghai Analytical Instrument Factory, the WFD-8 model by Tianjin Optical Instrument Factory, the Unican SP500 model from the UK, the Beckman DU from the USA, and the QV-50 from Shimadzu, Japan.

Additionally, simple spectrophotometers suitable only for the visible region, such as the 721, 722, and RX-727 models, also belong to single-beam spectrophotometers. These spectrophotometers have only one light source (tungsten lamp or halogen tungsten lamp), with a working wavelength range of 330 ~ 830 nm. The dispersive element is a glass prism, and the detector uses a phototube. The instrument structure is simple and inexpensive, but the monochromatic light purity is relatively poor [60].

Double-Beam Spectrophotometer

The double-beam optical path is widely used. Figure 5.34 shows the optical circuit of the Shimadzu UV-2100 double-beam spectrophotometer. The monochromatic light emitted from the monochromator is divided into two alternating single beams by a rotating sector mirror (also known as a chopper), passing through the blank solution and the sample solution, respectively. The same synchronous sector mirror then alternately projects the two beams onto the photomultiplier tube, generating an alternating pulse signal. After comparison and amplification, the display shows transmittance, absorbance, concentration, or performs wavelength scanning to

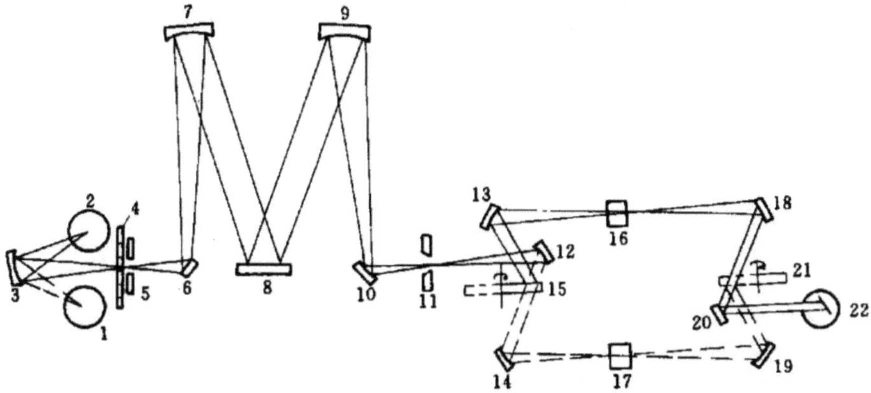

Fig. 5.34 Optical circuit diagram of Shimadzu UV-2100 Double Beam Spectrophotometer (1-tungsten lamp; 2-deuterium lamp; 3-concave mirror; 4-filter; 5-incidence slit; 6,10,20-flat mirror; 7,9-collimating mirror; 8-grating; 11- outgoing slit; 12–14,18,19-sector mirror; 15,21-sector mirror; 16- Reference cell; 17- Sample; 22 -photomultiplier tube)

record the absorption spectrum. The sector mirror rotates at a uniform speed of tens to hundreds of revolutions per second, allowing the monochromatic light to alternately pass through the blank and sample solutions in a very short time, reducing errors introduced by unstable light source intensity. The absorption cell does not need to be moved during measurement, allowing the photometric value to be recorded while changing the wavelength at will, facilitating the depiction of the absorption spectrum. Common instruments of this type include the 740 model from Shanghai Third Analytical Instrument Factory, the Unican SP 700, SP 1700, SP 1800 models from the UK, the Hitachi 200–20 from Japan, and the Shimadzu UV-2100 [61].

Dual-Wavelength Spectrophotometer

In the single-wavelength spectrophotometry measurement introduced above, two absorption cells are used, one containing the reference solution, with its transmittance adjusted to 100% (i.e., absorbance A = 0) at the selected wavelength, and then the absorbance of the sample solution is measured. When the test solution contains two components with overlapping absorption spectra, it becomes difficult to measure the absorbance of the component to be tested using this single-wavelength spectrophotometer alone, requiring extraction separation or the addition of masking agents to complete the measurement. On the other hand, since two absorption cells must be used, the difference between the absorption cells often affects the measurement accuracy, limiting the determination of more trace components. The dual-wavelength spectrophotometer overcomes these shortcomings, as shown in the block diagram in Fig. 5.35.

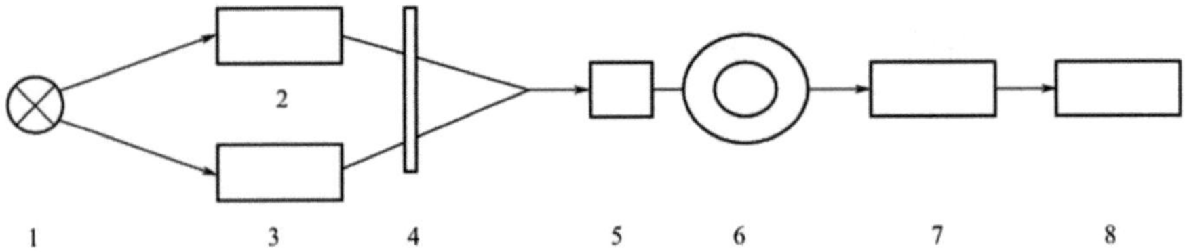

Fig. 5.35 Block diagram of a dual wavelength spectrophotometer 1. Light source 2, 3. Monochromator 4. Light cutter 5. Absorption cell 6. Detector 7. Electronic control system 8. Digital voltage

The spectral characteristics of pure egg yolk, Sudan IV, and spiked samples can be observed. The light emitted from the light source is split into two beams, which pass through their respective monochromators 2 and 3, resulting in two monochromatic beams with wavelengths λ1 and λ2. These beams are modulated by a light cutter and alternately illuminate the absorption cell containing the sample solution at certain time intervals. Through the photodetection and electronic control system, the digital voltmeter displays the transmittance difference ΔT or the absorbance difference ΔA between λ1 and λ2. According to Beer's law, the transmittance difference ΔA at the two wavelengths λ 1 and λ 2 is proportional to the concentration of the substance to be measured in the solution. The dual-wavelength spectrophotometry can set any wavelength of the absorption spectrum of the component to be measured as the zero point and measure the absorbance difference between it and any other wavelength [62].

Multi-Channel Spectrophotometer

The multi-channel spectrophotometer is based on a single-beam spectrophotometer and uses a multi-channel photon detector. It features rapid scanning, with the entire spectrum scan time being less than 1 second. This provides an extremely convenient means for tracking chemical reaction processes and studying rapid reactions, allowing for direct qualitative and quantitative analysis of samples separated by liquid chromatography columns and capillary electrophoresis columns. However, the resolution of this type of instrument is only 1 ~ 2 nm, and it is relatively expensive [63].

5.4.2.3 Application in the Detection in Foods

UV/Vis spectrophotometry is widely used in food testing due to its high sensitivity, accuracy, ease of operation, and wide applicability.

1. Application of nutritional component detection in foods.

   Protein and polysaccharides are important nutrients for the human, UV/VIS could be used for the detection of nutrients such as proteins and polysaccharides in foods. In addition, UV/VIS was also used for the detection of health foods such as flavonoids and saponins in foods.

2. Application in food additives detection.

   Food additives are used to improve the quality, color, aroma, and taste of food, as well as to prevent spoilage artificial or natural substances added to food for preservation and processing needs quality. UV/VIS was also used for quantitative and qualitative detection of common additives such as food coloring agents, preservatives, sweeteners, nutritional enhancers, antioxidants, etc.

3. Application harmful substances detection in food.

   UV/Vis spectrophotometry is an effective technique for detecting harmful substances in food, such as pesticides, heavy metals, and food additives. By analyzing the absorption spectra of food samples, this method can identify specific compounds based on their characteristic absorption peaks. The degree of absorption correlates with the concentration of these harmful substances, enabling rapid screening and quantification. This application is vital for ensuring food safety and compliance with regulatory standards, helping to protect consumer health and enhance food quality.

5.4.3 Fluorescence Detection Technology

5.4.3.1 Principle of Fluorescence Detection Technology

According to the different wavelengths emitted by fluorescent substances under excitation light, fluorescence can be divided into X-ray fluorescence, ultraviolet fluorescence, visible fluorescence, and infrared fluorescence. This section mainly introduces the principle and examples of using visible fluorescence excited by ultraviolet light for rapid and non-destructive detection of food and agricultural products.

Any substance capable of emitting fluorescence has two characteristic spectra: the excitation spectrum and the fluorescence spectrum. By using monochromatic light of different wavelengths to excite the fluorescent substance to emit light, and then measuring the fluorescence intensity produced by each wavelength of excitation light, plotting excitation light wavelength λ as the horizontal axis and fluorescence intensity I as the vertical axis, the excitation spectrum of the fluorescent substance can be obtained. In fact, the excitation spectrum is the apparent absorption spectrum of the fluorescent substance. The fluorescence excitation spectrum can be used to identify fluorescent substances and select suitable excitation light for fluorescence measurement. The excitation spectrum indicates the relative efficiency of fluorescence produced by excitation light of different wavelengths. The wavelength of the highest peak in the excitation spectrum can cause the fluorescent substance to emit the strongest fluorescence.

By fixing the excitation light wavelength and intensity, and allowing the fluorescence emitted by the substance to pass through a monochromator for dispersion, the fluorescence intensity at different wavelengths can be measured. Plotting the fluorescence wavelength λ as the horizontal axis and fluorescence intensity I as the vertical axis, the fluorescence spectrum, also known as the fluorescence emission spectrum, is obtained. The fluorescence spectrum represents the relative intensity of various wavelength components in the fluorescence produced by the substance, and the shape of the fluorescence spectrum is independent of the excitation light wavelength. Generally, the fluorescence of gaseous and vaporous substances appears as a line spectrum, while the fluorescence of liquid and solid substances appears as a band spectrum.

5.4.3.2 Application in Food

Fluorescence detection technology is a powerful analytical method that relies on the emission of light from a substance after it has absorbed photons, typically from ultraviolet or visible light sources. When molecules or ions absorb energy, their valence electrons are excited to higher energy states, and as these electrons return to their ground state, they release energy in the form of fluorescence. This emitted light can be measured to provide insights into the composition, concentration, and molecular structure of the substance being analyzed. Fluorescence detection is widely used in various fields, including biochemistry, environmental monitoring, and medical diagnostics, due to its sensitivity and ability to detect low concentrations of analytes [64].

5.4.4 Delayed Luminescence Detection Technology

It has been discovered that all photosynthetic organisms can exhibit a phenomenon known as "afterglow." For example, plants can continue to emit faint red light after being exposed to light for a period and then placed in the dark. This afterglow is called "delayed luminescence," also known as "delayed fluorescence," which is distinct from "prompt fluorescence" that stops once the excitation light is removed. Delayed luminescence is a fluorescence phenomenon closely related to various life activities of food and agricultural products, such as oxidative metabolism, cell division, apoptosis, growth regulation, and photosynthesis. Therefore, the chlorophyll content in agricultural products will directly affect the state of delayed luminescence. Although the mechanism of delayed luminescence is not yet fully understood, it is certain that it is closely related to the primary reactions of photosynthesis, electron transfer, and photophosphorylation processes, and it is also produced along with photosynthesis.

5.5 Image Detection Technology

5.5.1 *Visible Imaging Technology*

Visible light imaging technology refers to a technique that converts the target being captured into image signals and processes them within the electromagnetic wavelength range (400–780 nm) that can be directly perceived by the human eye. Initially, various calculations are used to extract the features of the target, such as area, quantity, position, and length. Then, based on preset tolerances and other limiting conditions, results are output, including size, angle, number, pass/fail, presence/absence, to achieve automatic recognition and control the actions of on-site equipment. Therefore, it is often used to replace human vision in large-scale industrial production processes, improving production efficiency and the degree of automation [65, 66].

A visible light imaging system typically consists of a light source, lens, imaging chip, image acquisition card (or simply acquisition card), and computer. The light source emitter emits light waves that illuminate the inspected product, which, after reflection, refraction, or transmission, enters the optical lens for spectral processing. The optical signal is converted into an analog image signal by the imaging chip and then into a digital signal by the acquisition card. The computer receives the digital signal of the image, stores it in the memory area, and analyzes the image information through corresponding image processing software [67].

Image processing technology can be basically divided into two categories: analog image processing and digital image processing, with digital image processing being the most commonly used method. Digital image processing generally uses real-time computer processing; hence it is also called computer image processing. Its advantages include high accuracy, rich processing content, the ability to perform complex nonlinear processing, and strong adaptability. Generally, changing the software can change the content. However, when performing complex image processing, the processing speed is relatively slow, especially since resolution and accuracy have a significant impact on processing speed; the higher the accuracy and resolution, the slower the processing speed [68].

5.5.2 *Applications in Quality Analysis of Food and Agricultural Products*

Currently, visible imaging technology has been widely applied in the quality inspection and evaluation of foods and agricultural products such as fruits, vegetables, meat, pizza, cheese, and bread, which can be summarized in the following aspects.

5.5.2.1 Application in Size and Shape Appearance Detection

Appearance size is an important basis for the grading of foods and agricultural products. By using visible light imaging technology combined with image processing methods, it is possible to determine the information of the tested object by detecting parameters such as perimeter, area, the orientation of dimples, and pore volume ratio. In addition, product shape is also an important aspect of appearance size detection, especially for fruits, where the quality of shape is a crucial grading indicator. Domestically and internationally, visible light imaging processing systems have been developed based on the shape characteristics of puffed foods, eggs, apples, and other foods and agricultural products, achieving automated detection of these products [69].

5.5.2.2 Application in Color Detection

Color is an important characteristic for evaluating the quality of foods and agricultural products. Using visible light imaging systems to evaluate their color can overcome human eye fatigue and differences. Differences in color characteristics can identify defects in fresh peaches and distinguish moldy soybeans, among others. Quality control of baked goods is a key step in the processing process, which previously relied solely on manual qualitative judgment. Now, visible light imaging technology can be used to detect the color of bread or other baked goods. Some food technicians are also attempting to use visible light imaging technology to detect the color of pizza to achieve quality control [70–73].

5.5.2.3 Application in Texture Feature

Detection color conventional imaging technology combined with image processing can obtain color images of the cross-section of puffed foods and perform texture analysis on the images, thereby evaluating the quality of puffed foods. There is a strong correlation between beef texture characteristics and tenderness, and the quality of beef can be analyzed using color conventional imaging technology combined with texture analysis methods [74].

5.5.2.4 Application in Surface Defect and Damage Detection

Detecting surface defects and damage has always been a challenge in the grading of food and agricultural products. By using visible light imaging technology combined with corresponding image processing methods, these morphological features can be quickly detected. For example, a stereoscopic color histogram based on a neural network classifier can analyze defects on the entire chicken carcass. Cracked eggs are susceptible to invasion by external microorganisms, leading to reduced freshness and even spoilage. The removal of cracked eggs is an important step in the egg product processing flow. Visible light imaging technology combined with morphological analysis methods can detect egg damage. Additionally, visible light imaging technology can also detect features such as the breakage rate and surface insect damage of rice grains [75].

5.5.2.5 Applications in Component Detection

Visible light imaging technology has many applications in the detection of food and agricultural product components. For instance, the lysine content can be determined based on the color characteristics of the digital image of a pizza base. Digital image processing technology can analyze features such as the uniformity of the cross-sectional size of bread, the thickness of the crust, and the brightness of the bread core to determine the role of emulsifiers in bread baking. Visible light imaging technology can also be used to evaluate cheese melting and browning. This new non-contact method is used to analyze the characteristics of cheddar and mozzarella cheese during cooking. The results show that this method can provide an objective and simple way to analyze the functional properties of cheese.

5.5.3 Spectral Imaging Technology

5.5.3.1 Basic Overview of Spectral Images

Spectral imaging technology (Spectral Imaging) is a new technology developed in the 1980s, integrating advanced technologies from fields such as optics, optoelectronics, electronics, information processing, and computer science, organically combining traditional two-dimensional imaging technology with spectral technology. Hyperspectral imaging technology (HIS) is characterized by an ultra-multi-band, high spectral resolution, and the integration of image and spectrum. In 1983, the first spectral imaging device, AIS-1, was successfully developed in this laboratory, demonstrating its great potential in quantitative research. Countries like Canada, Australia, France, and Germany have also invested heavily in the development and application research of hyperspectral imaging technology. Initially, this technology was mainly used in the field of space remote sensing, but it has now expanded to areas such as medical diagnosis, drug, and food analysis, and has seen significant development in aspects such as spectral image data acquisition, spectral calibration, three-dimensional data reconstruction, data processing analysis, and pattern recognition [76].

Spectral imaging technology combines image analysis and spectral analysis, offering unique advantages in the quality and safety detection of food and agricultural products. A spectral image is a three-dimensional data block, which refers to a

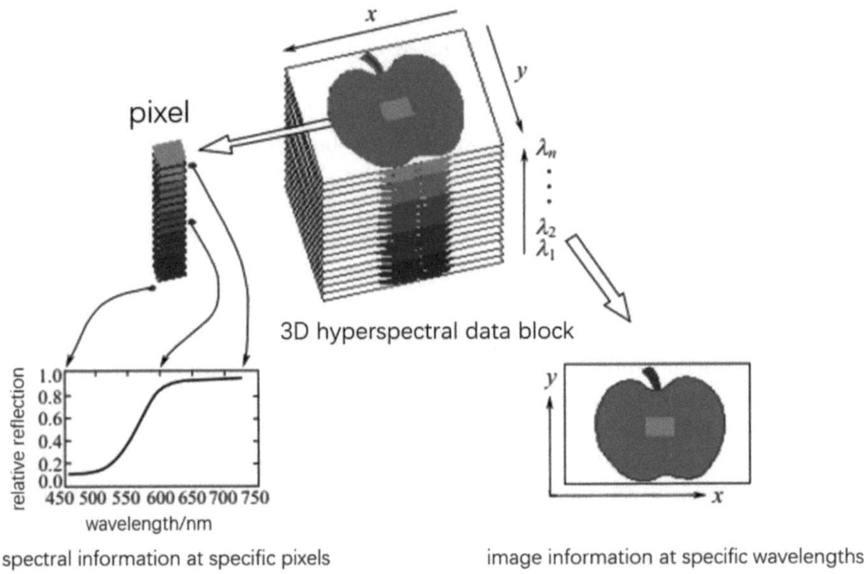

Fig. 5.36 Spectral image 3D data block format

series of two-dimensional images obtained simultaneously at continuous wavelengths within a specific spectral range using a spectroscopic system, as shown in Fig. 5.36. At each specific wavelength, spectral data can provide two-dimensional image information, while the grayscale of the same pixel at different wavelengths provides spectral information. Image information can characterize the size, shape, and color of agricultural products, while spectral information can reflect internal structure, component content, and other characteristic information of agricultural products. Thus, spectral imaging technology can perform visual analysis of the internal and external quality characteristics of agricultural products [77].

With the development of science and technology, the precision of spectral resolution in imaging spectroscopy is becoming increasingly higher. Based on different spectral resolutions, spectral images are divided into multi-spectral images, hyperspectral images, and ultra-spectral images. Generally, images with a spectral resolution in the order of $10^{-1}\ \lambda$ are called multi-spectral images, those in the order of $10^{-2}\ \lambda$ are called hyperspectral images, and those in the order of $10^{-3}\ \lambda$ are called ultra-spectral images. Depending on the precision and requirements of detection, different resolutions of spectral imaging technology are generally selected. Typically, for the detection of food and agricultural product quality and safety, hyperspectral imaging technology is generally chosen. Therefore, the spectral imaging technology characteristics listed in subsequent chapters of this book mainly refer to hyperspectral imaging technology.

As spectral resolution improves, the quality information of agricultural products that hyperspectral images can record becomes increasingly rich, which should be very beneficial for agricultural product quality detection. However, in practical applications, it is not so simple. Hyperspectral images have many unique characteristics that differ greatly from people's inherent prior knowledge about images. The lack of intuitive knowledge about hyperspectral images or spectra makes the effectiveness of using hyperspectral images for agricultural product quality detection fall short of people's expectations.

The advantage of hyperspectral image detection is that it can achieve comprehensive detection of the internal and external quality of agricultural products through rich spectral information [78]. In this regard, many different methods have been proposed, which can be mainly divided into two categories.

Detection Methods Based on Spectral Space

These methods are relatively traditional and commonly used in hyperspectral image detection. Common methods such as spectral matching and mixed spectral decomposition belong to this category. These methods require prior knowledge of the actual spectral parameters of the quality indicators of the detection object, which are then matched with the spectral parameters extracted from the hyperspectral image. Subsequently, the quality of the detection object is obtained through pattern recognition technology. These methods have issues such as spectral reconstruction and spectral variation uncertainties. Therefore, when using these methods for agricultural product quality detection, besides solving the spectral reconstruction problem of agricultural products, improving the spectral absorption characteristics of the detection object is also crucial.

Detection Methods Based on Feature Space

Another idea for hyperspectral image detection is to start from the distribution pattern of the feature space, extracting different distribution characteristics (such as different statistical parameters) presented by different quality indicators on the image of the same detection object, thereby achieving quality detection and recognition of the detection object. These methods first need to analyze the characteristics exhibited by different quality indicators of the detection object and their distribution differences with background characteristics. Then, through corresponding feature extraction methods, the structure of the detection object in the image is highlighted, and finally, the quality indicators of the detection object are extracted through image segmentation algorithms.

Based on the feature space, there are two main hyperspectral image detection schemes: The first is to perform detection operations directly in the original feature space, which can be represented by Fig. 5.43. Due to the many issues associated with analyzing data in high-dimensional space, this scheme struggles to fully leverage the advantage of the large information content in hyperspectral images. The second scheme is based on feature extraction and selection, which involves using feature extraction or selection methods to map the original data into a low-dimensional

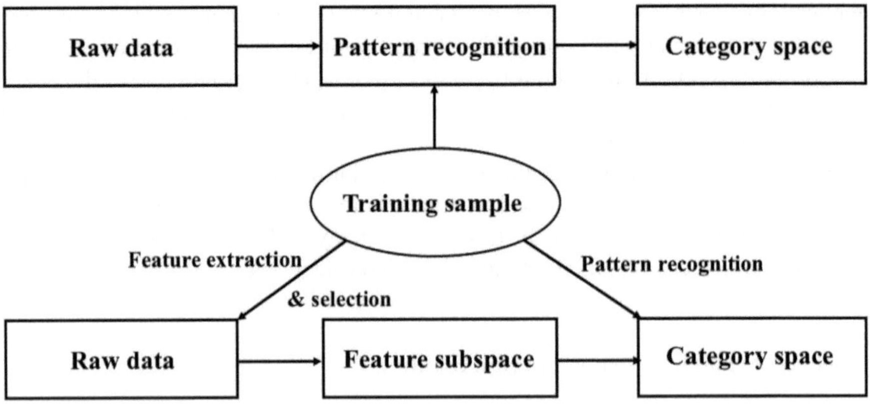

Fig. 5.37 Testing programs

feature subspace before conducting detection. This scheme can be represented by Fig. 5.37.

In the second scheme, the key to hyperspectral image detection depends on the following two aspects: The first is the feature extraction or selection algorithm. This is the most crucial aspect, as the ability to reflect the essential characteristics of detection indicators in the feature subspace is key to improving detection accuracy. Many studies have been conducted in this area. The second is the detection and classification algorithm. Theoretically, traditional machine learning algorithms (statistical pattern recognition, neural networks, etc.) should still be applicable in hyperspectral image analysis, but these methods require a large number of samples to train classifiers, which is often not feasible in hyperspectral image analysis. Therefore, researching machine learning methods suitable for small samples is an important research direction in hyperspectral image detection. For example, the support vector machine learning method developed in recent years has studied the issue of improving classification accuracy in small sample situations.

5.5.3.2 Spectral Imaging Methods

The hardware components of spectral imaging detection technology mainly include a light source, a CCD camera, a computer equipped with an image acquisition card, and a monochromator. The spectral range of the light source can be in the ultraviolet (200–400 nm), visible light (400–760 nm), near-infrared (760–2500 nm), and regions with wavelengths greater than 2500 nm. The camera can receive light reflected or transmitted from the surface of an object and convert the light signal into an electrical signal through a CCD sensor. CCD sensors are divided into line array (obtaining one-dimensional image signals per exposure) and area array (obtaining two-dimensional image signals per exposure), with the latter being more costly than the former. The image acquisition card converts the analog signal obtained by the CCD into a digital signal and displays it through a computer. The monochromator is used to obtain light of specific wavelengths, which can be achieved through two methods: filters (filter plates) and imaging spectrometers. Therefore, two different monochromators provide two methods for constructing hyperspectral image detection systems.

Filter-Based Spectral Imaging System

The filter-based hyperspectral image detection system is shown in Fig. 5.38. This method uses an imaging device mainly composed of a CCD camera and components for wavelength selection. Common wavelength selection components include narrowband filters, liquid crystal tunable filters, and acousto-optic tunable filters. The method for acquiring hyperspectral images is by continuously collecting a series of two-dimensional images of samples under different wavelength conditions (Fig. 5.39), for each wavelength K*i*($i = 1, 2, 3, \ldots, n$ where n is a positive integer), a two-dimensional image (with horizontal coordinate x and vertical coordinate y) is obtained, thus forming a three-dimensional image block (x, y, K).

Imaging Spectrometer-Based Spectral

The image spectrometer was capable of acquiring spectral information of objects in the visible and near-infrared spectral range. The hardware components include four 20 W halogen lamps (DECOSTAR51 MR16 type, OSRAM, Germany), an imaging spectrometer (ImSpector V10E, Specim, Finland), a CCD camera (C8484–05G, Hamamatsu, Japan), a displacement stage, a computer equipped with an image

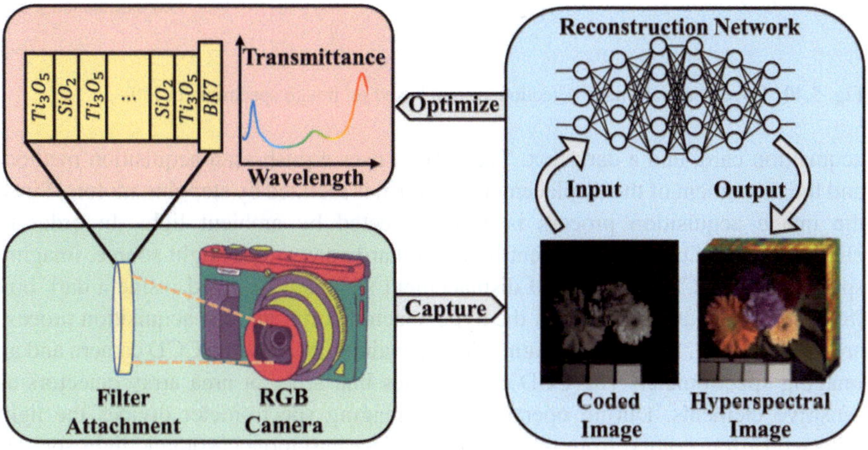

Fig. 5.38 Filter-based hyperspectral image system [79]

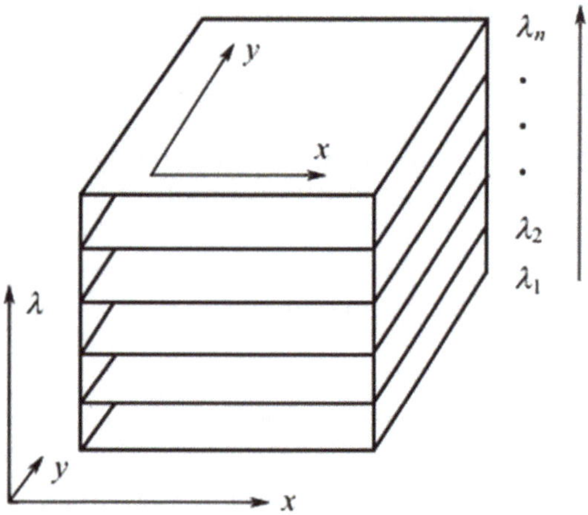

Fig. 5.39 Schematic of the image data obtained by the filter-based hyperspectral image system

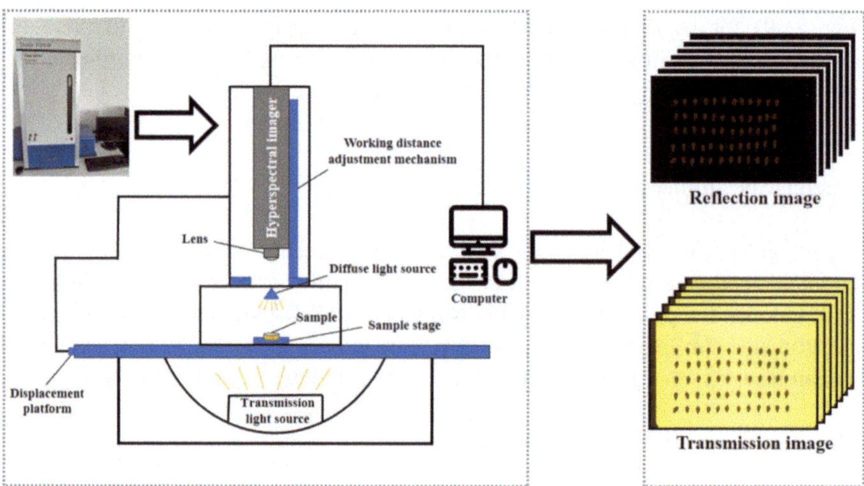

Fig. 5.40 Hyperspectral image detection system based on image spectrometer [80]

acquisition card, and a dark box. The system uses a push-scan acquisition method, and the movement of the displacement platform is realized by stepping motors. Since the image acquisition process is easily affected by ambient light. In order to eliminate the influence of ambient light, the hardware such as light source, imaging spectrometer, CCD camera, and displacement stage are installed inside a dark box [80]. The schematic structure of the hyperspectral sorter and the acquisition process are shown in Fig. 5.40. This imaging device mainly consists of a CCD camera and an imaging spectrometer. The CCD camera uses line array or area array detectors as sensitive elements. During operation, the imaging spectrometer divides the light reflected or transmitted from the detection object into monochromatic light sources before entering the CCD camera. This system uses a "broom" imaging method to obtain hyperspectral images: line array or area array detectors are arranged transversely in the optical focal plane to complete lateral scanning (x-axis direction), allowing the acquisition of image information for each pixel in the strip space at various wavelengths; simultaneously, during the advancement of the detection system's conveyor belt, the arranged detectors sweep out a strip-like trajectory like a brush sweeping the floor, thus completing longitudinal scanning (y-axis direction). By integrating horizontal and vertical scanning information, the three-dimensional hyperspectral image data of the sample can be obtained. In the detection of agricultural product quality and safety, the sample size in the direction of the production line is not limited by the size of the CCD camera's shooting range.

5.5.3.3 Application of Hyperspectral Imaging Technology in Foods

HSI is a potentially attractive process analysis technique for food applications, with the robust, non-destructive, and flexible characteristics are widely used in agricultural products, defect identification, component analysis, and food quality and safety detection.

Application in Agricultural Product Detection

HSI technology has been successfully applied to non-destructive testing of quality in vegetables and fruits. For example, by combining multiple algorithms, effectively extracting deep spectral features and establishing a robust heavy metal content prediction model for crop toxic stress, early detection to effectively monitor its growth status. By using spectral imaging and chemometric methods, it is possible to quickly distinguish sensory qualities such as tea appearance and aroma.

Application in Meat Quality Detection

During the storage and transportation of meat, the protein components gradually decompose, producing toxic small molecules such as histamine, tyramine, and serotonin, causing meat to spoil. HSI has been successfully applied to detect the freshness of pork and chicken meat, realizing meat quality control.

Application in Aquatic Product Detection

Fish bones often appear in fish products and are prone to ingestion, which may cause problems such as scratching the throat. HSI has been successfully applied to detect the bones in fish. Furthermore, HSI also used in freshness detection of aquatic products.

Application in Other Food Detection

In addition to the aforementioned applications, it is also widely used in many other fields. For example, HSI can extract image information from natural foods, detect changes in the quality of processed foods, thereby improving the level of food quality detection.

5.5.4 X-Ray Imaging Technology

In 1895, Rontgen discovered X-rays, in order to commemorate him, people will X-rays also called 'Rontgen rays." X-rays are a short wavelength electromagnetic waves, wavelength range of 0.001–10 nm, between ultraviolet and γ-rays, the energy of the photons in the 120–1.2×10^6 electron volts, much larger than the visible light, which shows obvious particles and strong penetration, is much larger than that of visible light, showing obvious particles and strong penetrability. With other electromagnetic waves, X-rays can produce reflection, refraction, scattering, interference, diffraction, polarization, and absorption phenomena. According to their energy size is divided into hard X-rays and soft X-rays. Hard X-rays are more energetic, more penetrating, and are often used to detect materials with large thickness and density; soft X-rays are less energetic, less penetrating, and are often used in agricultural quality inspection [81].

X-ray penetration ability is very strong, its image can intuitively reflect the internal defects of food and agricultural products, structural and organizational changes in the quality of the situation, the application of X-rays can be detected, such as bone residues in livestock products, apples, bruises, rot and water heart disease, potatoes, watermelon, the internal cavity, wrinkled skin in citrus, agricultural products and other defects of pests and diseases, etc. X-ray imaging technology in the detection of the quality of the internal products of agricultural and livestock products has a lot of potential for application. X-ray imaging technology has a great potential to be applied in the internal quality inspection of agricultural and livestock products, which is getting more and more attention.

5.5.4.1 X-Ray Inspection Devices

Figure 5.41 shows the experimental setup established for SG-DFXI studies. The setup primarily comprised of a microfocus X-ray tube (Oxford Instruments, UltraBright 96,000 Series) with a 20 μm focal spot, which was operated at 30 kVp, a focused linear X-ray grid (JPI Co, Grid 2100) with strip densities in the range of 6.99–13.16 lines/mm, and a complementary metal-oxide-semiconductor (CMOS) flat panel detector with a 49.5 μm pixel size (Rayence Co, Xmaru 1215). The average X-ray beam energy was evaluated to be approximately 10.7 keV using

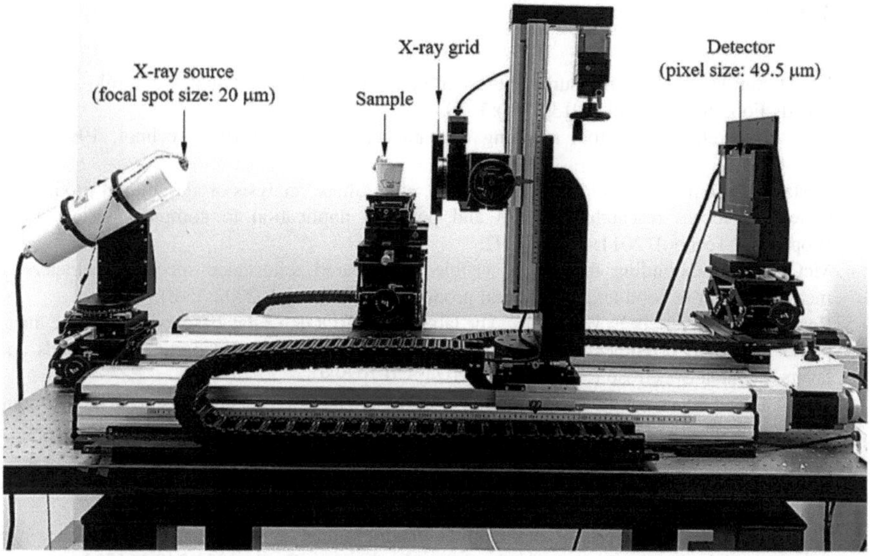

Fig. 5.41 Experimental setup established for SG-DFXI studies. The setup primarily comprised of a microfocus X-ray tube with a 20 μm focal spot, a focused linear X-ray grid with strip densities in the range of 6.99–13.16 lines/mm, and a CMOS flat panel detector with a 49.5 μm pixel size [82]

the SRS-78 program, which corresponds to an X-ray wavelength of $\lambda = 0.116$ nm [82].

The X-ray generator is used to generate and control the energy needed to detect materials, and it consists of several parts, including a high-voltage generator, an X-ray tube, and an X-ray controller. Among them, the X-ray tube is the core of the X-ray source, and its basic structure is a high-vacuum diode, consisting of a cathode, anode and a glass shell that maintains a high degree of vacuum. The detector receives X-rays that penetrate the inspected material, and converts the received X-rays into visible light through a fluorescent screen. The photodiode in the detector unit is irradiated by the visible light and produces a voltage signal, which is processed by an integrated circuit (ASIC) into a digital signal that is transmitted to a computer.

5.5.4.2 Application of X-Ray Detection in Foods

The penetration ability of X-rays is strong, and their attenuation degree is closely related to the density, thickness, and composition of the material to be tested. Therefore, the X-ray image can reflect the physical and chemical properties of the material to a certain extent. X-ray technology is used for detecting metal impurities in food, detecting insect infestations inside chestnut shells, and detecting internal defects in fruits [83].

References

1. Wu D, Sun D-W. Color measurements by computer vision for food quality control—a review. Trends Food Sci Technol. 2013;29(1):5–20.
2. Kays SJ. Preharvest factors affecting appearance. Postharvest Biol Technol. 1999;15(3): 233–47.
3. Costa C, Antonucci F, Pallottino F, Aguzzi J, et al. Shape analysis of agricultural products: a review of recent research advances and potential application to computer vision. Food Bioprocess Technol. 2011;4(5):673–92.
4. McCaig TN. Extending the use of visible/near-infrared reflectance spectrophotometers to measure colour of food and agricultural products. Food Res Int. 2002;35(8):731–6.
5. Renard J-B, Hadamcik E, Worms JC. The laboratory PROGRA2 database to interpret the linear polarization and brightness phase curves of light scattered by solid particles in clouds and layers. J Quant Spectrosc Radiat Transf. 2024;320:108980.
6. Ghosh T, Sen AK. Change in the direction of the polarization vector and redshift of an incoming light ray as observed from a rotating frame. Can J Phys. 2019;98(5):433–41.
7. Schaare PN, Fraser DG. Comparison of reflectance, interactance and transmission modes of visible-near infrared spectroscopy for measuring internal properties of kiwifruit (Actinidia chinensis). Postharvest Biol Technol. 2000;20(2):175–84.
8. Bhattacharjee PR. Discovery of total failure of the traditional laws of reflection and refraction of light to explain the phenomena of reflection and refraction. Optik. 2021;240:166923.
9. Salido EM, Servalli LN, Gomez JC, Verrastro C. Phototransduction early steps model based on Beer-Lambert optical law. Vis Res. 2017;131:75–81.
10. León K, Mery D, Pedreschi F, León J. Color measurement in L∗a∗b∗ units from RGB digital images. Food Res Int. 2006;39(10):1084–91.
11. Schweiggert R. Chapter 4 – Physical and biological fundamentals of color. In: Schweiggert R, editor. Handbook on natural pigments in food and beverages (Second Edition). Woodhead Publishing; 2024. p. 75–126.
12. Ling H. Understanding of the color space of standard chromaticity system[J]. Packaging Engineering. 2003;02:13–4,16. (In Chinese).
13. Blasco J, Aleixos N, Moltó E. Machine vision system for automatic quality grading of fruit. Biosyst Eng. 2003;85(4):415–23.
14. Chakraborty SK, Subeesh A, Dubey K, Jat D, et al. Development of an optimally designed real-time automatic citrus fruit grading–sorting machine leveraging computer vision-based adaptive deep learning model. Eng Appl Artif Intell. 2023;120:105826.
15. Huynh HX, Lam BH, Le HVC, Le TT, et al. Design of an IoT ultrasonic-vision based system for automatic fruit sorting utilizing size and color. Internet of Things. 2024;25:101017.
16. Spence C. Chapter 2 – On the psychological effects of food color. In: Schweiggert R, editor. Handbook on natural pigments in food and beverages (Second Edition). Woodhead Publishing; 2024. p. 33–60.
17. Sheng Q, Zheng J, Jing Z, Peng C, et al. Morphology reconstruction and defect detection of holes inner surface based on reflection characteristics. Opt Lasers Eng. 2024;179:108266.
18. Xu H, Wang S, Jin J, Duan Y, et al. A large-dynamic range conductivity-modulated PANI online analysis and sensing platform for ammonia gas detection using coupled optical probing mode. Sensors Actuators B Chem. 2024;418:136280.
19. Yin H, Xie B, Chen B, Ma J, et al. Detection of moisture content and size of pumpkin seeds based on hyperspectral reflection and transmission imaging techniques. J Food Compos Anal. 2023;124:105651.
20. Hu D, Guo T, Sun X, Lian K, et al. Internal quality evaluation of 'Fuji' apples during storage based on bulk optical properties or diffuse reflection and transmission spectra. LWT. 2024;200: 116202.

21. Jia Y, Cheng Z, Wang G, Shuang S, et al. Nitrogen doped biomass derived carbon dots as a fluorescence dual-mode sensing platform for detection of tetracyclines in biological and food samples. Food Chem. 2023;402:134245.
22. Yang J, Fan Y, Lu Z, Guo Y, et al. Positive profile of natural small molecule organic matters on emerging antivirus pharmaceutical elimination in advance reduction process: a deep dive into the photosensitive mechanism of triplet excited state compounds. Water Res. 2024;256:121611.
23. Sun F, Li P, Wu G, He F, et al. Carbon nanomaterials-based smart dual-mode sensors for colorimetric and fluorescence detection of foodborne hazards. Trends Food Sci Technol. 2024;152:104681.
24. Zhang M, Liu C, Li H, Teng S, et al. Scattering analysis about the visible diffraction orders of the reflective grating. Optik. 2013;124(14):1936–9.
25. Parker SF. A review of the theory of Fourier-transform Raman spectroscopy. Spectrochim Acta A: Mol Spectrosc. 1994;50(11):1841–56.
26. Ismail AA, Van de Voort FR, Sedman J. Chapter 4 Fourier transform infrared spectroscopy: principles and applications. In: Paré JRJ, Bélanger JMR, editors. Techniques and instrumentation in analytical chemistry, vol. 18. Elsevier; 1997. p. 93–139.
27. Li X, Gao M, Liu J, Li Y, et al. Design of MWIR hyperspectral imagers based on acousto-optic tunable filters. Optik. 2023;276:170636.
28. Xu R, Shen Z, Xiang Y, Huang J, et al. Portable electrochemiluminescence detection system based on silicon photomultiplier single photon detector and aptasensor for the detection of tetracycline in milk. Biosens Bioelectron. 2023;220:114785.
29. Caccia M, Nardo L, Santoro R, Schaffhauser D. Silicon photomultipliers and SPAD imagers in biophotonics: advances and perspectives. Nucl Instrum Methods Phys Res, Sect A. 2019;926: 101–17.
30. Grassi S, Alamprese C. Advances in NIR spectroscopy applied to process analytical technology in food industries. Curr Opin Food Sci. 2018;22:17–21.
31. Wang Y, Xing L, He H-J, Zhang J, et al. NIR sensors combined with chemometric algorithms in intelligent quality evaluation of sweet potato roots from 'Farm' to 'Table': progresses, challenges, trends, and prospects. Food Chem: X. 2024;22:101449.
32. Nagy MM, Wang S, Farag MA. Quality analysis and authentication of nutraceuticals using near IR (NIR) spectroscopy: a comprehensive review of novel trends and applications. Trends Food Sci Technol. 2022;123:290–309.
33. Xie L, Ye X, Liu D. Prediction of titratable acidity, malic acid, and citric acid in bayberry fruit by near-infrared spectroscopy. Food Res Int. 2011;44(7):2198–204.
34. Ebrahimi-Najafabadi H, Leardi R, Oliveri P. Detection of addition of barley to coffee using near infrared spectroscopy and chemometric techniques. Talanta. 2012;99:175–9.
35. Ding X, Guo Y, Ni Y. A novel NIR spectroscopic method for rapid analyses of lycopene, total acid, sugar, phenols and antioxidant activity in dehydrated tomato samples. Vib Spectrosc. 2016;82:1–9.
36. Lin H, Chen Q, Zhao J. Determination of free amino acid content in Radix Pseudostellariae using near infrared (NIR) spectroscopy and different multivariate calibrations. J Pharm Biomed Anal. 2009;50(5):803–8.
37. Borin A, Ferrao MF, Mello C. Least-squares support vector machines and near infrared spectroscopy for quantification of common adulterants in powdered milk. Anal Chim Acta. 2006;579(1):25–32.
38. Chen Q, Zhao J, Liu M. Determination of total polyphenols content in green tea using FT-NIR spectroscopy and different PLS algorithms. J Pharm Biomed Anal. 2008;46(3):568–73.
39. Tahir HE, Zou X, Shen T. Near-infrared (NIR) spectroscopy for rapid measurement of antioxidant properties and discrimination of Sudanese honeys from different botanical origin. Food Anal Methods. 2016;9(9):2631–41.
40. Berardo N, Pisacane V, Battilani P. Rapid detection of kernel rots and mycotoxins in maize by near-infrared reflectance spectroscopy. J Agric Food Chem. 2005;53(21):8128–34.

41. Bao Y, Liu F, Kong W. Measurement of soluble solid contents and pH of white vinegars using VIS/NIR spectroscopy and least squares support vector machine. Food Bioprocess Technol. 2014;7(1):54–61.
42. Chen Q, Zhao J, Huang X. Simultaneous determination of total polyphenols and caffeine contents of green tea by near-infrared reflectance spectroscopy. Microchem J. 2006;83(1):42–7.
43. Blanco-Díaz MT, Del Río-Celestino M, Martínez-Valdivieso D. Use of visible and near-infrared spectroscopy for predicting antioxidant compounds in summer squash (Cucurbita pepo ssp pepo). Food Chem. 2014;164:301–8.
44. Bellincontro A, Taticchi A, Servili M. Feasible application of a portable NIR-AOTF tool for on-field prediction of phenolic compounds during the ripening of olives for oil production. J Agric Food Chem. 2012;60(10):2665–73.
45. Dykes L, Hoffmann L Jr, Portillo-Rodriguez O. Prediction of total phenols, condensed tannins, and 3-deoxyanthocyanidins in sorghum grain using near-infrared (NIR) spectroscopy. J Cereal Sci. 2014;60(1):138–42.
46. Tahir HE, Zou X, Shi J. Rapid determination of antioxidant compounds and antioxidant activity of Sudanese Karkade (Hibiscus sabdariffa L.) using near infrared spectroscopy. Food Anal Methods. 2016;9(5):1228–36.
47. Arslan M, Zou X, Hu X. Near infrared spectroscopy coupled with chemometric algorithms for predicting chemical components in black goji berries (Lycium ruthenicum Murr.). J Near Infrared Spectrosc. 2018;26(5):275–86.
48. Arslan M, Zou X, Haroon ET. Near-infrared spectroscopy coupled chemometric algorithms for prediction of antioxidant activity of black goji berries (Lycium ruthenicum Murr.). J Food Measur Character. 2018;12(4):2366–76.
49. Zareef M, Chen Q, Ouyang Q. Prediction of amino acids, caffeine, theaflavins and water extract in black tea by FT-NIR spectroscopy coupled chemometrics algorithms. Anal Methods. 2018;10:3023.
50. Kutsanedzie FYH, Chen Q, Hassan MM. Near infrared system coupled chemometric algorithms for enumeration of total fungi count in cocoa beans neat solution. Food Chem. 2018;240:231–8.
51. Dong C, Zhu H, Wang J. Prediction of black tea fermentation quality indices using NIRS and nonlinear tools. Food Sci Biotechnol. 2017;26(4):853–60.
52. Magalhães LM, Machado S, Segundo MA. Rapid assessment of bioactive phenolics and methylxanthines in spent coffee grounds by FT-NIR spectroscopy. Talanta. 2016;147:460–7.
53. Teye E, Huang X. Novel prediction of total fat content in cocoa beans by FT-NIR spectroscopy based on effective spectral selection multivariate regression. Food Anal Methods. 2015;8(4):945–53.
54. Liu Y, Ying Y, Yu H. Comparison of the HPLC method and FT-NIR analysis for quantification of glucose, fructose, and sucrose in intact apple fruits. J Agric Food Chem. 2006;54(8):2810–5.
55. Xie L, Ye X, Liu D. Quantification of glucose, fructose and sucrose in bayberry juice by NIR and PLS. Food Chem. 2009;114(3):1135–40.
56. Tewari JC, Dixit V, Cho B. Determination of origin and sugars of citrus fruits using genetic algorithm, correspondence analysis and partial least square combined with fiber optic NIR spectroscopy. Spectrochim Acta A. 2008;71(3):1119–27.
57. Mabood F, Al-Harrasi A, Boqué R. Determination of sucrose in date fruits (Phoenix dactylifera L.) growing in the Sultanate of Oman by NIR spectroscopy and multivariate calibration. Spectrochim Acta A. 2015;150:170–4.
58. Zou X, Zhao J, Huang X. Use of FT-NIR spectrometry in non-invasive measurements of soluble solid contents (SSC) of 'Fuji' apple based on different PLS models. Chemom Intell Lab Syst. 2007;87(1):43–51.
59. Ameer K, Murtaza MA, Jiang G. Chapter six – Fluorescence and ultraviolet–visible spectroscopy in the honey analysis. In: Nayik GA, Uddin J, Nanda V, editors. Advanced techniques of honey analysis. Academic Press; 2024. p. 153–91.
60. Feng J, Khakipoor B, May J, Mulford M, et al. An open-source dual-beam spectrophotometer for citizen-science-based water quality monitoring. HardwareX. 2021;10:e00241.

61. Thulstrup EW, Spanget-Larsen J, Waluk J. Linear dichroism, instrumentation. In: Lindon JC, Tranter GE, Koppenaal DW, editors. Encyclopedia of spectroscopy and spectrometry (third edition). Oxford: Academic Press; 2017. p. 601–3.
62. Xu N, Xiao M, Yu Z, Jin B, et al. On-site quantitation of xanthine in fish and serum using a smartphone-based spectrophotometer integrated with a dual-readout nanosensing assay. Food Chem. 2024;431:137107.
63. Chance B, Legallais V, Sorge J, Graham N. A versatile time-sharing multichannel spectrophotometer, reflectometer, and fluorometer. Anal Biochem. 1975;66(2):498–514.
64. Zou Z, Li M, Wang Q, Wu Q, Zhen J, Yuan D, Yin S, Zhou M, Cui Q, Xu L. A non-destructive detection method of protein and TVB-N content changes in refrigerated and frozen-thawed salmon fillets using fluorescence hyperspectral technology. J Food Compos Anal. 2024;133: 106435.
65. Bao X, Zhang R, Zhong L. Research on apple recognition method based on artificial neural network and image processing. Trans Chin Soc Agric Eng. 2004;20(3):109–12.
66. Hu C, Li P. Research on identification of nitrogen deficiency and magnesium deficiency in cucumber based on image processing. J Jiangsu Univ (Nat Sci Ed). 2004;25(1):9–12.
67. Jia D, Jia X. Image color eigenvalue extraction and matching based on C#. J Hebei Softw Vocat Techn Coll. 2008;10(2):52–5.
68. Yao M. Digital image processing. Beijing: China Machine Press; 2006.
69. Zou X, Zhao J, Li Y. In-line detection of apple defects using three color cameras system. Comput Electron Agric. 2010;70(1):129–34.
70. Lv Q, Zhang J. Design of automatic fruit classification system based on neural network. J Anhui Agric Sci. 2009;37(35):17392–17394, 17439.
71. Li P. Research progress of fruit grading based on computer vision technology. Agric Mechan Res. 2007;11:10–5.
72. Zhao M, Hou W. Research progress of automatic fruit grading technology based on machine vision in China. Pack Food Machin. 2007;25(5):5–8.
73. Su B, Hua X. Research on strawberry image processing based on machine vision. J Cap Norm Univ (Nat Sci Ed). 2018;4:42–5.
74. Shi Z, Tong Y, Chen D. Electronic nose detection technology for beef freshness. J Agric Machin. 2009;40(11):184–8.
75. Xie W, Wei S, Wang F. Research on carrot surface defect recognition method based on machine vision. Trans Chin Soc Agric Machin. 2020;51:451–6.
76. Wu D, Sun D-W. Advanced applications of hyperspectral imaging technology for food quality and safety analysis and assessment: a review—part I: fundamentals. Innovative Food Sci Emerg Technol. 2013;19:1–14.
77. Shekarau Luka B, Mohammed Yunusa B, Msurshima Vihikwagh Q, Fanan Kuhwa K. Hyperspectral imaging systems for rapid assessment of moisture and chromaticity of foods undergoing drying: principles, applications, challenges, and future trends. Comput Electron Agric. 2024;224:109101.
78. Marín-Méndez J-J, Luri Esplandiú P, Alonso-Santamaría M, Remirez-Moreno B, et al. Hyperspectral imaging as a non-destructive technique for estimating the nutritional value of food. Curr Res Food Sci. 2024;9:100799.
79. Zhang M, Wang L, Zhu L, Huang H. Learning a physics-based filter attachment for hyperspectral imaging with RGB cameras. Neurocomputing. 2024;580:127474.
80. Ou-yang S-T, Yao C, Wan Y-R, Zou J-P, et al. Detection of moisture and size of winter melon seeds based on hyperspectral imaging and convex polygon size measurement. J Food Compos Anal. 2024;136:106789.
81. Bull CR, Zwiggelaar R, Speller RD. Review of inspection techniques based on the elastic and inelastic scattering of X-rays and their potential in the food and agricultural industry. J Food Eng. 1997;33(1):167–79.

82. Lim H, Lee J, Lee S, Cho H, et al. Low-density foreign body detection in food products using single-shot grid-based dark-field X-ray imaging. J Food Eng. 2022;335:111189.
83. Musielak M, Musielak E, Sitko R. Ultrasensitive determination of selenium in food samples and its speciation in water and beverages using thiosemicarbazide-incorporated graphene and total-reflection X-ray fluorescence spectrometry. Food Chem. 2024;439:138156.

Chapter 6
Thermal Characteristics in Foods

Man Zhou

6.1 Introduction

Thermal processing operations are critical for ensuring food quality and safety. Various processing techniques—such as drying, rapid freezing, refrigeration, dehydration, heat treatment, baking, steaming, intelligent cold chain transportation, and more—are all closely tied to the thermal properties of food materials. These thermal characteristics include specific heat, thermal conductivity, thermal diffusivity, water states, and the temperature and enthalpy of phase transitions. They are essential for designing processes and determining appropriate thermal treatments for specific food products. Reliable estimation of these thermal properties is crucial for kinetic modeling of nutrient degradation and microbial inactivation.

Common methods for analyzing thermal characteristics include differential scanning calorimetry (DSC) and thermogravimetric analysis (TGA), along with its derivative form, derivative thermogravimetric (DTG). These techniques are rapid and reliable, requiring only small sample sizes and minimal preparation. DSC and TGA/DTG are widely used to assess the thermophysical properties of food materials, which in turn influence processing decisions and the quality of the final product.

This chapter provides an overview of food thermal characteristics and the methods for measuring them. It also highlights recent advancements in the use of DSC and TGA/DTG for estimating the thermophysical, nutritional, structural, and quality attributes of food products, particularly during both traditional and novel processing methods. Additionally, it explores new perspectives on assessing food structure and their applications in intelligent cold chain logistics. The technologies and applications discussed are summarized in Fig. 6.1.

M. Zhou
School of Food and Biological Engineering, Jiangsu University, Zhenjiang, Jiangsu, China
e-mail: manzhou@ujs.edu.cn

H. Lin et al., *Advanced Technologies for Physical Properties Measurement of Food and its Processing*, https://doi.org/10.1007/978-981-96-2347-1_6

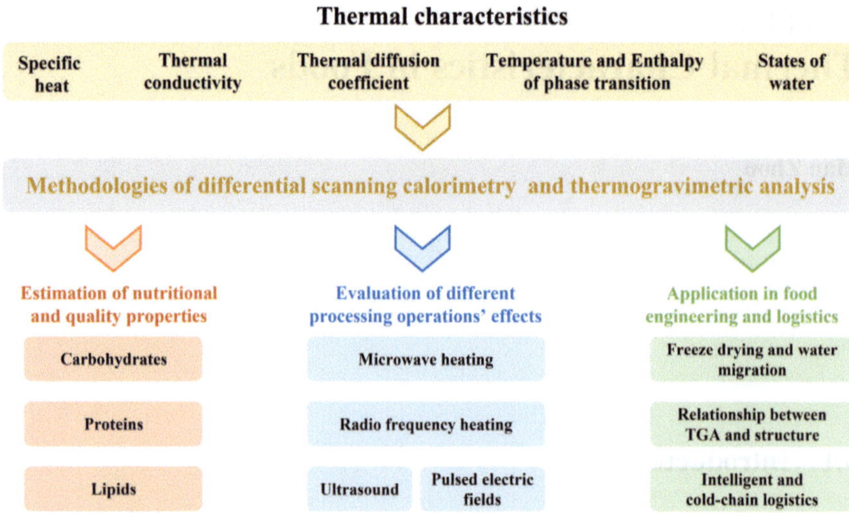

Fig. 6.1 The related technologies and applications

The thermal characteristics of foods are essential for conducting various heat transfer calculations that inform the design of processing equipment and the estimation of processing times for a wide range of operations, including drying, rapid freezing, refrigeration, cooling, dehydration, heat treatment, baking, steaming, pasteurization, evaporation, and frying. Key thermal properties include specific heat, thermal conductivity, thermal diffusivity, temperature, enthalpy of phase transition, and the states of water [1]. These properties are vital to ensuring the production efficiency, safety, and quality of food products. Thermal characteristics can be measured directly using equipment such as DSC, or they can be estimated through predictive models that utilize advanced techniques like artificial neural networks, fuzzy logic, and other data analysis tools. These predictive models allow for sophisticated analysis and are critical for optimizing food processing operations, especially when direct measurement is impractical or when developing new food processing technologies.

6.2 Thermal Characteristics in Foods

6.2.1 Specific Heat Capacity

Specific heat (C) is an important parameter in thermal processes, playing a fundamental role in energy cost estimation, machinery selection, and equipment design [2]. It refers to the amount of energy needed to raise the temperature of a sample by 1 °C and indicates a food's capacity to retain and store heat.

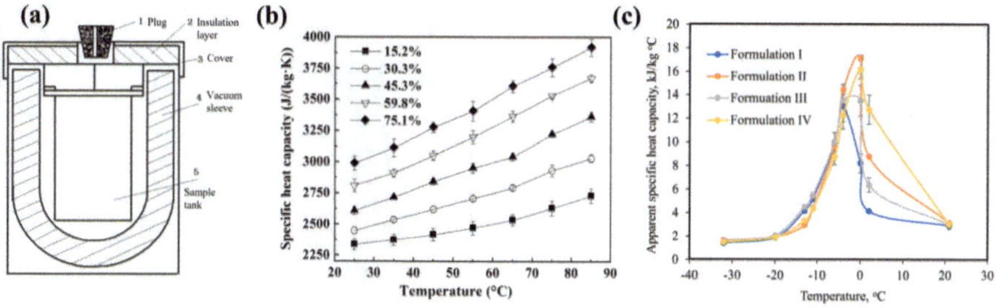

Fig. 6.2 Schematic of vacuum sleeve calorimeter apparatus (**a**), moisture content and temperature-dependent specific heat capacity of wolfberries (**b**), and apparent specific heat capacity of pancakes with different compositions (**c**). (Adapted from Refs. [3–5])

Two common methods are used for measuring specific heat: the method of mixtures and DSC. In the method of mixtures, a vacuum sleeve calorimeter is used to calculate the specific heat of an unknown sample by balancing the heat exchanged between the liquid, the calorimeter, and the sample, as depicted in Fig. 6.2a [3]. For example, using this method, the specific heat of red lentil seeds ranges from 1.08 to 2.03 kJ/kg·K [2].

In the DSC method, deviations from a baseline obtained during the transition from a scanning regime to an isothermal regime are used to calculate specific heat [6]. The process involves three measurements: (1) an empty container in the reference cell, (2) a reference substance with known C_p in the reference cell, (3) the investigated sample in the operating cell. Specific heat (C_p) is then calculated using Eq. 6.1:

$$C = C_r[(Y_s - Y_e)m_r] / [(Y_r - Y_e)m_s] \tag{6.1}$$

Where C_r represents the specific heat of reference material, Y_e, Y_r, and Y_s represent the respective deviations of DSC curves from the baselines for three separate experiments, m_s and m_r represent the masses of the sample and reference materials, respectively.

Specific heat is influenced by factors such as moisture content, temperature, thermal transitions, physical state, and chemical composition. Many studies have shown that the specific heat of materials like corn flour increases with moisture content [7]. As shown in Fig. 6.2b, the specific heat capacity of wolfberries increases with both temperature and moisture content [4]. Similarly, Fig. 6.2c demonstrates that the specific heat capacity of frozen pancakes is closely related to the phase state changes of water and the amount of freezable water [5]. These insights into specific heat provide valuable information for optimizing food processing and ensuring product quality.

6.2.2 *Thermal Conductivity*

Thermal conductivity, λ, signifies the ease with which heat moves through a material. According to Fourier's law of heat conduction, it represents the steady-

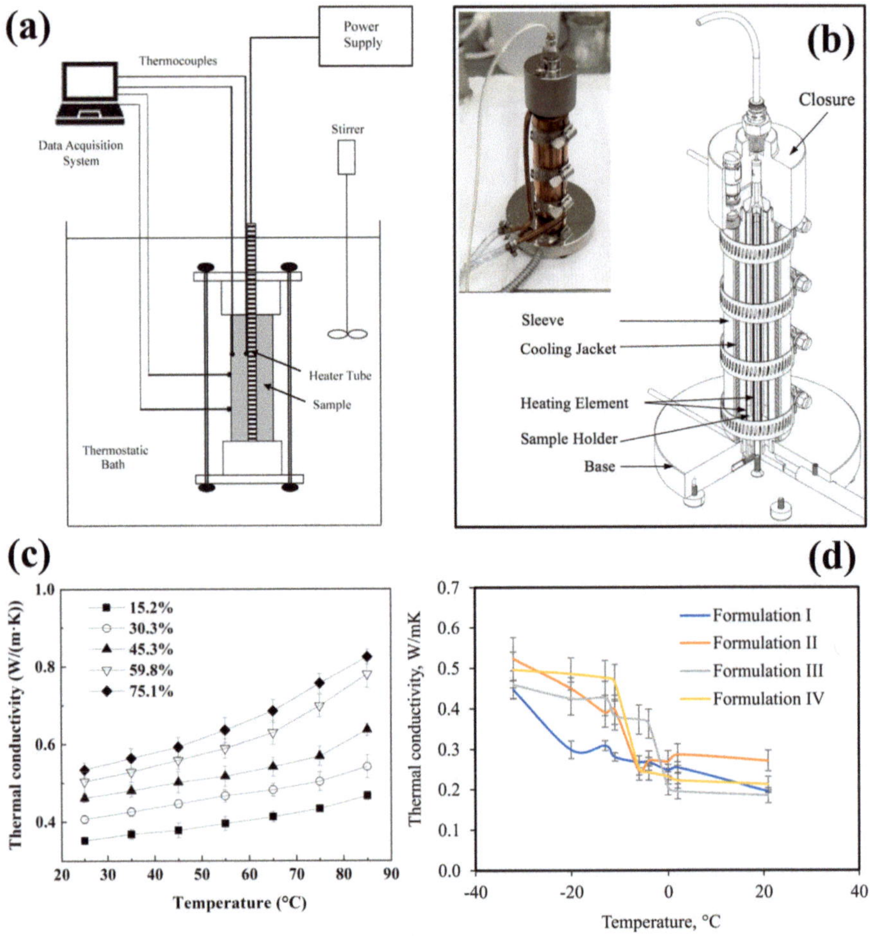

Fig. 6.3 Apparatus used for measuring effective thermal conductivity (**a**), the full assembly of TPCell equipment (**b**), moisture content and temperature-dependent thermal conductivity of wolfberries (**c**), and thermal conductivity of pancakes of with various formulations (**d**). (Adapted from Refs. [4, 5, 8, 9])

state rate of heat transfer (dQ) along the direction of a temperature gradient (dT/dx), as shown in Eq. 6.2 [3]:

$$dQ = -\lambda \cdot dA \cdot \frac{dT}{dx} \tag{6.2}$$

Where dA is the area perpendicular to the direction of heat flow; dQ is the amount of heat passing through dA per unit time starting; dT/dx denotes the temperature gradient.

Thermal conductivity can be measured using various methods: the steady-state method, the transient hot-wire method, and the inverse method using sequential parameter estimation. Figure 6.3a exemplifies a device for determining effective thermal conductivity via the steady-state technique [8]. This device consists of a jacketed cylindrical cell, two Teflon discs to prevent axial heat transfer, thermal resistance, Type J thermocouples, a laboratory power supply, a data acquisition system, a thermostatic bath, and a sample cell. A key drawback of these methods is the long duration (at least 45 minutes) required to reach the desired temperature range.

To reduce the detection time, an apparatus called the Thermal Properties Cell (TPCell) has been developed. This tool utilizes a rapid non-isothermal technique and consists of a specially designed sample holder and fittings that allow the heater to be installed in a pressurized environment (Fig. 6.3b). The experiment is completed in under 1 minute [9]. This cutting-edge tool enables food producers to quickly gather thermal property data, allowing them to design processes that optimize product quality while maintaining safety.

Different materials exhibit varying thermal conductivities (λ) due to differences in their chemical and physical compositions. Factors such as temperature, porosity, nutritional content, component phase distribution, and moisture content all affect thermal conductivity [2, 7, 10]. For instance, the thermal conductivity of mango peel differs significantly from that of mango pulp because of differences in moisture content. Additionally, the λ values of both mango peel and pulp show a linear increase with temperature [11]. Figure 6.3c illustrates how, as moisture content increases, the thermal conductivity of wolfberries also rises with both moisture level and temperature [4]. Similarly, Fig. 6.3d shows that the λ values of frozen pancakes are much higher than those of unfrozen pancakes [5]. This is because the thermal conductivity of ice is approximately four times higher than that of water. The structure of food components also influences thermal conductivity, with most foods generally exhibiting poor thermal conductivity. For example, the thermal conductivity values of green and yellow bananas fall within the range of 0.302–0.338 W/m·K [12]. In contrast, the λ values of red lentil seeds range from 0.191 to 0.241 W/m·K, depending on moisture content [2]. These examples demonstrate the diverse factors that influence thermal conductivity across different food materials.

6.2.3 *Thermal Diffusivity*

When heating or cooling a material in an unstable condition, the rate at which heat moves through it is indicated by its thermal diffusivity (α). Thermal diffusivity is critical for forecasting the nutritional and microbiological changes in food during thermal processing, as well as for modeling and computing transient heat transfer in fundamental food processing operations [13]. It represents the combined effect of thermo-physical properties, and as shown in Eq. 6.3, it can be calculated using experimentally measured values of thermal conductivity (λ), specific heat capacity (C_p), and density (ρ):

$$\alpha = \frac{\lambda}{\rho \times C_p} \tag{6.3}$$

Here, α represents thermal diffusivity, λ is thermal conductivity, ρ is density, and C_p is specific heat capacity.

Thermal diffusivity can also be determined using the inverse technique by applying analytical and numerical solutions to the diffusion equation. For example, the average α value of lean beef during the heating stage was found to be approximately $0.970 \times 10^{-7} \pm 0.0074 \times 10^{-7}$ m^2/s, closely matching the true value. Similar studies by Muramatsu et al. have used sequential estimation and ordinary least squares to calculate the thermal diffusivity of low-moisture food samples, which ranged from 9.8×10^{-8} to 1.3×10^{-7} m^2/s [13]. One of the main advantages of this method is its ability to account for significant spatial temperature variations. The inverse method was also employed to estimate the thermal diffusivity of cashew pulp using analytical and numerical solutions of the diffusion equation [14].

Thermal diffusivity is influenced by factors such as moisture content, polymer structure, and food formulation. For instance, the α values of red lentil seeds increase with higher moisture content [2], as do the values for corn flour [7] and mango [11]. During ripening, water moves from the peel to the flesh of bananas due to osmotic pressure changes, causing the α values of green and yellow bananas to shift from 1.442×10^{-7} to 1.500×10^{-7} m^2/s [12]. Lyng and colleagues studied how ingredients affect the thermal diffusivity of mashed potatoes [10]. They observed that adding butter decreased the thermal diffusivity, while adding salt caused the diffusivity to increase quadratically. These findings highlight the complex relationship between thermal diffusivity and food composition, emphasizing the importance of understanding these properties to optimize thermal processing techniques.

6.2.4 Temperature and Enthalpy of Phase Transition

The components of food, such as protein and starch, undergo various phase transitions during processes like heating, cooling, freezing, and storage. These transitions are particularly significant in amorphous materials, where temperature changes cause the material to pass through several states. Figure 6.4a illustrates the phase transitions an amorphous material experience when subjected to temperature variations. As the material is heated, it gradually transitions from a glassy state to a rubbery state, eventually leading to crystallization, and further heating causes melting. Rapid cooling from the molten state leads to the formation of a disordered glass structure.

First-order phase transitions, such as crystallization, melting, condensation, gelatinization, retrogradation, denaturation, and evaporation, are common in food systems. These transitions are marked by a sudden change in specific volume, as shown in Fig. 6.4b. In contrast, second-order transitions, like the glass transition, are characterized by a continuous change in slope with temperature (Fig. 6.4c).

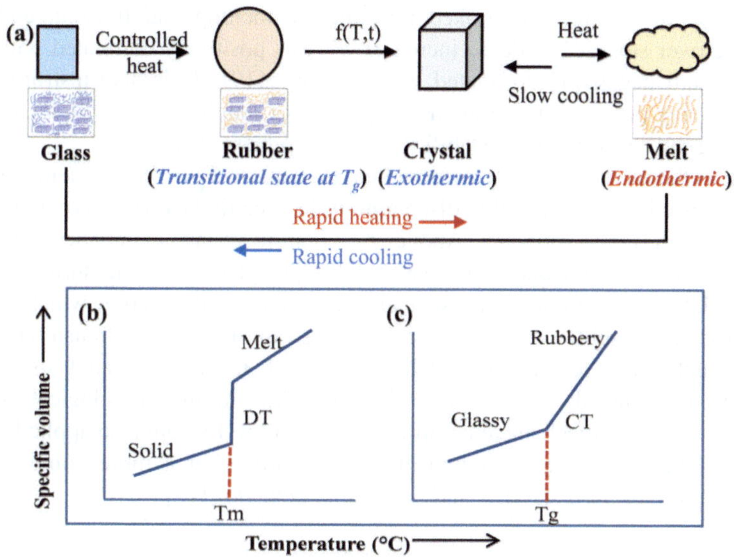

Fig. 6.4 An overview of the physical state transitions of an amorphous glass from a rubbery to a crystalline state (**a**), a schematic depiction of first-order transition (**b**) and second-order transition (**c**). (Adapted from Refs. [15, 16])

The glass transition, often observed in food systems, occurs as a result of changes in relaxation time and molecular mobility in amorphous solids. The temperature at which this inflection point occurs is known as the glass transition temperature (T_g). This transition reflects the shift in the material's physical properties, marking a change from a rigid, glassy state to a more pliable, rubbery state. Understanding these transitions is crucial for food processing and storage, as they affect the stability, texture, and quality of food products.

The temperature and enthalpy of phase transitions are critical factors in controlling food processing and storage conditions [6]. For example, changes in the enthalpy of food can be used to calculate the energy required to raise or lower the temperature. In confectionery products, phase behavior changes during storage can negatively impact shelf-life. The onset temperature (T_o) or peak transition temperature (T_p) is used to estimate the stability of samples, while the glass transition temperature (T_g) is essential for assessing the storage stability of dried fruit products. T_g provides insights into how much energy is needed to detect a temperature change based on the alteration in enthalpy.

One of the most critical physicochemical properties of amorphous food solids is the glass transition [17]. Glass transition is the most significant state transition influencing the processing, stability, and quality of food materials [18]. This phenomenon has been widely applied to food products to assess their stability [15]. The glass transition is also connected to various changes in food systems, such as microbiological stability, oxidation processes, caking, sticking, collapse, agglomeration crystallization, and non-enzymatic browning. T_g notably impacts the internal mass transfer mechanism, rheological behavior, and transport characteristics of food materials. For instance, in cooked rice, the ice melting and freezing enthalpies decrease over storage, while T_g increases. T_g was positively correlated with adhesiveness and negatively correlated with hardness [19]. T_g is also instrumental in characterizing porosity during the frying process of starchy foods, with increased T_g indicating improved physical stability in fried products [17].

DSC is the most commonly used method to measure T_g in food systems. In DSC, T_g is indicated by a measurable discontinuity in specific heat (C_p), typically calculated through DSC software. Above T_g, food materials experience significant changes in free volume and molecular mobility [18]. However, measuring T_g solely through DSC can be complex, especially in materials like starch, which undergo extensive and subtle endothermic changes during glass transition. Other thermomechanical techniques—such as thermal mechanical analysis (TMA), dynamic mechanical thermal analysis (DMTA), thermal rheological analysis (TRA), and thermomechanical compression test (TMCT)—are also applied to measure T_g. These methods complement DSC in providing a more comprehensive understanding of food stability and processing behavior [17].

6.2.5 Status, Amount, and Migration of Water

Water is a vital component of food materials, significantly influencing various food attributes. The state of water (bound, immobilized/unfreezable, and free/freezable), its amount, and its migration are key factors in food stability and can be analyzed using differential scanning calorimetry (DSC) and low-field nuclear magnetic resonance (LF-NMR). DSC provides a detailed analysis of macroscopic water changes, making it suitable for evaluating the thermal properties and freezing behaviors of water. On the other hand, LF-NMR is more appropriate for examining microscopic modifications within cell compartments, offering insights into water mobility and distribution at a cellular level [6]. Free/freezable water is loosely bound to the food matrix, displaying high fluidity and freezing ability. In contrast, bound or unfreezable water interacts more intensely with other components in the food matrix, contributing to food texture, chemical stability, and microbiological stability. Bound water plays a crucial role in food stability, especially in unfrozen biological systems, where water activity is more closely associated with stability than the total moisture content.

The amount of free water m^f and bound water m^u in a food sample can be experimentally calculated using the following equations:

$$m^f = m_s \times \frac{\Delta H_s}{\Delta H_r} \tag{6.4}$$

$$m^u = m - m^f \tag{6.5}$$

$$m = m_s \times W_s \tag{6.6}$$

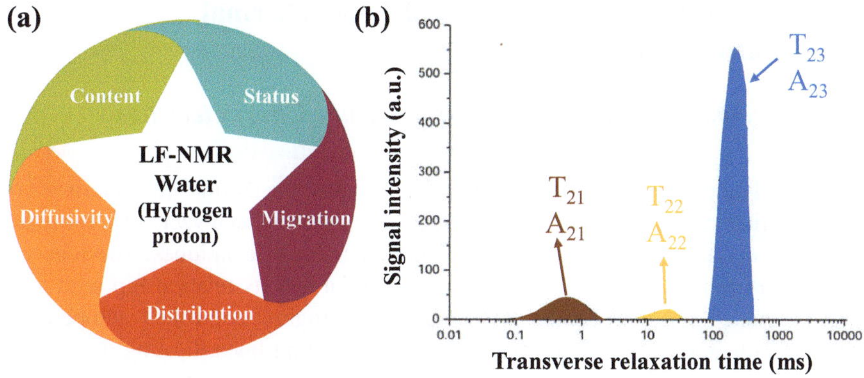

Fig. 6.5 Applications of LF-NMR in the study of water (**a**) and typical distribution of T_2 relaxation (**b**)

Where ΔH_s and ΔH_r represent the melting enthalpies of the sample and reference, respectively; m_s represents the total mass of the sample, m denotes the total mass of water in the sample; W_s is the moisture content of the sample. These equations enable the determination of the proportions of free and bound water in food materials, which is crucial for understanding their thermal behavior, texture, and overall stability.

LF-NMR provides a rapid, non-invasive, and real-time method for measuring the moisture status and migration patterns in food materials [20]. The LF-NMR spectra provide indirect information on the content, status, migration, distribution, and diffusivity of water within a sample, as depicted in Fig. 6.5a. According to the NMR principle, the relaxation time (T_2) of protons is influenced by the water's mobility and its surrounding chemical environment. The number of peaks and the area under these peaks in the T_2 relaxation graph reflect the condition and quantity of water present in the sample.

Figure 6.5b shows that the relaxation time T_2 typically ranges from 0.01 to 1000 ms. T_2 is divided into three main categories. T_{21}, ranging from 0.01 to 1 ms, represents bound water. T_{22}, spanning 1 to 10 ms, indicates immobilized water, while T_{23}, from 10 to 1000 ms, corresponds to free water. The areas under the peaks, labeled A_{21}, A_{22}, and A_{23}, represent the amounts of these three types of water. Larger peak areas suggest a higher content of the respective water category. Additionally, changes in T_2 values provide insights into water migration and redistribution within the sample over time [21]. This makes LF-NMR a highly effective tool for tracking moisture dynamics and understanding water behavior in food systems, particularly during processing methods such as drying, freezing, and storage.

6.3 Measurement Technologies of Food Thermal Characteristics

6.3.1 *Measurement Methods for Thermal Characteristics*

With the advancement of thermal analysis instruments, the thermal characteristics of food materials can now be directly measured using thermal analyzers [3]. Table 6.1 provides a classification and definition of various thermal techniques. These instruments measure the physical properties of a material and its relationship with temperature under program-controlled conditions. Program-controlled temperature refers to a range of temperature manipulations, including linear heating or cooling, constant temperature maintenance, cyclic, or nonlinear heating processes.

The term "substance" in this context refers to the reaction products generated either from the sample itself or during interactions within the sample. Physical properties that are measured include mass, temperature, specific heat capacity, thermal conductivity, enthalpy change, size, mechanical properties, acoustic properties, as well as optical, electrical, and magnetic properties. These techniques allow for a detailed analysis of how temperature influences the thermal behavior of food materials, facilitating precise control in food processing and storage.

The commonly used methods for determining the thermal characteristics of food materials include techniques such as differential scanning calorimetry (DSC), thermogravimetric analysis (TGA) and derivative thermogravimetry (DTG), thermomechanical analysis (TMA), nuclear magnetic resonance (NMR), and dynamic mechanical analysis (DMA). Among these, DSC and TGA/DTG are particularly favored due to their ability to handle minimal amounts of material, their capacity to detect thermal transitions across a wide range of temperatures, and their straightforward quantitative and qualitative transition analysis [6, 16]. The following sections will provide a detailed introduction to DSC and TGA/DTG, outlining their significance and methodologies in the determination of thermal properties in food materials.

6.3.2 *Differential Scanning Calorimetry*

6.3.2.1 Principle and Structure of DSC

The principle of DSC involves maintaining equal temperatures between samples and references, placed in separate micro-ovens, throughout a programmed heating or cooling process. DSC assumes that the sample and reference are exposed to similar conditions at a pre-programmed heating or cooling rate and calculates the amount of energy required to maintain a zero temperature difference (ΔT) between them [22]. The energy required is applied as heat, which varies with temperature or time. DSC hardware operates on two main principles: the ΔT-measuring system

Table 6.1 The classification and definitions of thermal techniques

Measured physical parameters	Name of technique	Definition
Mass	Thermogravimetric analysis (TGA)	Measures the relationship between the quantity of substances and temperature under controlled, programmed heating conditions.
	Derivative thermogravimetry (DTG)	The method of the first derivative of the thermogravimetric curve obtained by the thermogravimetric method concerning time or temperature.
Temperature	Differential thermal analysis (DTA)	Measures the temperature difference (ΔT) and its relationship between a substance and a reference substance under programmed temperature control.
Enthalpy	Differential scanning calorimetry (DSC)	Measures the power difference between the input material and the reference material under program-controlled temperature and their temperature relationship.
Size	Thermodilatometry (TD)	Measures the size and temperature relationship of a substance under negligible load under program-controlled temperature
Mechanical properties	Thermomechanical analysis (TMA)	Measures the relationship between deformation and temperature of a substance under non-vibrating loads under programmed temperature control
	Dynamic mechanical analysis (DMA)	Measures the dynamic modulus and mechanical loss of a substance under vibration load under program-controlled temperature, about temperature
Electrical properties	Thermoelectrometry	Measures the electrical properties of a substance and its temperature relationship controlled, programmed heating conditions
	Dynamic dielectric analysis (DDA)	Measures the dielectric constant and loss of a substance under alternating electromagnetic field under program-controlled temperature, as well as the relationship with temperature
Optical properties	Thermophotometry	Measures the optical properties of a substance and its temperature relationship under programmed temperature control
	Thermospectrometry	Measures the relationship between the transmittance and absorption coefficient of a substance at a certain characteristic wavelength and temperature under programmed temperature control
	Thermoluminescence	Measures the relationship between the luminous intensity of a substance and temperature under controlled, programmed heating conditions
Acoustic properties	Thermosonimetry	Measures the relationship between sound emitted by substances and temperature under programmed temperature control
	Thermoascoustimetry	Measures the relationship between sound wave characteristics and temperature after passing through a substance under programmed temperature control
Magnetic properties	Thermomagnetometry	Measures the relationship between magnetic susceptibility and temperature of materials under the programmed control temperature

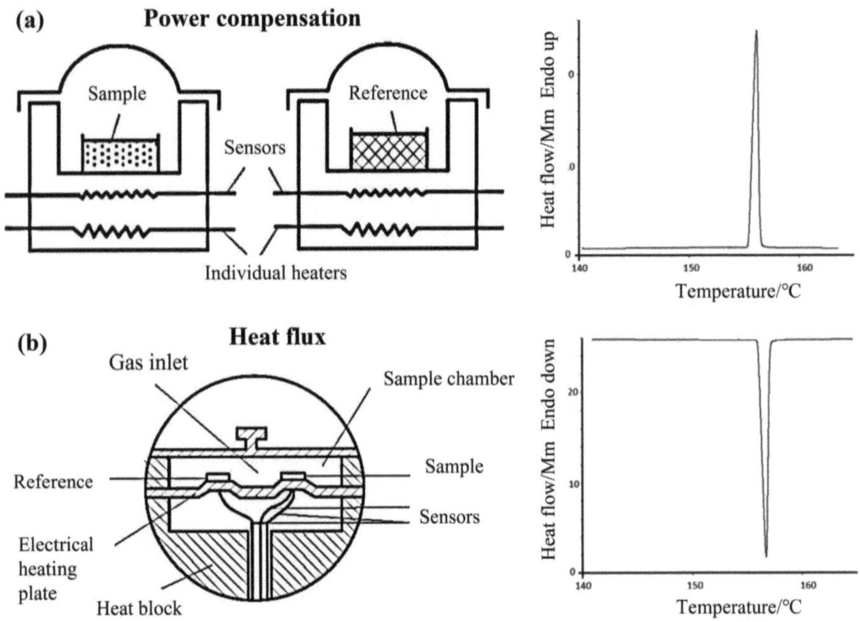

Fig. 6.6 Schematic structure diagrams and comparison of the power compensation (**a**) and heat flux (**b**) DSC instruments

and the power compensation system [1]. Based on the measurement principle, DSC instrumentation can be classified into two types: heat flux DSC and power compensation DSC. Figure 6.6 illustrates the schematic structure and key differences between these two DSC types.

In power compensation DSC, two isolated heaters are used—one for the sample pan and one for the reference pan. These independent heaters maintain equal temperatures between the sample and reference pans by regulating the heat flow between them. Throughout the experiment, the temperature difference (ΔT) remains zero. The power required to achieve identical temperatures in both pans is quantitatively measured in watts, and the heat flow ($\mathrm{d}H/\mathrm{d}t$) is determined using Eq. 6.7:

$$\Delta W = \frac{\mathrm{d}Q_S - \mathrm{d}Q_r}{\mathrm{d}t} = \frac{\mathrm{d}H}{\mathrm{d}t} \tag{6.7}$$

Where $\mathrm{d}Q_s/\mathrm{d}t$ is the amount of heat applied to the sample per unit time, $\mathrm{d}Q_r/\mathrm{d}t$ is the amount of heat applied to the reference per unit time, and $\mathrm{d}H/\mathrm{d}t$ is the rate of change in enthalpy or heat flow.

In contrast, heat flux DSC uses two crucibles, one for the sample and one for the reference, placed symmetrically on an electrical heating plate surrounded by a single heat block. When the temperature of the sample crucible differs from that of the reference crucible, the resulting heat flow is recorded by thermosensors and is described by the thermal equivalent of Ohm's law [22], as shown in Eq. 6.8:

$$\Delta T = \mathrm{R} \cdot \mathrm{d}Q_s s / \mathrm{d}t \tag{6.8}$$

Where ΔT is the temperature difference between the sample and reference, R is the thermoelectric disk resistance, and $\mathrm{d}Q_s/\mathrm{d}t$ is sample heat flow.

A significant difference between these two DSC instruments is the form of the DSC curves: in power compensation DSC, the curve points upward, while in heat flux DSC, the curve points downward (as illustrated in Fig. 6.6). These distinctions affect how energy input and thermal transitions are represented and analyzed during the experiment.

6.3.2.2 Typical DSC Curves

The DSC method generates a curve that plots heat flow against temperature or time. Figure 6.7 illustrates a typical power compensation DSC curve, where heat flow (milliwatts, mW) is represented on the vertical axis, and temperature (°C) or time (minutes) is plotted on the horizontal axis. The curve reflects significant thermal events such as glass transitions, exothermic reactions, and endothermic peaks.

A typical DSC curve consists of three key points: the onset of a peak (point **a**), the peak itself (point **b**), and the end of the peak (point **c**). Point **a**, marking the start of the peak, indicates the beginning of crystal melting, a crucial point for determining the initial melting temperature of crystals. Point **b** represents the majority of the crystal melting process, where the most significant portion of the material undergoes melting. Point **c** signals the end of the crystallization or melting process.

The area under the curve is proportional to the change in enthalpy and the amount of sample deposited in the DSC pan. The heat flow $\mathrm{d}H/\mathrm{d}t$, is calculated by measuring the temperature difference (ΔT) between the reference material and the sample. The shape and size of the peaks provide valuable information about the thermal

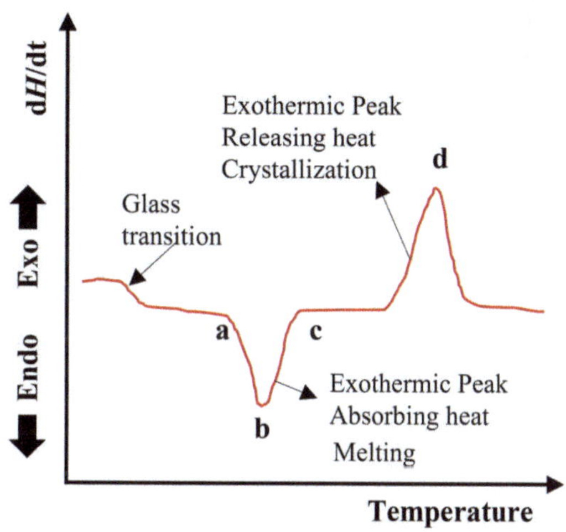

Fig. 6.7 Standard points on a DSC curve are identified as follows: (**a**) represents the start of an endothermic peak; (**b**) signifies the peak itself; (**c**) denotes the end of the endothermic peak; (**d**) indicates an exothermic peak

properties of the material, such as melting points, phase transitions, and heat capacities, making DSC a powerful tool for analyzing food materials and other substances.

6.3.2.3 Analysis of DSC

DSC can be applied to measure several important thermal properties of materials, including the temperature of phase transitions, enthalpy, specific heat capacity, and crystal quantity.

Temperature of Phase Transition

The transformation temperatures detected by DSC include the glass transition temperature (T_g), crystallization temperature (T_c), and melting temperature (T_m). Both crystallization and melting exhibit clear exothermic and endothermic peaks, making it relatively easy to determine these transition temperatures. Generally, the baseline before and after crystallization or melting is connected, and the temperature at the intersection (T_e) of the tangent line of the starting edge and the baseline is considered the transition temperature. The transition peak temperature (T_p) is also used as an alternative transition temperature (Fig. 6.8a).

Several methods are used to determine T_g, which is the temperature corresponding to the beginning, middle, or end of the glass transition process. Since the glass transition occurs over a range of temperatures, there may be some variation in the reported T_g. One common method involves determining the slope of the transition curve, with T_g being the temperature corresponding to the key point on the curve (Fig. 6.8b). For curves with subtle changes, auxiliary methods such as extending the tangent of the baseline before and after the change are often used to more accurately determine T_g (Fig. 6.8c).

In food materials, the glass transition temperature varies depending on factors such as molecular weight, component content, and individual component properties. The T_g of different food components can vary significantly, leading to a broad range of T_g values during thermal processing. When reporting T_g, it is important to include

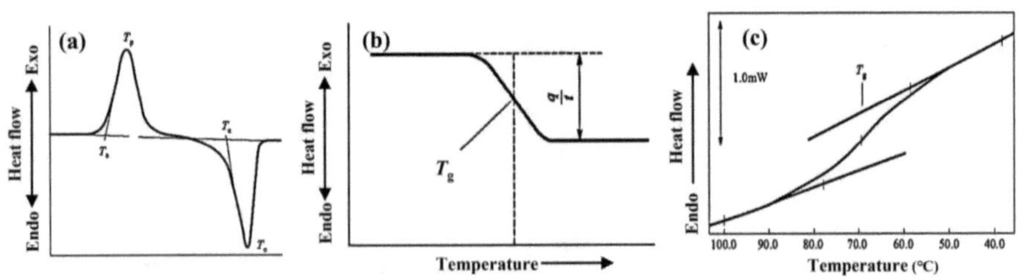

Fig. 6.8 Determination of the start and end temperatures of glass transition from DSC curves

accurate experimental conditions such as the thermal history before testing, the DSC heating or cooling rate, and the time held at a constant temperature, as these factors affect the validity and reproducibility of the data.

Enthalpy

Enthalpy is a key thermodynamic parameter that relates to both physical (e.g., phase transitions) and chemical changes (e.g., decomposition or bond cleavage) in a sample. The enthalpy change (ΔH) measured by DSC is the difference in enthalpy before and after the thermal transformation of the sample. For processes occurring at constant pressure, ΔH is equivalent to the heat (Q) absorbed or released during the change. To compare enthalpy changes across different substances, ΔH is normalized by calculating the enthalpy change for 1 mole of the sample's molecules.

DSC records a curve of heat flow rate over time, where the peak area formed by the baseline corresponds linearly to the amount of heat absorbed or released during the sample's thermal transformation. The proportionality between the peak area and the enthalpy can be determined by comparing the sample size (amount of substance) to the standard material with a known phase transition enthalpy. The area of the DSC phase transition peak of the sample is used to determine its enthalpy. Modern DSC data processing software can accurately calculate the peak area, making it easier to determine the enthalpy of the phase transition for a given sample.

Specific Heat Capacity

Compared to the conventional calorimeter hot melt method for measuring specific heat capacity (C_p), the DSC method offers significant advantages. It requires fewer sample quantities, ensures faster measurement speeds, and simplifies the operational process. In DSC, the sample undergoes controlled linear temperature programming, and the heat flow rate into the sample is continuously measured and is proportional to the sample's specific heat capacity. The specific heat capacity can be calculated using Eq. 6.9:

$$\frac{\mathrm{d}Q}{\mathrm{d}t} = mC_{\mathrm{p}}\frac{\mathrm{d}T}{\mathrm{d}t} \tag{6.9}$$

Where Q is the heat, m is the mass of sample, C_p is the specific heat capacity.

The specific steps for measuring C_p using DSC have been described earlier (refer to Sect. 6.2.1 for details).

State of Water and Crystal Quantity

Many food materials contain varying amounts of crystals and glass structures, and the ratio of these two phases directly affects the physical properties of food. During storage and processing, the ratio of crystals and glasses changes continuously, necessitating the measurement of crystal content in food materials. The crystal content can be determined using the DSC curve by calculating the areas of the melting peak A_{m} and crystallization peak A_{c}. The formulas for calculating the areas are as follows:

$$A_{\mathrm{m}} = \frac{H_{\mathrm{m}}T}{tm} \tag{6.10}$$

$$A_{\mathrm{c}} = \frac{H_{\mathrm{c}}T}{tm} \tag{6.11}$$

Where H_{m} is the melting heat absorption per unit time and unit mass; H_{c} is the crystallization heat release per unit time and per unit mass; T is the temperature, t is the unit time, and m is the unit mass.

To determine the heat absorbed and released per gram of sample, divide the area of the melting or crystallization peak by the heating rate. Then multiply by the actual sample mass used in the experiment to obtain the total absorbed heat $H_{\mathrm{m,total}}$ and total released heat $H_{\mathrm{c,total}}$ for the sample material. The number of crystals (m_{c}) that existed before the sample reached the crystallization phase transition is calculated using the difference between the total absorbed and released heat, divided by the heat released during crystallization for a unit mass sample, as shown in Eq. 6.12:

$$m_{\mathrm{c}} = \frac{H_{\mathrm{m,\ total}} - H_{\mathrm{c,total}}}{H'_{\mathrm{c}}} \tag{6.12}$$

Where H_{c}' is the heat released during the crystallization of a unit mass sample. This method allows the determination of the proportion of crystals in food materials, providing insights into their physical properties and their transformations during thermal processing and storage.

6.3.3 *Thermogravimetric Analysis*

Thermogravimetric analysis/derivative thermogravimetry (TGA/DTG) is primarily used to calculate mass loss under controlled environmental conditions as a sample undergoes heating, cooling, or is maintained at a constant temperature [23]. These techniques are applied in various processes, including the measurement of volatile compounds, material degradation, combustion, and the analysis of residual matter. TGA/DTG is also useful for estimating the thermal stability of polymers, such as modified starches, by determining the decomposition temperature. This method tracks the evolution of individual components and is used to study the breakdown of organic materials, including the thermal properties of food packaging composites and encapsulated materials [24]. In a single experiment, simultaneous TGA-DSC can measure both mass loss and heat flow, providing two types of data [25].

In a single experiment, simultaneous TGA-DSC can measure both mass loss and heat flow, providing two types of data [16]. This technique is particularly useful in cases where processes such as melting and degradation occur within the same temperature range, as is common in systems high in sugar content. Simultaneous TGA-DSC can effectively differentiate between melting and degradation events, as the temperatures at which they occur are significantly influenced by the sample's moisture content. However, accurately determining these properties can be challenging due to the potential for event overlap or weak signals. These factors make it difficult to fully capture the distinct thermal transitions, requiring careful analysis to separate the effects of overlapping phenomena, especially in complex samples.

6.3.3.1 Principle and Structure of TGA/DTG

TGA/DTG is an instrument designed to measure the continuous change in the mass of a sample exposed to varying temperatures or over time, under program-controlled temperature and a specific atmosphere. In this setup, the crucible containing the sample is placed in a sample holder connected to the weight measurement system. The system conducts experimental measurements and collects real-time data according to pre-set temperature programs and atmospheric conditions. The TGA/DTG equipment consists of several key components, including the instrument host, which features a programmed temperature controller, heating furnace, support frame, gas controller, temperature measurement system, and weight measurement system. It also includes auxiliary equipment like an automatic sampler, pressure control device, lighting, and cooling device, alongside control and data acquisition modules that manage the instrument's operation and process the collected data in real-time [23]. TGA/DTG systems are categorized into three types based on the relative positioning of the sample and the balance knife line. These are the bottom dish type, top dish type, and horizontal type. Each configuration is designed to meet specific experimental requirements, ensuring accurate and reliable measurement of mass variations during thermal analysis.

6.3.3.2 Typical TGA/DTG Curves

The mass change in a TGA curve reflects how the properties of a sample vary with temperature. For any thermal process, both temperature and mass are typically described together to provide a comprehensive understanding of the material's behavior. Several characteristic temperatures are commonly used to analyze these changes. These include the initial temperature (T_i), the extrapolated onset temperature (T_{onset}), the final temperature (T_f), the extrapolated end temperature (T_{endset}), the

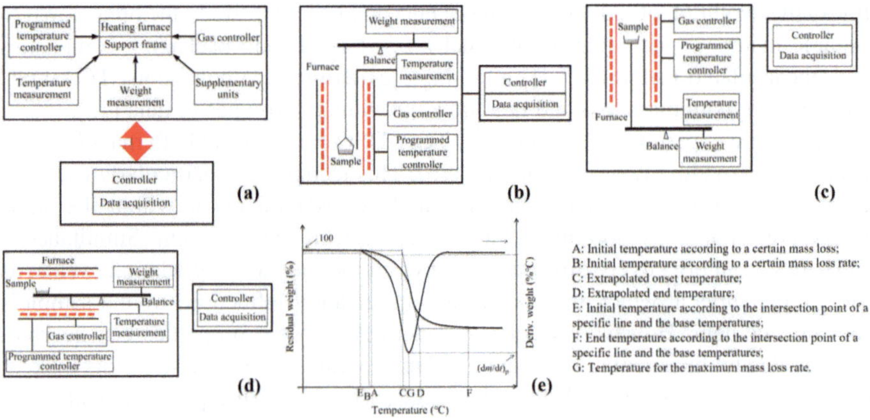

Fig. 6.9 Schematic of typical TGA equipment with the sample in a heating furnace (**a**), whose temperature is controlled with a program. Schematic of TGA equipment with the crucible at low (**b**) and high (**c**) positions of the vertical heating furnace and with the horizontal heating furnace (**d**). Typical TGA/DTG curve points (**e**). (Adapted from Ref. [23])

n% decomposition temperature ($T_n\%$), and the peak temperature of the fastest mass change (T_p), which corresponds to the DTG peak temperature. Using specialized analysis software, these characteristic temperatures can be easily identified and marked on the TGA/DTG curve. Figure 6.9e provides an example of a typical TGA/DTG curve, illustrating where these characteristic temperatures appear, offering a visual representation of the material's thermal stability, decomposition, and other related processes as a function of temperature.

6.4 Application of DSC and TGA/DTG in Food Industry

6.4.1 Application of DSC and TGA/DTG in Food Nutritional Properties and Quality Assessments

Natural food matrices primarily consist of carbohydrates, proteins, lipids, and water, all of which have thermal properties that significantly influence the nutrition, safety, quality, and processability of foods. These thermal characteristics include denaturation temperature, enthalpy, conformational transitions, and thermal stability. Such properties play crucial roles in determining how food behaves during processing, potentially affecting cooking loss, water-holding capacity, volume shrinkage, odor, shelf-life, and other key quality attributes. DSC is a valuable and sensitive tool for analyzing the thermal characteristics of these nutrients. DSC analysis provides insights into how carbohydrates, proteins, lipids, and water behave under different thermal conditions, making it an important method for assessing food quality and stability during processing. This section highlights the application of DSC in examining the changes that occur in these macronutrients during various processing steps, offering a deeper understanding of how their thermal behavior impacts food quality.

6.4.1.1 Carbohydrates

Starch

Gelatinization and retrogradation are two key thermal properties of starch that play a critical role in food processing. Gelatinization occurs when water penetrates starch granules, disrupting their structure. Starch granules that are highly cross-linked or substituted typically have lower gelatinization temperatures. Retrogradation, on the other hand, involves the reassociation of disaggregated amylose and amylopectin chains into an ordered structure during the processing and storage of gelatinized starch. While retrogradation negatively affects the sensory and storage qualities of starch-based foods, such as causing staleness or hardening, it can improve the textural and nutritional properties of these foods, making it a double-edged sword in terms of food quality.

Starch gelatinization and retrogradation are particularly important in starch-based foods, as they directly influence texture and quality. These characteristics vary depending on the type of starch and its modifications. For example, Jeong and colleagues studied four types of rice starches with varying amylose content and found that both the length of amylopectin chains and the amount of amylose significantly impacted their gelatinization and retrogradation behaviors [26]. Figure 6.10a illustrates the thermal behaviors of rice starch-water systems, highlighting these differences. Physical treatments like heat moisture treatment and microwave processing can elevate the temperature at which gelatinization [27]. Additionally, Kumar and his team measured the retrogradation enthalpy (ΔH_R) of frozen cooked rice and discovered that rice subjected to temperature fluctuations that reached the thawing stage exhibited the greatest increase in ΔH_R [28].

TGA/DTG has been extensively utilized to assess the thermal stability of starch modified through various techniques, including pulsed electric field (PEF) modifications and microwave-ultrasound-assisted cross-linking. PEF-oxidized starch, for example, demonstrates reduced thermal stability compared to native starch due to the decrease in molecular weight caused by oxidation [29]. In contrast, cassava starch modified through microwave-ultrasound-assisted cross-linking exhibits improved thermal stability, making it particularly suitable for applications in the food and pharmaceutical industries [30]. This increased stability enhances its potential for use in processes requiring elevated temperatures.

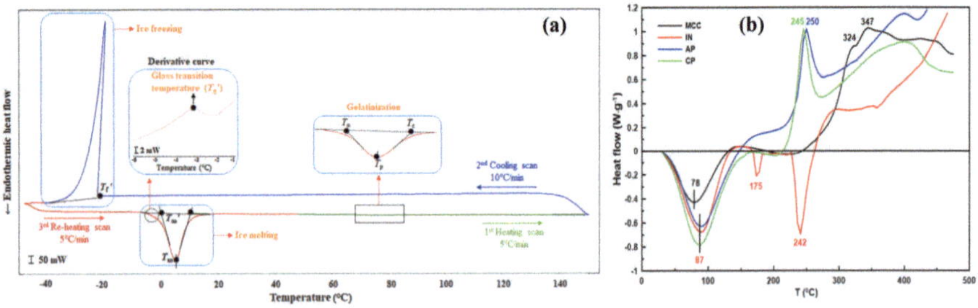

Fig. 6.10 DSC thermograms of rice starch-water system during heating and cooling scan processes (**a**) [26] and four polysaccharides (**b**) [25]

Polysaccharide's Thermal Behavior

Polysaccharides, such as pectin, cellulose, carrageenan, cyclodextrin, chitosan, and gum, are crucial components in many foods, and their thermal behaviors—including denaturation peak temperature, enthalpy, conformational transitions, and thermal stability—can be measured using DSC and TGA techniques. For example, DSC thermograms for four polysaccharides are presented in Fig. 6.10b [25], which allow for the determination of their denaturation peak temperature (T_p). In one case, the Tp of inulin was recorded at 87, 172, and 242 °C. Similarly, the DSC curve for polysaccharides from *Musa sapientum* L. showed a glass transition at 40.8 °C and an onset temperature of weight loss at 240 °C [31].

The thermal behavior of pectin is influenced by structural and functional group variations. Ultrasound-assisted extraction of pectin results in greater structural stability and enhanced thermal stability compared to pectin obtained through conventional heating extraction, as shown by DSC analysis [32]. Einhorn-Stoll and colleagues examined the impact of relative humidity on pectin de-methoxylation and observed a shift in DSC curves to lower temperatures after storage at reduced humidity. Furthermore, the height of the initial endothermic peak decreased after 28 days of storage due to partial hydrogen bond cleavage, with the glass transition temperature (T_g) and melting temperature (T_m) of commercial citrus peel pectin recorded at 139 °C and 158 °C, respectively. The thermal breakdown of flaxseed gum, subjected to continuous microwave exposure, was found to be primarily influenced by the accumulation of phenolic compounds rather than by the minor depolymerization of polysaccharides [33].

6.4.1.2 Proteins

Denaturation and Aggregation of Proteins

DSC is an invaluable tool for investigating thermal denaturation, conformational changes, phase transitions, gelation, and the thermodynamic behavior of proteins. As shown in Fig. 6.11a, the DSC thermograms of beef muscle illustrate the denaturation

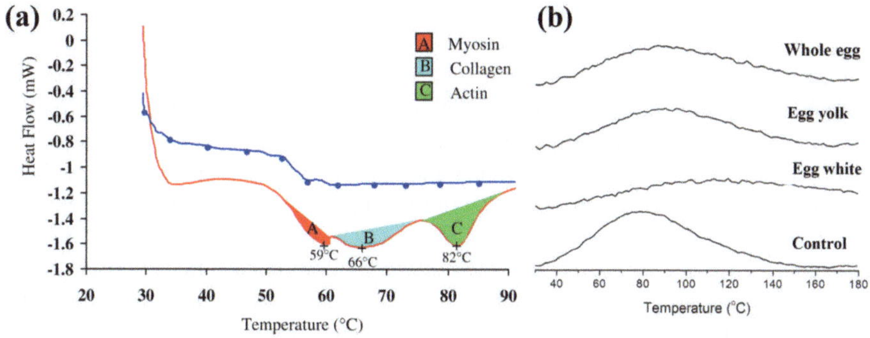

Fig. 6.11 DSC thermogram of uncooked and cooked beef muscle (**a**) [34], and DTG plot of dried pasta (**b**) [36]

of proteins such as actin, collagen, and myosin, with endothermic peaks observed between 56 and 81 °C [34]. Such data provide insights into the selection of appropriate formulations and processing conditions to improve food functionality, particularly in terms of texture and sensory attributes. For instance, the thermal denaturation of meat proteins can serve as an indicator for determining optimal cooking times, ensuring desirable meat quality [35].

Additionally, TGA/DTG techniques can be employed to investigate the effect of proteins (such as gluten and egg proteins) on the quality of dried noodles or pasta. Figure 6.11b shows DTG data that reveal a strong interaction between water molecules and egg proteins in pasta, suggesting that the broader DTG curves of pasta with egg proteins indicate more difficulty in water release compared to control pasta [36]. TGA analysis has also been used to demonstrate that dielectric drying methods can enhance the stability of protein networks in Chinese dried noodles [37].

In meat science, understanding the thermal properties of muscles is crucial. For example, García-Sánchez and colleagues studied the effects of freezing storage on the thermal transitions of jumbo squid muscle using DSC. They found no significant differences in T_{max} and ΔH values between fresh and frozen muscle, indicating that freezing storage did not cause protein denaturation [38]. Vaskoska and co-workers discovered that the lower melting temperature (T_m) of myosin and the higher cumulative ΔH in the *cutaneous trunci* muscle contributed to greater cooking loss and volume shrinkage compared to the masseter muscle, providing insights into the physicochemical properties of meat [39].

Whey protein aggregation, influenced by pH and protein concentration, has also been studied using DSC. Dissanayake and colleagues found that the denaturation temperatures and ΔH of whey proteins remained independent of concentration, offering valuable information on protein stability under different processing conditions [40]. Moreover, PEF treatments had no significant effect on the denaturation temperature (T_d) of xanthine oxidase, whereas thermal treatments increased T_d by 2–3 °C. DSC analysis revealed a decrease in ΔH in treated milk, with higher denaturation rates corresponding to increased treatment intensities, further illustrating the usefulness of DSC in understanding protein behavior in food systems [41].

Interaction of Proteins with Other Components

Natural foods are complex systems consisting of multiple macronutrients, such as carbohydrates, proteins, and lipids. Under heat, these components interact with each other, altering the system's thermal properties and heat transfer behavior. For example, Li and colleagues examined the thermal characteristics of a soybean protein isolate (SPI) and corn starch mixture using DSC [42]. Their results revealed that SPI hindered the gelatinization of corn starch, while the corn starch, in turn, protected SPI from denaturation, highlighting the reciprocal interactions between these macronutrients. Nawrocka and co-workers investigated how different polysaccharides affect the thermal properties of gluten proteins using both DSC and TGA/DTG [25]. Their findings showed that dietary fiber polysaccharides structurally altered gluten proteins by competing for water during dough mixing, which led to the dehydration of the gluten network. This indicates that the addition of polysaccharides can significantly influence the thermal behavior and stability of gluten proteins. In another study, Sobanwa and colleagues used micro-DSC to examine the effects of cellulose and starch on the denaturation of collagen [43]. Their results indicated that while cellulose and starch had no effect on the collagen denaturation temperature, the denaturation temperature shifted to lower values when reheating collagen pastes containing starch granules. These studies illustrate the intricate interactions between macronutrients in natural food systems and how these interactions modify the food's thermal properties during processing.

6.4.1.3 Lipids

DSC and TGA/DTG have been widely utilized to investigate the thermal properties of lipids in various edible oils and fats, such as olive oil, palm oil, confectionery fats, and butter [44]. These techniques are valuable for monitoring oxidative deterioration, lipid oxidation kinetics, and antioxidant activity, as well as evaluating how formulations or process technologies impact lipid stability. DSC is recommended as a prime method for objectively assessing the oxidative stability of oils [45–48]. Key parameters such as oxidation induction time (OIT), oxidation onset temperature (OOT), and activation energy (E_a) can be derived from DSC data, as shown in Fig. 6.12a. The OIT is determined from the curve at the intersection of the baseline and the tangent to the exothermal peak [45], while the OOT is calculated in an oxygen atmosphere [6]. E_a is derived from the slope and intercept of the linear regression model of $\ln k_{OIT,i}$ versus the inverse of the absolute temperature ($1/T_i$, E_a). The TGA signal can also be used to assess the relative oxidative stability of oils, with key indicators including the initial weight loss temperature (T_{on}) and the temperature corresponding to the maximum increase in weight loss (TM), as shown in Fig. 6.12b. For instance, the TGA curve of palm oil indicates a T_{on} of 225.32 °C, which provides insight into its thermal stability [44, 49, 50].

These techniques have been applied to study a wide range of oils, including non-conventional plant and animal-derived oils like blackberry and raspberry seed [45], flaxseed oil [44, 46], *Terminalia catappa* fruit oils [50], and microalgae oil

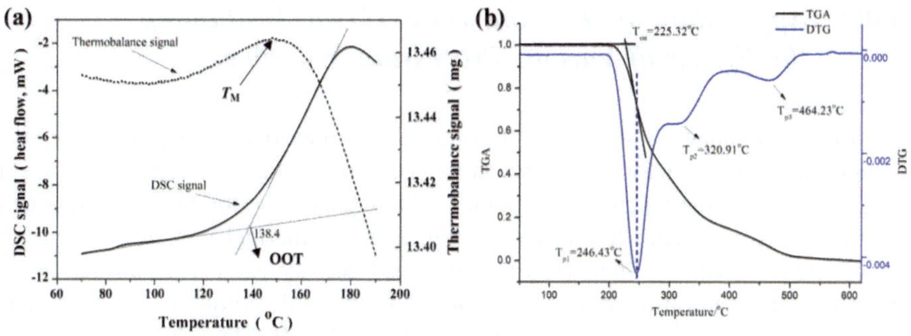

Fig. 6.12 DSC (line) and TG (dots) profiles of grape seed oil (**a**) (Adapted from Ref. [49]; and the specific principles and derivation process of the T_{on} based on TGA curve (**b**) (Adapted from Ref. [44])

[51]. For example, the thermal stability of *Terminalia catappa* oils was found to differ between pulp and nut oils, with the pulp oil exhibiting better stability due to its higher saturated fatty acid content. The T_{on} values for the nut and pulp oils were 280 °C and 230 °C, respectively [50].

Moreover, DSC and TGA/DTG have been employed to explore the effects of nutrient enrichment and physical treatments such as pulsed electric fields (PEF) and ultrasound on lipid stability. Lee and co-workers found that riboflavin increases the crystallization enthalpy of vegetable oil in salad dressings, delaying oxidation [47]. Similarly, polyphenolic compounds in grape seed oil have been found to enhance its antioxidative properties, with ultrasound-assisted extraction showing a correlation between longer extraction times and greater oxidative stability [49]. OOT and T_M values from DSC and TGA analyses indicate a possible correlation between longer ultrasound-assisted extraction time and higher oxidative stability. The oxidative stability of *Schizochytrium* microalgae oil was determined by an accelerated DSC technique [51]. Kinetic analyses of microalgae oil provided E_a, activation enthalpy, and entropy, based on secondary and tertiary oxidation products. The crystallization and melting profiles of flaxseed oil during thermal-induced oxidation were studied by DSC [46]. This information provided a new method for evaluating oxidation levels and chemical composition changes in flaxseed oils using DSC.

Additionally, the effects of nutrient enrichment and physical treatment (e.g., PEF and ultrasound) on the thermal stability of fats and oils can be studied using DSC and TGA/DTG. The polymorphism and crystallization behavior of milk fat were measured by DSC after PEF treatment, and the final temperatures of fat melting were found to be unaffected by PEF treatments. However, thermal treatments enhanced the final melting temperatures of milk fat [41]. The impact of ultrasound on the crystallization and melting properties of virgin olive oil was also studied using DSC [48]. Results revealed that ultrasound generally did not affect the oxidation state or thermal properties of virgin olive oils, further underscoring the versatility of DSC and TGA/DTG in assessing lipid thermal behavior and stability under various conditions and treatment.

6.4.2 Application of DSC and TGA/DTG in Evaluating the Effects of Various Processing Procedures

Food products undergo significant transformations due to various processing procedures. In the modern food industry, physical treatments that don't require chemical reagents are becoming increasingly popular. Among these are thermal methods, such as microwave and radio frequency heating, and non-thermal methods, like ultrasound and pulsed electric field (PEF) treatments. Recently, extensive analyses on the impact of different physical treatments on the nutritional and qualitative characteristics of various food products have been published [6]. Food products subjected to non-conventional processing have had their quality and nutritional characteristics evaluated using DSC and TGA/DTG methods. Below are some of the most noteworthy findings.

6.4.2.1 Microwave Heating

In food processing, microwaves are a form of electromagnetic radiation known for their high penetration and heating efficiency. Microwave treatment offers several advantages, including improved energy efficiency, simplified automatic control, and enhanced food product quality. The development of high-power microwave sources has led to its widespread industrial use.

Microwave heating is commonly used in the preparation and heating of polysaccharide-rich foods. It has been shown to modify starch crystallinity and promote the formation of amylose-lipid complexes, which influence the thermal properties of food. Microwaves play a critical role in the early phases of gelatinization, while thermal effects dominate later. Lyng and colleagues evaluated the impact of formulation changes on the thermal properties of reconstituted potato flakes after microwave warming [10]. The results from DSC revealed the specific heat capacity (C_p) values for the different formulations, though the impact of composition on specific heat capacity was not conclusively determined. The DSC analysis indicated that microwave-treated potato starches exhibited a higher gelatinization temperature compared to native potato starches [27]. Moreover, microwave treatment of potato starch resulted in a decrease in onset T_o, T_p, and T_e temperatures [52], with these values decreasing as microwave radiation intensity increased. Higher microwave power led to greater disruption of starch's crystalline structure, as well as changes in its rheology, as indicated by DSC measurements. DSC and TGA curves (Fig. 6.13a and b) demonstrated that after flaxseed gum was exposed to microwave radiation for 5 minutes, the values of T_o, T_p, T_e, and ΔH increased, and the first weight loss (intrinsic water) disappeared due to microstructural changes [33]. These changes were attributed to the microwave's modification of the polysaccharides' crystal structure.

Microwave-assisted thawing methods are also well-known for reducing protein denaturation and preserving the quality of frozen meat. DSC was used to assess the time-dependent degrees of protein denaturation in fish samples during microwave

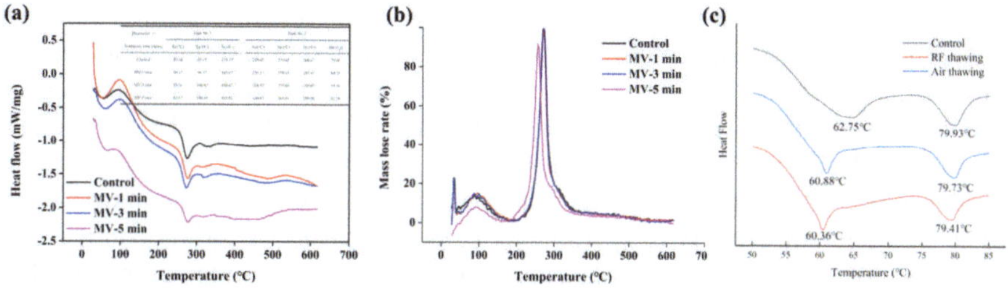

Fig. 6.13 DSC (**a**) and DTG (**b**) curves of flaxseed gum treated by microwave (Adapted from Ref. [33]), and DSC curves (**c**) of frozen mutton after thawing (Adapted from Ref. [53])

irradiation [54]. Additionally, Liu and colleagues examined the gelation properties of surimi treated with microwave heating and water bath heating [55]. According to the DSC results, the natural structures of myosin and actin were less damaged by microwave heating compared to water bath heating, suggesting that microwave heating is a promising technique for enhancing surimi's gelation properties. Furthermore, TGA/DTG curves revealed variations in the weight loss rates of Chinese dried noodles when subjected to hot air, microwave, and radio frequency drying methods [37]. Microwave drying enhanced the cross-linking of gluten proteins, increasing the stability of the protein network.

6.4.2.2 Radio Frequency Heating

Radio frequency (RF) treatments have been used in many food process operations such as drying [37], pasteurization [56], blanching [57], tempering [58], and thawing [53]. RF treatment generates heat within the product, resulting in uniform heating, which is particularly advantageous for low-moisture foods that typically have low thermal diffusivities. RF heating addresses the limitations of conventional heating methods by providing volumetric heating that can penetrate deeply into foods. Its rapid heating rate helps maintain product quality while significantly reducing the heat treatment time. Understanding the thermal characteristics of foods is essential for optimizing RF treatment and improving heating uniformity [59]. Several factors, including frequency, electrode gaps, sample geometry, moisture content, superimposed arrangement, and bulk density, affect the heating uniformity of RF treatment.

RF-treated starch has a lower gelatinization temperature compared to native and microwave-treated starches [27]. When the water content was held constant, the ΔH values of microwave-treated starches were higher than those of RF-treated starches, indicating that RF treatments penetrate deeper and cause greater destruction in starch granules. The thermal characteristics of potatoes subjected to RF blanching were studied using DSC [57], showing that RF-treated potatoes required less thermal energy to gelatinize, as evidenced by lower ΔH values.

The DSC curves of RF-treated powdered infant formula milk revealed that protein denaturation temperature increased with decreasing moisture content and RF treatment resulted in more stable protein structures [56]. Zhang and colleagues also investigated the effects of RF tempering on shrimp [58]. SC results showed that RF-tempered shrimp exhibited significantly higher ΔH_1 and ΔH_2 compared to shrimp tempered using water or refrigerator methods after all freeze-thaw cycles. Moreover, RF tempering took less time to temper six layers of shrimp compared to the other methods. Figure 6.13c presents the DSC curves of mutton before and after two thawing methods, showing that both air thawing and RF thawing reduced the denaturation temperature of mutton proteins, indicating a reduction in their thermal stability [53].

6.4.2.3 Ultrasound

Ultrasound is a novel, safe, eco-friendly, and efficient method used in food production. It is widely recognized for its role in enhancing processes such as extraction, heating, marination, thawing, and freezing. For instance, ultrasound-assisted freezing promotes ice formation, reduces the width of the metastable zone, shortens the induction period, and increases the nucleation rate [60]. Additionally, ultrasound has shown potential in improving the efficiency of the malaxation process in the production of virgin olive oil [48]. Ultrasound treatments (150 W, 40 KHz, for durations of 0, 15, 30, and 60 minutes) did not alter the thermal properties of virgin olive oil, highlighting its promising applications in this industry.

McDonnell and co-workers studied the effects of ultrasound power on myosin and actin in pork [61]. Their DSC thermographs of salted pork muscle after various ultrasound treatments showed a reduction in myosin T_d from 54.2 to 52.7 °C. Higher power ultrasound treatments induced myosin denaturation at the meat's surface. Similarly, Frydenberg et al. analyzed the thermal properties of whey proteins treated with high-density ultrasound using DSC [62]. They found that changes in denaturation enthalpy were linked to alterations in bonding patterns; fewer or weaker bonds resulted in reduced energy for protein unfolding, leading to lower enthalpy values. In ultrasound-assisted marination, T_m and ΔH values were significantly lower than in the control group, indicating that ultrasound-induced conformational changes in actomyosin [63].

Low-intensity ultrasound is also effective in assisting the freezing and thawing of food. DSC curves of mushrooms subjected to ultrasound-assisted freezing displayed an endothermic peak [60]. Gambuteanu and Alexe evaluated the influence of five conventional thawing methods on the maximum transition temperatures and ΔH values of pork muscle using DSC and found no significant differences between ultrasound-assisted thawing and other conventional methods, suggesting that ultrasound thawing did not significantly alter the protein structure [64]. A study on ultrasound-assisted thawing of mushrooms revealed that bath-type ultrasound caused the least protein denaturation without significant structural disruption, as confirmed by DSC data [65].

6.4.2.4 Pulsed Electric Fields

Pulsed electric fields (PEF) technology applies short bursts of electric fields and has been primarily used in food preservation and pre-treatment techniques for processes like freeze-thawing, drying, peeling, cooking, and modifying food structure, such as meat tenderization. PEF induces structural changes in food components, influencing their processing and nutritional characteristics. Understanding the effects of PEF on the thermal stability of nutritional components is essential for developing effective PEF protocols to improve food quality.

PEF-assisted acetylation of cassava and potato starches was investigated using DSC and TGA/DTG [24]. The results showed that PEF treatment was more efficient for acetylation than conventional methods. However, TGA/DTG analysis indicated that the thermal stability of PEF-treated acetylated cassava starches was lower than that of native starch. Similarly, Zeng and co-workers examined the effects of PEF treatment on the thermal properties of waxy rice starch [66]. DSC data revealed significant changes in the digestibility of waxy rice starch after PEF treatment. In another study, PEF-assisted oxidation of starch led to a T_p that was 10.3 °C lower than native starch, with TGA/DTG data indicating decreased thermal stability due to oxidation [29].

The thermal characteristics of PEF-treated milk proteins were extensively studied using DSC [41]. The denaturation of α-lactalbumin, β-casein, β-lactoglobulin, and xanthine oxidase increased with higher treatment intensity. Additionally, PEF treatment of connective tissue isolated from beef pectoralis muscle resulted in a lower denaturation temperature compared to untreated samples [67]. As the specific energy and electric field strength increased, the Ringer-soluble collagen fraction also rose, indicating enhanced solubility of collagen. This suggests that PEF could be used to reduce cooking times for tough beef cuts by breaking down connective tissue. Moreover, DSC analysis of α-amylase pre- and post-PEF treatment showed a decrease in enthalpy and an increase in the denatured peak, indicating that PEF enhanced the intramolecular interactions of α-amylase [68].

6.4.3 Application of Thermal Characteristics in Food Engineering and Logistics

6.4.3.1 Freeze Drying and Water Migration of Fruits and Vegetables

Freeze drying is a widely used dehydration technology that preserves the structure and nutritional value of food by removing water through sublimation. This method effectively maintains the cellular structure of fruits and vegetables, allowing them to retain their shape and texture even after rehydration. Freeze drying has been applied in the processing of various fruits and vegetables, such as stem lettuce cubes [69], pumpkin [70], *Dioscorea opposite* Thunb [71], garlic slices [72], mushroom [60, 73], mango [74], strawberry [75, 76], blueberry [77], raspberry [20], and carambola [78].

During freeze drying, water migrates from the interior of the food to the surface, where it sublimates. Water migration occurs through various processes such as sublimation, absorption, and drip loss. This migration is influenced by factors like temperature gradients, water gradients, and the food's structure [79]. The water gradient promotes the movement of immobile water to bound water [70]. Due to the non-uniform structure of food, water migration may vary across different areas, leading to varying degrees of drying in freeze-dried products.

The efficiency and uniformity of water migration during freeze drying directly affect the final product's quality, including appearance, taste, texture, juiciness, weight, rehydration capacity, and nutrient retention. LF-NMR can non-destructively monitor the content, status, diffusivity, distribution, and migration of water in various foods [77]. For instance, Wang and colleagues used LF-NMR to study the distribution and movement of unfrozen water in lettuce stem pieces during microwave freeze drying [69]. The content of un-freezable water and its distribution significantly affected microwave energy input. Zhang et al. examined how water mobility in frozen mangoes was influenced by state/phase transitions [74]. Frozen mangoes in their glassy state retained more vitamin C due to reduced water mobility, fewer ice crystals, and less freezable water.

One challenge in freeze drying is ensuring uniform water migration throughout the food, preventing issues like excessive or incomplete drying. Current research explores improving freeze drying technology to enhance water migration control, efficiency, and product quality. Various auxiliary technologies, including microwave, ultrasound, and high hydrostatic pressure, are being coupled with freeze drying to achieve more uniform water migration and higher drying efficiency. As microwave power density increased, both the effective water diffusion coefficient and water migration velocity rose, leading to a gradual conversion of unbound water to bound water [71]. Pulse-spouted microwave freeze drying improved the drying uniformity of raspberries compared to conventional microwave freeze drying [20]. Pre-ultrasound treatment before vacuum freeze drying enhanced moisture migration and quality characteristics in strawberry slices [75]. High hydrostatic pressure pretreatment also promoted water migration during freeze drying in both strawberry slices [76] and *carambola* slices [78], resulting in more uniform water distribution and shorter drying duration.

6.4.3.2 Relationship Between TGA/DTG Technology and Internal Structure of Food Materials

TGA and DTG are analytical techniques extensively applied to investigate the thermal stability and composition of food materials. The relationship between TGA/DTG technology and the internal structure of food materials can be understood through the insights these methods provide into thermal decomposition and mass loss behavior. Thermal transitions, such as glass transitions, melting, and gelatinization, are linked to the internal structure of the food material. Changes in the internal structure of food significantly affect its texture and stability.

TGA/DTG can be used to study protein denaturation, starch swelling, and the crystallization of vegetable oil, all of which contribute to structural changes in food. Researchers have used TGA/DTG to examine the effects of enrichment compounds on the internal structure of food products. For example, the impact of four types of dietary fiber polysaccharides on the structural and thermal characteristics of gluten proteins was revealed by TGA analysis [25]. The shape of the TGA/DTG curves indicated that MCC and inulin caused the dehydration of the gluten network. Similarly, Witek and co-workers used DTG analysis to study the effect of egg protein enrichment on gluten-free rice pasta [36]. The addition of egg white resulted in a distinctly different vaporization curve in the dough, suggesting a modification of the starch-protein matrix. TGA/DTG analysis of rice kernels further confirmed the hydrophobic structural transition of protein during cooking [80].

Additionally, TGA/DTG has been employed to investigate the effects of processing technologies on the internal structure of food products. Hong and co-workers studied the effect of PEF conditions on the acetylation of cassava starch, with TGA/DTG revealing that the thermal stability of starch deteriorated after acetylation [24]. Native starches exhibited reduced freeze-thaw stability and increased syneresis compared to acetylated starches, attributed to the re-crystallization of amylose molecules. Yu and colleagues explored how microwave exposure affected the structure of flaxseed gum polysaccharides [33]. Their study showed that after microwave treatment for 5 minutes, 63.07% of intrinsic water in flaxseed gum was lost due to polysaccharide depolymerization, leading to changes in water absorption and retention capacity. Moreover, TGA/DTG results demonstrated that microwave and radio frequency drying increased the cross-linking of gluten proteins in Chinese dried noodles, enhancing the stability of the protein network and improving textural and cooking properties [37].

Overall, TGA/DTG technology proves to be a valuable tool for understanding the complex relationships between thermal properties and the internal structure of food materials. By analyzing thermal decomposition behavior and identifying component interactions, TGA/DTG offers crucial insights into the quality, stability, and safety of food products. These insights are essential for optimizing food processing techniques, contributing to the production of high-quality and safe food for consumers.

6.4.3.3 Application of Thermal Characteristics of Foods in Intelligent and Cold Chain Logistics

Understanding the thermal characteristics of food is essential for designing and optimizing intelligent and cold chain logistics systems. The specific heat capacity of food helps calculate the energy required to maintain a constant temperature during transportation, optimizing energy efficiency in refrigeration or heating systems. Selecting materials with appropriate thermal conductivity can minimize heat loss or gain in intelligent transportation systems, maintaining the required food temperature, which is particularly crucial for insulation in cold chain logistics. For instance, the specific heat and thermal conductivity of sugar solutions correlate with the

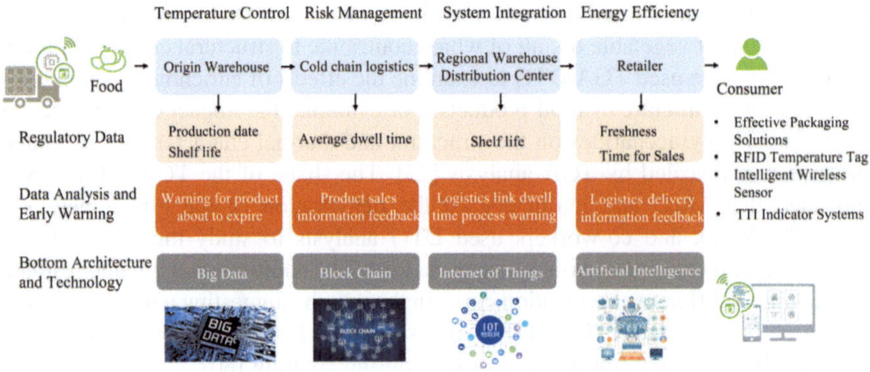

Fig. 6.14 The intelligent and cold chain logistics of food materials

Prandtl number, which is vital for engineering and transport analysis [8]. Additionally, knowledge of thermal diffusivity is crucial in designing efficient pasteurization systems [14]. State and phase transitions in frozen foods, such as ice melting or crystallization, affect quality and storage. Phase change materials (PCMs) can help stabilize temperatures during transportation, minimizing fluctuations—particularly important for temperature-sensitive products [81].

Intelligent and cold chain logistics systems, through the integration of sensors, real-time monitoring, and advanced control algorithms, can accurately control and manage these thermal characteristics. Figure 6.14 illustrates the intelligent and cold chain logistics of food materials. The application of food thermal properties in such systems is crucial for maintaining the quality and safety of perishable goods during transportation and storage.

Here are some key points outlining the application of food thermal properties in intelligent and cold chain logistics:

Temperature Control

Many foods, especially perishable items like fruits, vegetables, prepared dishes, and frozen foods, are highly sensitive to temperature changes. Fluctuations or unsuitable temperature conditions can lead to spoilage, compromising the quality and safety of the food. By understanding and leveraging the thermal properties of food, more effective cooling and heating technologies can be developed to maintain optimal temperatures throughout the supply chain. These thermal properties are also critical in designing packaging solutions that protect food from temperature variations. Intelligent and cold chain logistics systems can utilize food's thermal characteristics to regulate temperatures more accurately, ensuring that products remain within safe temperature ranges during transportation. For example, chilling injuries can occur when bananas are stored below 11 °C. Knowledge of thermal diffusivity and thermal conductivity is vital for the design and optimization of cooling units

[12]. Additionally, Staniszewska and colleagues found that glass transition temperature information was useful for optimizing storage conditions for dried cranberry powders [82].

Risk Management

Thermal property data can help identify and manage potential risk points in cold chain logistics, such as temperature abuse or breaks in the cold chain. In recent years, time-temperature-integrator (TTI) indicator systems have significantly improved the management of frozen food distribution and safety. Advances in TTI technology have enabled the delivery of safe, affordable, and high-quality frozen foods to consumers. The integrated application of RFID temperature tags, intelligent wireless sensors, GPS, GPRS, and GIS allows for real-time monitoring of the entire process—from the raw material origin, through the food processing plant, preservation and storage, to the final sales point—ensuring better control and safety throughout the supply chain.

System Integration

Intelligent transportation systems can integrate thermal property data into logistics planning and execution software to optimize route planning, scheduling, and loading of food products. This information allows for more efficient planning of transportation routes and storage conditions, reducing food loss during transit and storage. Researchers are increasingly using advanced technologies such as big data, blockchain, the Internet of Things, artificial intelligence, and fuzzy logic to explore the non-linear relationships between thermal characteristics and variables like temperature, pressure, and formulation [83]. Accurate thermophysical property measurements are essential for refining and validating predictive models. Innovations in instrumentation, such as three-dimensional imaging techniques and improvements to calorimetric and line heat source methods, have led to more precise quantification of specific heat and thermal conductivity.

Energy Efficiency

A comprehensive knowledge of the thermal properties of food also aids in designing more energy-efficient refrigeration and freezing systems, reducing energy consumption and operational costs, leading to more environmentally friendly transportation practices. Applying thermal properties in intelligent and cold chain logistics enhances food quality and safety while improving the overall efficiency and reliability of the food supply chain. Optimizing temperature control in cold chain logistics contributes to reducing food waste and supports sustainable development goals.

6.5 Conclusion

Understanding the thermal properties of food is fundamental and essential for every aspect of the food industry. It plays a pivotal role in enhancing the overall quality and safety of food while meeting consumer needs and market expectations. With advances in instrumentation, food manufacturers and researchers now have the ability to identify a wide range of food attributes. Techniques like DSC and TGA/DTG are key measurement tools for determining the thermal characteristics of food. These technologies provide both quantitative and qualitative data that are crucial for assessing nutritional properties and evaluating processing methods. This insight aids in designing formulations and optimizing processing conditions to achieve the desired qualities and nutritional value of food products. Moreover, the thermal properties of food are vital in the design and operation of intelligent and cold chain logistics systems, which are essential for maintaining food quality and safety during storage and transportation. By leveraging these thermal characteristics, the industry can ensure that food products remain stable and safe throughout the supply chain, thereby improving consumer satisfaction and reducing waste.

References

1. Figura LO, Teixeira AA. Food physics-physical properties measurement and applications. 2nd ed. Berlin: Springer; 2023. p. 75.
2. Gharibzahedi SMT, Ghahderijani M, Lajevardi ZS. Specific heat, thermal conductivity and thermal diffusivity of red lentil seed as a function of moisture content. J Food Process Preserv. 2014;38(4):1807–11.
3. Jiang S, Zhao J. Thermal characteristics of food. Food Phys. 2015;
4. Bai S, Liu L, Yu H, Guan X, et al. Thermal and dielectric properties of wolfberries as affected by moisture content and temperature associated with radio frequency and microwave dehydrations. Food Secur. 2022;11(23)
5. Kumar PK, Parhi A, Sablani SS. Development of high-fiber and sugar-free frozen pancakes: influence of state and phase transitions on the instrumental textural quality of pancakes during storage. LWT-Food Sci Technol. 2021;146:111454.
6. Parniakov O, Bals O, Barba FJ, Mykhailyk V, Lebovka N, et al. Application of differential scanning calorimetry to estimate quality and nutritional properties of food products. Crit Rev Food Sci Nutr. 2018;58(3):362–85.
7. Ozturk S, Kong F, Singh RK, Kuzy JD, Li C. Radio frequency heating of corn flour: heating rate and uniformity. Innovative Food Sci Emerg Technol. 2017;44:191–201.
8. Martins MJN, Guimarães B, Polachini TC, Telis-Romero J. Thermophysical properties of carbohydrate solutions: correlation between thermal and transport properties. J Food Process Eng. 2020;43(9)
9. Mehta HN, Benyathiar P, Mishra DK, Varney M. Complementary experiments for parameter estimation in heat transfer model. Food Bioprod Process. 2021;128:240–6.
10. Lyng JG, Arimi JM, Scully M, Marra F. The influence of compositional changes in reconstituted potato flakes on thermal and dielectric properties and temperatures following microwave heating. J Food Eng. 2014;124:133–42.
11. Pierre-Noel-Gilles B, Iribe-Salazar R, Vazquez-Lopez Y, Caro-Corrales J. Heat transfer analysis during hydrothermal treatment of mango. J Food Sci. 2022;87(3):1161–73.

12. Erdoğdu F, Linke M, Praeger U, Geyer M, et al. Experimental determination of thermal conductivity and thermal diffusivity of whole green (unripe) and yellow (ripe) Cavendish bananas under cooling conditions. J Food Eng. 2014;128:46–52.
13. Muramatsu Y, Greiby I, Mishra DK, Dolan KD. Rapid inverse method to measure thermal diffusivity of low-moisture foods. J Food Sci. 2017;82(2):420–8.
14. da Silva WP, da Silva AP, de Souto LM, da Silva Jr AF, Ferreira JP d L, et al. Determination of constant and variable thermal diffusivity of cashew pulp during heating: experimentation, optimizations and simulations. Case Stud Therm Eng. 2022;39
15. Balasubramanian S, Devi A, Singh KK, Bosco SJ, et al. Application of glass transition in food processing. Crit Rev Food Sci Nutr. 2016;56(6):919–36.
16. Leyva-Porras C, Cruz-Alcantar P, Espinosa-Solis V, Martinez-Guerra E, Balderrama CIP, et al. Application of differential scanning calorimetry (DSC) and modulated differential scanning calorimetry (MDSC) in food and drug industries. Polymers (Basel). 2019;12(1)
17. Dehghannya J, Ngadi M. The application of glass transition temperature in the frying of starchy foods: a review. Food Rev Intl. 2023;40:1–19.
18. Li R, Lin D, Roos YH, Miao S. Glass transition, structural relaxation and stability of spray-dried amorphous food solids: a review. Dry Technol. 2018;37(3):287–300.
19. Jeong D, Zhang C, Ha M, Chung H-J. Relationships between phase transition and quality characteristics of cooked rice during storage. Food Hydrocoll. 2022;132:107874.
20. Sun Y, Zhang M, Mujumdar AS, Yu D. Pulse-spouted microwave freeze drying of raspberry: control of moisture using ANN model aided by LF-NMR. J Food Eng. 2021;292
21. Sun X, Hu J, Xiao H, Liu C, Yang F, et al. Effect of pectin addition on the drying characteristics of freeze-dried restructured strawberry blocks. LWT. 2024;192
22. Islam M, Bełkowska L, Konieczny P, Fornal E, et al. Differential scanning calorimetry for authentication of edible fats and oils - what can we learn from the past to face the current challenges? J Food Drug Anal. 2022;30(2):185–201.
23. Xie QY, Chen DD, Ding YW. Thermogravimetric analysis and its applications in polymer characterization. Acta Polym Sin. 2022;53(2):193–210.
24. Hong J, Zeng X-A, Buckow R, Han Z, et al. Nanostructure, morphology and functionality of cassava starch after pulsed electric fields assisted acetylation. Food Hydrocoll. 2016;54:139–50.
25. Nawrocka A, Szymańska-Chargot M, Miś A, Wilczewska AZ, et al. Effect of dietary fibre polysaccharides on structure and thermal properties of gluten proteins—a study on gluten dough with application of FT-Raman spectroscopy, TGA and DSC. Food Hydrocoll. 2017;69:410–21.
26. Jeong D, Lee JH, Chung HJ. Effect of molecular structure on phase transition behavior of rice starch with different amylose contents. Carbohydr Polym. 2021;259:117712.
27. Xia T, Gou M, Zhang G, Li W, et al. Physical and structural properties of potato starch modified by dielectric treatment with different moisture content. Int J Biol Macromol. 2018;118 (Pt B):1455–62.
28. Kumar PK, Joyner HS, Tang J, Rasco BA, et al. Kinetics of starch retrogradation in rice (Oryza sativa) subjected to state/phase transitions. Food Bioprocess Technol. 2020;13(9):1491–504.
29. Li Y, Wang JH, Wang EC, Tang ZS, et al. The microstructure and thermal properties of pulsed electric field pretreated oxidized starch. Int J Biol Macromol. 2023;235:123721.
30. Sun J, He RM, Gao FY, Kou ZL, et al. High-efficient preparation of cross-linked cassava starch by microwave-ultrasound-assisted and its physicochemical properties. Starch—Stärke. 2019;71 (7–8)
31. Suvakanta D, Narsimha MP, Pulak D, Joshabir C, Biswajit D. Optimization and characterization of purified polysaccharide from Musa sapientum L. as a pharmaceutical excipient. Food Chem. 2014;149:76–83.
32. Wang W, Ma X, Jiang P, Hu L, et al. Characterization of pectin from grapefruit peel: a comparison of ultrasound-assisted and conventional heating extractions. Food Hydrocoll. 2016;61:730–9.

33. Yu X, Huang S, Yang F, Qin X, et al. Effect of microwave exposure to flaxseed on the composition, structure and techno-functionality of gum polysaccharides. Food Hydrocoll. 2022;125:107447.
34. Brunton NP, Lyng JG, Zhang L, Jacquier JC. The use of dielectric properties and other physical analyses for assessing protein denaturation in beef biceps femoris muscle during cooking from 5 to 85 degrees C. Meat Sci. 2006;72(2):236–44.
35. Chian FM, Kaur L, Oey I, Astruc T, Hodgkinson S, et al. Effect of pulsed electric fields (PEF) on the ultrastructure and in vitro protein digestibility of bovine longissimus thoracis. LWT Food Sci Technol. 2019;103:253–9.
36. Witek M, Maciejaszek I, Surowka K. Impact of enrichment with egg constituents on water status in gluten-free rice pasta - nuclear magnetic resonance and thermogravimetric approach. Food Chem. 2020;304:125417.
37. Lin Q, Shen H, Ma S, Zhang Q, Yu X, et al. Morphological distribution and structure transition of gluten induced by various drying technologies and its effects on Chinese dried noodle quality characteristics. Food Bioprocess Technol. 2023;16(6):1374–87.
38. García-Sánchez G, Sotelo-Romero CR, Pacheco-Aguilar R, Ramírez-Suárez JC, et al. Effect of freezing on protein denaturation and gelling capacity of jumbo squid (Dosidicus gigas) mantle muscle. LWT Food Sci Technol. 2015;60(2):737–42.
39. Vaskoska R, Ha M, Ong L, Chen G, et al. Myosin sensitivity to thermal denaturation explains differences in water loss and shrinkage during cooking in muscles of distinct fibre types. Meat Sci. 2021;179:108521.
40. Dissanayake M, Ramchandran L, Donkor ON, Vasiljevic T. Denaturation of whey proteins as a function of heat, pH and protein concentration. Int Dairy J. 2013;31(2):93–9.
41. Sharma P, Oey I, Everett DW. Thermal properties of milk fat, xanthine oxidase, caseins and whey proteins in pulsed electric field-treated bovine whole milk. Food Chem. 2016;207:34–42.
42. Li S, Wei Y, Fang Y, Zhang W, et al. DSC study on the thermal properties of soybean protein isolates/corn starch mixture. J Therm Anal Calorim. 2014;115(2):1633–8.
43. Sobanwa M, Foster TJ, Watson NJ. The effect of cellulose and starch on the viscoelastic and thermal properties of acid-swollen collagen paste. Food Hydrocoll. 2020;101:105460.
44. Li J, Liu J, Sun X, Liu Y. The mathematical prediction model for the oxidative stability of vegetable oils by the main fatty acids composition and thermogravimetric analysis. LWT- Food Sci Technol. 2018;96:51–7.
45. Micić DM, Ostojić SB, Simonović MB, Krstić G, et al. Kinetics of blackberry and raspberry seed oils oxidation by DSC. Thermochim Acta. 2015;601:39–44.
46. Zhang Z, Wei Y, Guan N, Li B, et al. Changes in chemical composition of flaxseed oil during thermal-induced oxidation and resultant effect on DSC thermal properties. Molecules. 2022;27 (20)
47. Lee YH, Lee J, Min DB, Pascall MA. Effect of riboflavin on the photo-oxidative stability of vegetable oil in salad dressing. Food Chem. 2014;152:349–54.
48. Gila A, Sanchez-Ortiz A, Jimenez A, Beltran G. The ultrasound application does not affect to the thermal properties and chemical composition of virgin olive oils. Ultrason Sonochem. 2021;70:105320.
49. Malićanin M, Rac V, Antić V, Antić M, et al. Content of antioxidants, antioxidant capacity and oxidative stability of grape seed oil obtained by ultra sound assisted extraction. J Am Oil Chem Soc. 2014;91(6):989–99.
50. Santos OV, Lorenzo ND, Souza ALG, Costa CEF, et al. CO(2) supercritical fluid extraction of pulp and nut oils from Terminalia catappa fruits: thermogravimetric behavior, spectroscopic and fatty acid profiles. Food Res Int. 2021;139:109814.
51. Bañares C, Chabni A, Reglero G, Torres CF. Oxidative stability of microalgae oil and its acylglycerol mixture obtained by enzymatic glycerolysis and the antioxidant effect of supercritical rosemary extract. LWT. 2022;171:171.

52. Przetaczek-Roznowska I, Fortuna T, Wodniak M, Labanowska M, et al. Properties of potato starch treated with microwave radiation and enriched with mineral additives. Int J Biol Macromol. 2019;124:229–34.
53. Sun Y, Jia Y, Song M, Liu Y, et al. Effects of radio frequency thawing on the quality characteristics of frozen mutton. Food Bioprod Process. 2023;139:24–33.
54. Liu S, Yu X, Fukuoka M, Sakai N. Modeling of fish boiling under microwave irradiation. J Food Eng. 2014;140:9–18.
55. Liu X, Feng D, Ji L, Zhang T, Xue Y, et al. Effects of microwave heating on the gelation properties of heat-induced Alaska Pollock (Theragra chalcogramma) surimi. Food Sci Technol Int. 2018;24(6):497–506.
56. Zhang Y, Zhu G, Xie Y, Liu Y. Effects of radio frequency on physicochemical properties of powdered infant formula milk as compared with conventional thermal treatment. LWT. 2020;134
57. Zhang Z, Yao Y, Shi Q, Zhao J, et al. Effects of radio-frequency-assisted blanching on the polyphenol oxidase, microstructure, physical characteristics, and starch content of potato. LWT. 2020;125
58. Zhang Y, Li F, Yao Y, He J, et al. Effects of freeze-thaw cycles of Pacific white shrimp (Litopenaeus vannamei) subjected to radio frequency tempering on melanosis and quality. Innovative Food Sci Emerg Technol. 2021;74:102860.
59. Altemimi A, Aziz SN, Al-Hilphy ARS, Lakhssassi N, Watson DG, et al. Critical review of radio-frequency (RF) heating applications in food processing. Food Qual Saf. 2019;3(2):81–91.
60. Islam MN, Zhang M, Fang Z, Sun J. Direct contact ultrasound assisted freezing of mushroom (Agaricus bisporus): growth and size distribution of ice crystals. Int J Refrig. 2015;57:46–53.
61. McDonnell CK, Allen P, Morin C, Lyng JG. The effect of ultrasonic salting on protein and water-protein interactions in meat. Food Chem. 2014;147:245–51.
62. Frydenberg RP, Hammershoj M, Andersen U, Greve MT, et al. Protein denaturation of whey protein isolates (WPIs) induced by high intensity ultrasound during heat gelation. Food Chem. 2016;192:415–23.
63. Zou Y, Shi H, Xu P, Jiang D, et al. Combined effect of ultrasound and sodium bicarbonate marination on chicken breast tenderness and its molecular mechanism. Ultrason Sonochem. 2019;59:104735.
64. Gambuteanu C, Alexe P. Comparison of thawing assisted by low-intensity ultrasound on technological properties of pork Longissimus dorsi muscle. J Food Sci Technol. 2015;52(4): 2130–8.
65. Farahnak R, Nourani M, Riahi E. Ultrasound thawing of mushroom (Agaricus bisporus): effects on thawing rate, protein denaturation and some physical properties. LWT. 2021;151
66. Zeng F, Gao QY, Han Z, Zeng XA, et al. Structural properties and digestibility of pulsed electric field treated waxy rice starch. Food Chem. 2016;194:1313–9.
67. Alahakoon AU, Oey I, Silcock P, Bremer P. Understanding the effect of pulsed electric fields on thermostability of connective tissue isolated from beef pectoralis muscle using a model system. Food Res Int. 2017;100(Pt 2):261–7.
68. Jin W, Wang Z, Peng D, Shen W, et al. Effect of pulsed electric field on assembly structure of α-amylase and pectin electrostatic complexes. Food Hydrocoll. 2020;101:105547.
69. Wang Y, Zhang M, Mujumdar AS, Mothibe KJ. Experimental investigation and mechanism analysis on microwave freeze drying of stem lettuce cubes in a circular conduit. Dry Technol. 2012;30(11–12):1377–86.
70. Chao E, Li J, Fan L. Influence of combined freeze-drying and far-infrared drying technologies on physicochemical properties of seed-used pumpkin. Food Chem. 2023;398:133849.
71. Wang H, Duan X, Duan L, Ren G. Mutual transformation of the water binding state and moisture diffusion characteristics of Chinese yams during microwave freeze drying. Dry Technol. 2019;39(1):66–76.

72. Feng Y, Ping Tan C, Zhou C, Yagoub AEA, et al. Effect of freeze-thaw cycles pretreatment on the vacuum freeze-drying process and physicochemical properties of the dried garlic slices. Food Chem. 2020;324:126883.
73. Younas S, Mao Y, Liu C, Murtaza MA, et al. Measurement of water fractions in freeze-dried shiitake mushroom by means of multispectral imaging (MSI) and low-field nuclear magnetic resonance (LF-NMR). J Food Compos Anal. 2021;96:103694.
74. Zhang Y, Zhao J-H, Ding Y, Xiao H-W, Sablani SS, et al. Changes in the vitamin C content of mango with water state and ice crystals under state/phase transitions during frozen storage. J Food Eng. 2018;222:49–53.
75. Xu B, Chen J, Sylvain Tiliwa E, Yan W, et al. Effect of multi-mode dual-frequency ultrasound pretreatment on the vacuum freeze-drying process and quality attributes of the strawberry slices. Ultrason Sonochem. 2021;78:105714.
76. Zhang L, Qiao Y, Wang C, Liao L, Shi D, et al. Influence of high hydrostatic pressure pretreatment on properties of vacuum-freeze dried strawberry slices. Food Chem. 2020;331: 127203.
77. Liu W, Zhang M, Bhandari B, Yu D. A novel combination of LF-NMR and NIR to intelligent control in pulse-spouted microwave freeze drying of blueberry. LWT. 2021;137
78. Zhang S, Zhao Y, Yao X, Zheng Z, et al. Effect of high hydrostatic pressure pretreatment on flavour and physicochemical properties of freeze-dried carambola slices. Int J Food Sci Technol. 2022;57(7):4245–53.
79. Kumar PK, Rasco BA, Tang J, Sablani SS. State/phase transitions, ice recrystallization, and quality changes in frozen foods subjected to temperature fluctuations. Food Eng Rev. 2020;12 (4):421–51.
80. Zhu L, Wu G, Cheng L, Zhang H, et al. Investigation on molecular and morphology changes of protein and starch in rice kernel during cooking. Food Chem. 2020;316:126262.
81. Qi T, Ji J, Zhang X, Liu L, Xu X, et al. Research progress of cold chain transport technology for storage fruits and vegetables. J Energy Storage. 2022;56
82. Staniszewska I, Dzadz Ł, Xiao H-W, Zielinska M. Evaluation of storage stability of dried cranberry powders based on the moisture sorption isotherms and glass transition temperatures. Dry Technol. 2020;40(1):89–99.
83. Paluri S, Phinney DM, Heldman DR. Recent advances in thermophysical properties—measurements, prediction, and importance. Curr Opin Food Sci. 2018;23:142–8.

Chapter 7
Acoustical Characteristics in Foods

Hao Lin

The dynamic vibration characteristics of food and agricultural products refer to the vibration signal characteristics generated when they are subjected to external excitation. By analyzing these vibration signals, the internal quality of food and agricultural products, such as hardness, maturity, and cracks, can be understood. The reflection, scattering, transmission, and absorption characteristics of vibration signals, as well as parameters like attenuation coefficient and propagation speed, can provide important information about the quality of agricultural products [1].

The vibrations of sound waves can be perceived by the human ears, and the propagation characteristics of vibration signals can be analyzed through acoustic detection technology. This technology can not only detect the mechanical properties of foods but also evaluate the quality structure characteristics of agricultural products. This chapter will introduce the overview of acoustic detection technology, system and sensing principles, signal processing, and analysis methods and explore their applications in the detection of food and agricultural products.

7.1 Basic Theory of Acoustic Detection Technology

7.1.1 Basic Theory of Sound

Sound is produced by the vibration of objects, and the fluctuations that cause the auditory organs of humans or animals to have a sense of sound are called sound waves. Sound is the phenomenon of fluctuations that can be perceived by the auditory organs of humans or animals as sound waves propagate through a medium

H. Lin (✉)
School of Food and Biological Engineering, Jiangsu University, Zhenjiang, Jiangsu, China
e-mail: linhao@ujs.edu.cn

H. Lin et al., *Advanced Technologies for Physical Properties Measurement of Food and its Processing*, https://doi.org/10.1007/978-981-96-2347-1_7

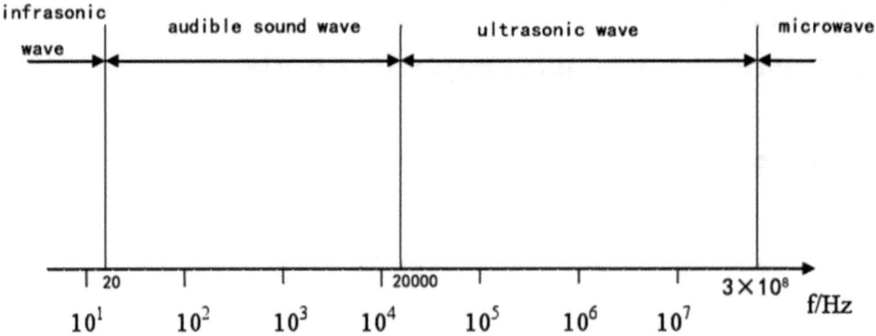

Fig. 7.1 Acoustic phenomena (properties of sound)

(gas, solid, or liquid) [2]. The object that initially emits the vibration is called the sound source. Sound propagates through the medium in the form of waves. As shown in Fig. 7.1, the number of vibrations an object makes in 1 s is called frequency. Based on the frequency of vibration, sound can be divided into three types: infrasound, audible sound, and ultrasound. Due to different vibration frequencies, only audible sound can be perceived by the human ears.

Infrasound refers to sound waves with a frequency below 20 Hz, which cannot be perceived by the human ears. It has advantages such as long propagation distance and strong penetration ability and is widely used in military, natural disaster prediction, medical fields, etc. Audible sound waves refer to sound waves with a frequency between 20 Hz and 20 kHz, which can be perceived by the human ears and are widely used in industrial detection, scientific research, music, and language fields. Ultrasound refers to sound waves with a frequency above 20 kHz, which cannot be perceived by the human ears. It has advantages, such as good directionality, strong breaking ability, and low energy loss during propagation, and is therefore used in industrial detection, food processing, military, and security fields.

The essence of sound is the transmission process of a series of mechanical vibrations produced by the particles of the sound-transmitting medium. One is the medium, and the other is the vibration [3]. On one hand, the vibration of sound waves is mechanical vibration, which is transmitted by the vibration of medium particles driving the vibration of adjacent medium particles. The propagation of sound waves (vibration waves) is not the propagation of the vibrating particles themselves but the propagation of the form of particle vibration, such as electronic instruments using electronic oscillators as the vibration source, but the sound heard by the human ears is transmitted through air vibration or through the human body to reach the ears [4]. On the other hand, the propagation of sound waves cannot be separated from the medium. Without a medium, sound waves cannot propagate, and sound disappears accordingly. For example, when two astronauts successfully landed on the moon, although they were eager to share their joy, they could not hear each other's voices and could only convey emotions through gestures.

Additionally, if we place an alarm clock that can continuously sound in a glass cover connected to a vacuum pump and gradually remove the air from the cover, we will find that the sound of the alarm clock becomes weaker and weaker. This phenomenon indicates that sound waves cannot propagate in a vacuum. Meanwhile, in different media, such as solids, gases, liquids, or plasmas, the propagation speed of sound waves will also vary.

7.1.2 Sound Wave Propagation Methods

Sound waves can be divided into longitudinal waves (L waves), transverse waves (TV, TH waves or SV, SH waves), surface waves (R waves), and plate waves (Lamb waves and Love waves) based on wave type [5].

Longitudinal waves refer to waves where the direction of particle vibration in the medium or physical quantity is parallel to the propagation direction, also known as compression waves or density waves [4], as shown in Fig. 7.2. The propagation process of longitudinal waves appears as parts with different densities along the direction of wave advancement. Essentially, the propagation of longitudinal waves is due to the medium. In each element, compression and stretching deformations occur, and an elastic force is generated to restore the element to its original state. Therefore, longitudinal waves can only propagate in elastic media that undergo tensile and compressive deformation. Generally, solids, liquids, and gases all possess tensile and compressive elasticity, so they can all transmit longitudinal waves.

Transverse waves refer to waves where the direction of particle vibration in the medium or physical quantity is perpendicular to the direction of propagation. As shown in Fig. 7.3, when a rope is shaken in one direction, the wave travels in the x-direction, while the particles of the rope move up and down along the y-axis.

British physicist Lord Rayleigh first confirmed the existence of surface waves (as shown in Fig. 7.4) in 1987. Surface waves are a type of wave that propagates

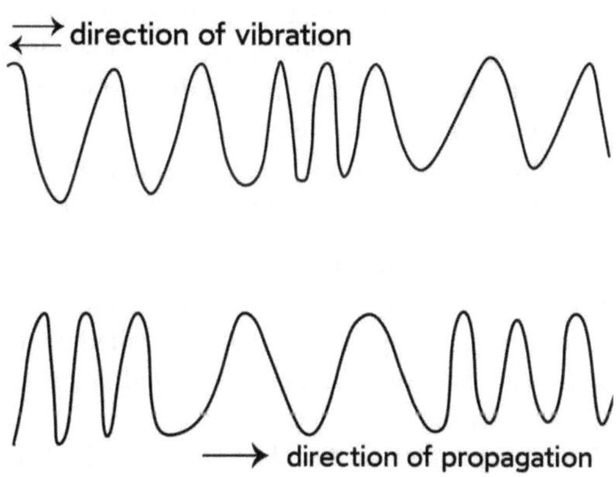

Fig. 7.2 Longitudinal wave schematic

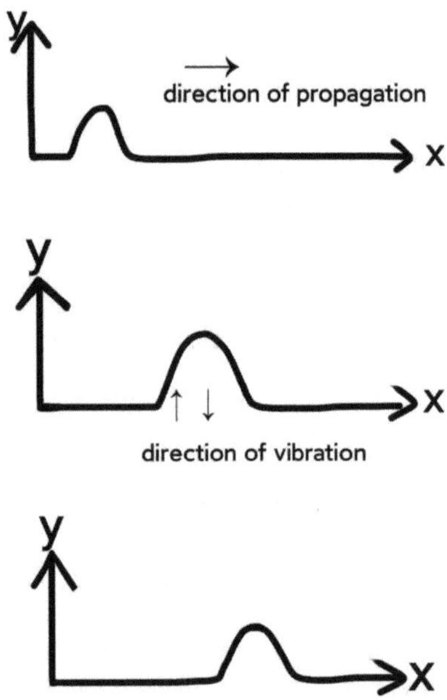

Fig. 7.3 Schematic of rope vibration propagation

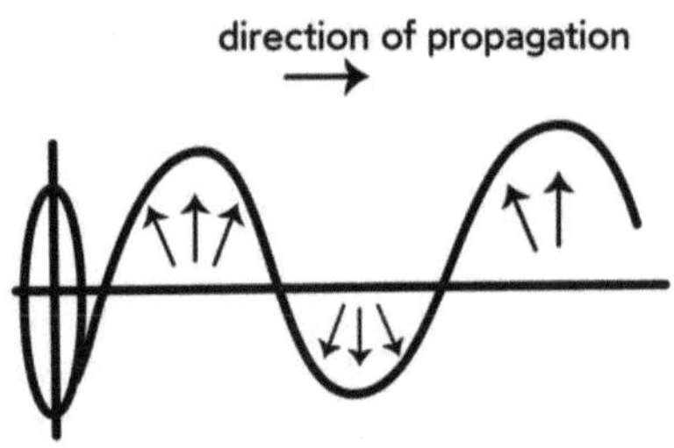

Fig. 7.4 Surface wave schematic

along the surface of a medium and are a special kind of transverse wave. The characteristics of surface waves include:

1. Having a lower propagation speed and shorter wavelength
2. Propagating along the surface of solids with a slower speed
3. Elastic waves propagating on the surface of crystals, not involving the migration of electrons inside the crystal
4. Manufactured using single-crystal materials and planar processes, with good repeatability and consistency, making them easy for mass production

Plate waves are a type of mechanical wave (as opposed to electromagnetic waves), and the transmission of mechanical waves requires a medium. Mechanical waves propagating in a flat plate are called plate waves. Plate waves are a mixture of transverse and longitudinal waves, and in a given plate, three different polarizations of plate waves can exist: the first is a pure transverse wave, with the polarization direction parallel to the plate surface, which is significant in material testing; the second is a symmetric plate wave, with the particle polarization direction parallel to the propagation direction on the plate's central plane; and the third is an asymmetric plate wave, with the particle polarization direction perpendicular to the propagation direction on the plate's central plane.

Audible sound waves are generally longitudinal waves. On one hand, this is because the sound we hear with our ears is the compression and expansion of air stimulating the eardrum. On the other hand, when audible sound waves propagate in the air, the vibration direction of air particles is consistent with the wave's propagation direction, so audible sound waves are longitudinal waves.

The wave type of ultrasound propagating in a medium is determined by the inherent properties of the medium and boundary conditions. For fluid media, since ultrasound only causes volumetric deformation and not shear deformation during propagation, only ultrasonic longitudinal waves exist, and there are no ultrasonic transverse waves (also known as shear waves). In solid media, because ultrasound causes deformation and shear, ultrasonic transverse waves also exist.

7.1.3 Interaction Between Sound Waves and Matter

7.1.3.1 Interaction in Sound Wave Detection

The interaction between sound waves and matter can be used to evaluate and detect the state of objects or materials. Common interactions include propagation, reflection, refraction, scattering, and resonance:

1. Propagation: The speed and characteristics of sound wave propagation vary in different media. In gases, sound waves propagate more slowly, while in liquids and solids, they propagate faster. This is due to differences in molecular density and elasticity.
2. Reflection and Refraction: When sound waves encounter the interface of different media, reflection and refraction occur. Reflection causes sound waves to return to the original medium, while refraction allows sound waves to enter a new medium and change their propagation direction.
3. Scattering: When sound waves encounter irregular material interfaces or particles, scattering occurs, leading to changes in the direction and intensity of the sound waves.
4. Resonance: When the frequency of sound waves matches the natural frequency of a material, resonance occurs, resulting in an increase in the amplitude of the material's vibration.
5. Nonlinear Effects: Under certain conditions, the propagation of sound waves may exhibit nonlinear characteristics, such as changes in wave shape and frequency multiplication. This has important applications in ultrasound technology.

These interactions are often used in acoustic detection technology, having a positive impact on the detection of food quality, internal structure, and other aspects.

7.1.3.2 Interaction in Ultrasonic Processing in Ultrasound Applications

The interaction between ultrasound and matter holds unique significance, specifically manifested in two aspects. On one hand, the effect of objects on ultrasound is reflected in the influence on the ultrasound propagation process. On the other hand, the effect of ultrasound on objects is due to the high intensity, high power, and high frequency generated during the propagation process, which alters the physical properties of the object. The interaction between ultrasound and objects can be summarized as thermal effects, mechanical effects, and cavitation effects:

1. Thermal effects: During the propagation of ultrasound waves, due to the internal friction of the propagation medium, part of the sound wave energy is absorbed by the medium and converted into heat energy, raising the temperature of the medium itself. This method of heating achieves the same effect as other heating methods, and this effect of raising the medium's temperature is called the thermal effect of ultrasound.
2. Mechanical effects: For ultrasound waves, even at low power, they can cause particles in the medium to undergo periodic compression and expansion, producing linear or nonlinear alternating vibrations. These vibrations can excite interaction forces such as the Bernoulli effect and viscous forces within the medium, thereby promoting the movement of particles within the medium and accelerating the material transfer process. This rapid and intense mechanical motion is referred to as the mechanical effect of ultrasound.
3. Cavitation effects: Under the influence of the sound pressure generated by ultrasound waves, tiny gas bubble nuclei in the liquid will vibrate, and the intermolecular distance in the medium will change with the vibration of the molecules. When the sound pressure reaches a certain value, the bubbles will rapidly expand and then quickly collapse, simultaneously generating shock waves. This dynamic process of a series of expansions, closures, and oscillations is called the cavitation effect of ultrasound waves.

Ultrasound technology has significant applications in various food-processing operations such as thawing, marinating, disinfecting, and cutting of meat, fruits, vegetables, and liquids [6], which can help shorten processing time and improve industrial efficiency.

7.2 Acoustic Detection Technology

7.2.1 Principles of Acoustic Detection Technology

7.2.1.1 Principles of Acoustic Detection

Acoustic detection is a dynamic non-destructive testing method that can predict upcoming changes by detecting signals emitted by the material itself. By using external conditions (such as force, heat, electricity, magnetism, etc.), the object to be tested generates sound waves, which are influenced by the internal structure of the object during propagation, resulting in phenomena such as acoustic impedance and amplitude attenuation. Based on the reflection and resonance of sound waves, the state or internal structure changes of the object can be inferred.

7.2.1.2 Principles of Ultrasonic Detection

Ultrasonic detection is a method that uses the propagation characteristics of low-intensity ultrasound in different media to detect internal defects in materials. Low-intensity ultrasound, also known as high-frequency ultrasound, operates on the basic principle that when ultrasound propagates through a medium, if it encounters defects such as pores and cracks, the difference in acoustic impedance between the defect and the surrounding medium will cause the ultrasound to reflect [7]. The reflected ultrasound is received by a probe and forms a waveform of the reflected wave on a display screen [8]. Based on the position and amplitude of the reflected wave, the presence, size, and location of defects can be determined.

7.2.2 Acoustic Detection Sensors

Microphones are a type of electrical component that can convert sound signals into electrical signals and are an important component of acoustic detection. Currently, microphones on the market are mainly classified based on the transduction mechanism into several types: dynamic microphones, electrostatic microphones, electromagnetic microphones, piezoelectric microphones, carbon microphones, semiconductor microphones, fiber optic microphones, and ultrasonic sensors [9].

7.2.2.1 Dynamic Microphones

The working principle of dynamic microphones is to place a conductor in a magnetic field, and when sound causes the vibration system to vibrate, an electromotive force is generated on the conductor through electromagnetic induction. Dynamic microphones are mainly divided into moving coil and ribbon types.

Moving coil microphones are the most common type of dynamic microphone. In these microphones, the coil is installed in the gap of the magnetic circuit, and the diaphragm is made of synthetic materials such as metal aluminum film or polyester resin, responsible for converting sound signals into electrical signals [10]. The frequency response, output voltage, and maximum output power of moving coil microphones are mainly influenced by the size of the coil and diaphragm, the magnetic flux of the magnetic circuit, and the vibration speed of the diaphragm. Moving coil sensors typically have a robust structure, perform well in frequency response, and are relatively low-cost. However, they are not very precise in capturing sound details and have lower sensitivity. A ribbon microphone flattens the conductor to give it a certain area, replacing the diaphragm and coil. It is named a ribbon microphone because its shape resembles a ribbon. In fact, it is a variation of the dynamic microphone. Additionally, from the perspective of vibration velocity, the ribbon microphone is also known as a velocity microphone, and the conductor is relatively thin, possessing excellent performance characteristics.

7.2.2.2 Electrostatic Microphone

Electrostatic microphones come in two types: condenser and electret, and their basic principles are entirely the same. They both obtain changes in the electrostatic charge between the diaphragm and the backplate through a load resistor. A typical condenser microphone structure is shown in Fig. 7.5. The condenser microphone consists of a thin diaphragm and a backplate close to the diaphragm, forming a capacitor. The diaphragm uses a stretched metal film or a plastic film coated with metal, with a polarization voltage of about 150–200 V (for battery-powered condenser microphones, the polarization voltage is 28 V). Condenser microphones have an advantage in quality, providing a flat sensitivity response over a relatively wide frequency range and possessing good stability. Therefore, they are often used as laboratory standards to calibrate other types of microphones (Table 7.1).

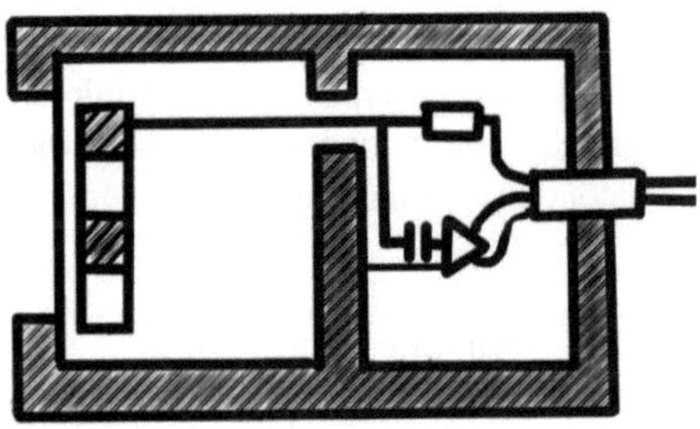

Fig. 7.5 Condenser microphone structure

Table 7.1 Typical technical characteristics of microphones

Name	Frequency characteristic	Sensitivity (mV/Pa)	Direction property
Condenser microphone	40–14,000 Hz, ±6 dB	>5	Line of heart; curve eight; all-directional form
Electret microphone	50–18,000 Hz, ±6 dB	>3	All-directional form; unidirectional form
Moving-coil microphone	50–10,000 Hz, <12 dB	<2.2	Line of heart; all-directional form
Ribbon-type microphone	50–10,000 Hz, 12 dB	<2	Curve eight

7.2.2.3 Electromagnetic Microphone

The electromagnetic microphone, also known as a magnetic microphone, has a magnetic circuit and diaphragm combination that is basically the same as the dynamic microphone. Its working principle is that a part of the magnetic circuit, the armature (directly connected to the diaphragm), vibrates, causing changes in the magnetic flux through the armature, which in turn induces voltage in the coil wound around the magnetic flux. This type of sound sensor is characterized by its small size, robustness, and high sensitivity, making it suitable for use in hearing aids and microphones for special purposes.

7.2.2.4 Piezoelectric Microphone

A piezoelectric microphone is a type of microphone that utilizes the piezoelectric effect of crystals and ceramics. Crystal microphones generally use an aluminum diaphragm, fixed to the crystal piezoelectric material with an armature. Its high-frequency characteristics depend on the resonant frequency of the vibration system. Ceramic microphones replace the crystal piezoelectric material in crystal microphones with ceramic piezoelectric material, connected to the diaphragm through a linkage. Compared to crystals, ceramic components are more resistant to high temperatures and less prone to damage, but they have lower sensitivity. The common characteristic of these two types of microphones is high electrical impedance, so the lead wires should not be too long. Although their sensitivity and reproduction bandwidth are not as good as condenser microphones, crystal and ceramic microphones have the advantage of being compact and lightweight.

7.2.2.5 Semiconductor Microphone

Semiconductor microphones come in forms such as piezoresistive microphones and transistor microphones. Piezoresistive microphones utilize the piezoresistive effect, where applying pressure to single crystals or diodes of germanium, silicon, etc. causes a change in resistance. Based on this principle, piezoresistive microphones can output voltage like carbon particle transducers. Transistor microphones, on the other hand, reduce the current amplification rate of the transistor by applying pressure, thereby exhibiting their unique sound transmission characteristics.

7.2.2.6 Fiber Optic Microphone

The fiber optic microphone, as a key application area of fiber optic sensing technology, has demonstrated significant advantages compared to traditional electrical microphones: high sensitivity, strong electromagnetic interference resistance, low

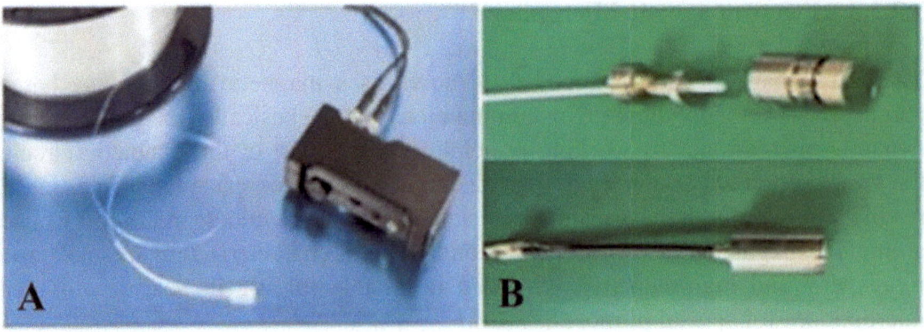

Fig. 7.6 Fiber optic sensors: (**a**) intensity; (**b**) F-P interferometric

transmission loss, and high convenience. Additionally, they possess characteristics of corrosion resistance and high temperature and pressure resistance [11]. By employing technologies such as time-division multiplexing and wavelength-division multiplexing, fiber optic microphones can be flexibly constructed into large-scale microphone arrays, which not only improve the array's gain but also enhance the signal processing capability [12]. With continuous advancements in acoustics, optics, materials science, and signal processing technology, fiber optic microphone technology has gradually moved towards commercialization, providing strong technical support for the development of related fields [13]. According to different working principles, fiber optic microphones are mainly divided into intensity-type fiber optic microphones (as shown in Fig. 7.6a), fiber Bragg grating microphones, and Fabry-Perot (F-P) interferometric microphones (as shown in Fig. 7.6b).

7.2.2.7 Ultrasonic Sensors

An electronic instrument that uses ultrasound (through air) to measure the distance to a target object and convert the reflected sound into an electrical signal. It determines the distance, position, shape, and other information of an object by emitting ultrasound and receiving its reflected echo, based on the time difference and direction of the echo. Ultrasonic sensors are widely used in medical diagnostics, industrial non-destructive testing, robotic navigation, automotive collision avoidance systems, and other fields [14]. Ultrasonic sensors can be divided into straight probe sensors, angled probe sensors, surface wave probe sensors, Lamb wave probe sensors, and dual probe sensors based on their probes. The main performance indicators of ultrasonic sensors include operating frequency, operating temperature, sensitivity, and directivity. These performances are affected by environmental factors such as temperature, humidity, and airflow. When in use, it is advisable to avoid using them outdoors, in high temperatures, or in strong wind environments. The working principle and application fields of ultrasonic sensors are very broad, playing an important role in modern technology.

7.2.3 Acoustic Detection Systems

In practical applications, acoustic detection systems typically include sound wave generators, sensors (microphones), signal amplifiers, dynamic signal analyzers, and computers. Through these devices, non-destructive testing of food and agricultural products can be conducted. Acoustic detection is a technology that uses vibration response signals to detect agricultural products, as shown in Fig. 7.7, which uses it to test the quality of eggshells. By collecting and analyzing the response signals generated by tapping agricultural products, different quality eggshells can be distinguished. The schematic diagram of the system's structural principle usually shows its main components (as shown in Fig. 7.8). These components work together, collecting vibration response signals through microphones, which are then amplified, filtered, and converted from analog to digital (A/D) before being input into a computer for processing.

Acoustic detection systems are typically composed of the following key parts:

1. *Tapping Module*

The tapping module provides pulse excitation to agricultural products, causing them to produce free vibrations. This excitation can be mechanical tapping or other forms of pulse force, with the aim of stimulating the vibration characteristics of agricultural products.

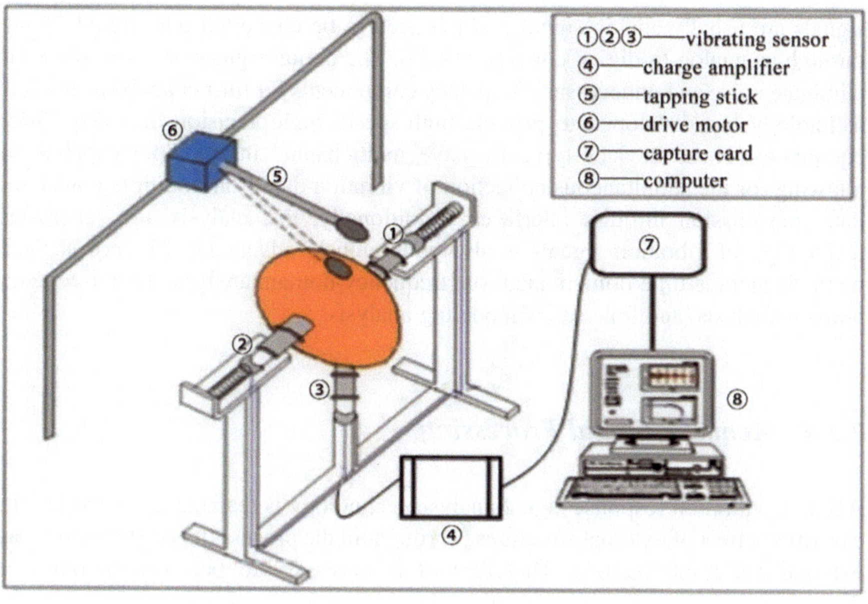

Fig. 7.7 Schematic diagram of eggshell quality testing of eggs

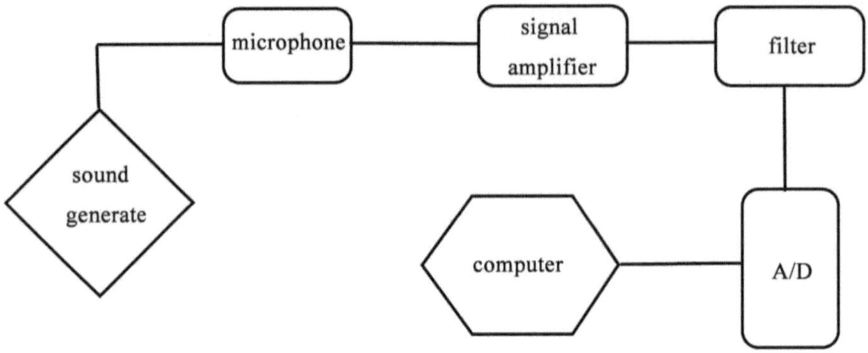

Fig. 7.8 Schematic structure of acoustic vibration detection system for agricultural products

2. *Support Module*

The support module is used to securely support the agricultural products to be tested, ensuring that they can produce effective vibrations under the action of the tapping module without external interference. The design of this module needs to consider the size, shape, and material of the agricultural products to ensure their stability during the testing process.

3. *Signal Acquisition and Processing Module*

This module is the core of the entire system, responsible for collecting the response signals generated by the vibrations of agricultural products. The collected signals are usually analog signals, which need to be converted into digital signals through an analog-to-digital converter (A/D). The digital signals are then filtered to eliminate noise and unnecessary frequency components for further analysis. Modern technology has developed to provide high-speed, high-precision data acquisition equipment. These devices typically have multichannel input/output capabilities, allowing for the simultaneous collection of vibration data from multiple points and data transmission through interfaces. Additionally, the analysis and processing technology of vibration signals is also continuously advancing. Modern analysis methods include time-domain analysis, frequency-domain analysis, time-frequency domain analysis, and time series modeling analysis.

7.2.4 Acoustic Signal Processing

Acoustic vibration response signal analysis technology is realized by analyzing the excitation effect of external structures. Throughout the process, it is difficult to avoid external noise interference. Therefore, it is necessary to perform filtering and denoising on the collected acoustic vibration response signals to improve the resolution and sensitivity of the signals. Adaptive filtering algorithms can continuously update parameters recursively based on statistical rules to maintain the optimal filtering state [15]. There are two common adaptive filtering algorithms: the least mean square (LMS) algorithm and the recursive least squares (RLS) algorithm. Independent component analysis (ICA) is also often used in signal processing.

7.2.4.1 LMS (Least Mean Square) Adaptive Filtering Algorithm

The LMS (least mean square) algorithm is a simple and widely used adaptive filtering algorithm, extensively applied in the noise reduction processing of digital audio signals. It is based on the gradient descent method to minimize the mean square value of the error signal, responding to the error between the input signal and the desired response.

The basic principle of the LMS algorithm is as follows:

Assume there is a filter w[n] of length N, and the relationship between its output signal y[n] and the input signal x[n] can be expressed using convolution operation:

$$y[n] = \sum_{k=0}^{N-1} w[k]x[n-k] \quad (7.1)$$

where $w[k]$ are the coefficients of the filter, n is the current time, and k is the delay of the filter. The goal of the LMS algorithm is to continuously adjust the filter coefficients $w[k]$ so that the filter output $y[n]$ approximates the desired signal $d[n]$ as closely as possible. Therefore, the LMS algorithm can be described by the following iterative update rule:

$$w[n+1] = w[n] + 2\mu e[n]x[n] \quad (7.2)$$

where $w[n]$ is the coefficient vector of the filter, μ is the step size parameter, $e[n] = d[n] - y[n]$ is the estimated error signal, and $d[n]$ is the desired signal.

The LMS algorithm is a method that continuously optimizes filter coefficients through an iterative process, relying on the error signal to adjust parameters in order to achieve optimal results. This algorithm is favored for its simplicity and low computational burden, making it suitable for applications requiring quick response. However, the performance of the LMS algorithm may be challenged when facing nonstationary noise, which can slow down the convergence speed or make it difficult to reach a stable state. As a result, variant algorithms based on the fundamental principles of the least mean squares algorithm have been developed.

Variants of the LMS algorithm include:

Normalized LMS: Adjusts the step size by normalizing the power of the input signal, making the algorithm insensitive to changes in signal amplitude.

Leaky LMS: Introduces a forgetting factor during the iteration process, so that the weight update depends not only on the current error but also considers past errors.

Affine Projection Algorithm: Combines features of LMS and RLS (recursive least squares) algorithms, offering better convergence performance.
Transform Domain LMS: Applies the LMS algorithm in the transform domain (such as the frequency domain) to improve convergence performance.

7.2.4.2 RLS (Recursive Least Square) Adaptive Filtering Algorithm

The recursive least square (RLS) algorithm is an adaptive filtering algorithm that uses the least squares principle to estimate filter parameters recursively, minimizing the sum of squared errors. The RLS algorithm is widely used in signal processing fields such as system identification, prediction, noise cancellation, and inverse modeling.

The core idea of the RLS algorithm is to minimize a weighted sum of squared errors, where the weights are determined by a forgetting factor λ, typically ranging between 0 and 1. The forgetting factor is used to determine the weight of historical data in the current estimate; the closer λ is to 1, the greater the weight of historical data, resulting in slower adaptability but stronger noise interference resistance. In RLS, the relationship between the desired response $d(n)$ and the tap input vector $u(n)$ can be represented by a multiple linear regression model:

$$d(n) = W_0^T u(n) + e_0(n) \tag{7.3}$$

In this equation, $W_0 = [W_0, 0, W_0, 1, \ldots, W_0, M\text{-}1]$, T is the regression parameter vector, and $e_0(n)$ has a mean of 0. The tap input vector $u(n)$ is generated by a random process with an autocorrelation function that is ergodic, allowing time averages to replace statistical averages.

The advantage of the RLS algorithm is its fast convergence speed and strong adaptability to changes in the statistical characteristics of the input signal. However, its computational complexity is relatively high, especially when dealing with long filters, as it requires storing and updating the covariance matrix, which can lead to large computational and storage demands. To address this issue, researchers have proposed some improved algorithms, such as the fast recursive least squares (fast RLS) algorithm and the RLS lattice algorithm based on lattice structures, which aim to reduce computational complexity while retaining the fast convergence characteristics of the RLS algorithm. In practical applications, the RLS algorithm can be adjusted and optimized according to different application scenarios to achieve the best filtering effect. For example, in noise cancellation and echo cancellation applications, the RLS algorithm can effectively reduce background noise, improving the clarity and quality of the signal.

7.2.4.3 ICA (Independent Component Algorithm) Analysis

Independent component analysis (ICA) is a multivariate statistical data analysis technique that originated in the field of blind source separation (BSS). This method first constructs an objective function for multidimensional detection data based on the principle of statistical independence; then, it uses optimization techniques to identify features of non-Gaussian distributions, thereby decomposing the detection data into multiple independent components; and finally, by selectively removing unwanted target components, it extracts the desired signals [16].

The main purpose of ICA is to extract mutually independent components from mixed signals. In acoustic detection, mixed signals are usually a superposition of multiple sound signals, such as the mixture of background noise and target signals. ICA achieves denoising through the following steps:

1. Data Modeling

   Set the observed signal $=[x_1, x_2, \ldots, x_n]^T$, where x_i is the mixed observed signal.

   Assume these observed signals are generated by several independent source signals $s = [s_1, s_2, \ldots, s_m]^T$ through an unknown linear transformation A. That is:

$$x = As \tag{7.4}$$

2. Unmixing Matrix Estimation

   ICA attempts to estimate the inverse matrix W of this linear transformation, so that the signals transformed by W can be as independent as possible. This step is completed by optimizing an objective function, which aims to maximize the independence of the signals. The main objective functions of independent analysis algorithms include maximizing non-Gaussianity, minimizing mutual information, and maximum likelihood estimation. This section will illustrate with maximizing non-Gaussianity:

$$y = Wx \tag{7.5}$$

   To estimate an independent component, consider the linear combination of $y = W^T x = \sum W_i x_i$, where W is an undetermined variable, if W is the inverse matrix of A, this linear combination estimates the variable y as an estimate of an original signal. Define $Z = A^T W$, then $y = W^T x = W^T As$; y is a linear combination of s_i, with weights given by z_i. W can be considered as a vector that maximizes Gaussian $W^T x$, such a vector corresponds to z, and then $W^T x = Z^T s$ equals one of the independent components.

3. Signal Separation

   Once the inverse matrix W of the mixing matrix is estimated, it can be used to separate the observed signal x into various independent components s. These independent components contain the original signal and noise. By selecting and processing these components, the clean target signal can be extracted.

7.3 Ultrasonic Processing Technology

7.3.1 Principles of Ultrasonic Processing Technology

Food processing plays a key role in addressing food security issues, and ultrasonic technology, as an environmentally friendly and economical solution, has shown its importance in various stages of food processing [17]. This noninvasive technology has multiple advantages, including accelerating processing, enhancing production efficiency, simplifying processes, improving product quality, and maintaining the original characteristics of the product (such as texture, nutritional value, and sensory quality), while also extending the product's shelf life.

The combination of ultrasonic technology and food processing represents an innovative and sustainable approach, enhancing speed and efficiency in several key stages of the food industry. This technology can also be combined with temperature (thermosonication) or pressure (sonication) to produce synergistic effects, further enhancing its effectiveness [18].

High-intensity ultrasound, also known as low-frequency ultrasound, has high energy and intensity, capable of producing mechanical and thermal effects, disrupting the structure and organization of the processed object, and altering the physical and chemical properties of food. Therefore, high-intensity ultrasound is used in various operations in food processing, such as emulsification, extraction, cutting, sterilization, and cleaning.

7.3.2 Ultrasonic Generators

An ultrasonic generator consists of a switching power supply, feedback system, power amplification device, time controller, protector, and power regulation device, generating ultrasonic signals to allow ultrasound to propagate in the medium. It is widely used in cleaning, welding, atomization, emulsification, and other fields [19]. According to the working principle and usage scenarios, ultrasonic generators can be divided into the following types:

7.3.2.1 Self-Excited Ultrasonic Generator

A self-excited ultrasonic generator is a device that utilizes the resonant characteristics of the ultrasonic transducer itself to automatically generate oscillations when the power is turned on. It does not require an external oscillator or starting circuit but achieves oscillation and amplification through a high-power switching tube. The circuit structure of this generator is relatively simple and cost-effective, but the output stability is relatively poor, mainly suitable for low-power cleaning applications. A self-excited ultrasonic generator typically consists of an inverter circuit, control circuit, and feedback circuit. The inverter circuit converts the DC power supply into an AC power supply, the control circuit controls the working state of the inverter circuit, and the feedback circuit feeds the output signal of the ultrasonic generator back to the control circuit to achieve closed-loop control.

7.3.2.2 Externally Excited Ultrasonic Generator

An externally excited ultrasonic transducer is a high-performance ultrasonic device. Its working principle is to generate high-frequency signals through an external oscillator, which are then amplified by a power amplifier and converted into mechanical vibrations by the transducer to produce ultrasound. Quartz crystal oscillators or LC oscillators are usually used to ensure high precision and stability of the frequency.

7.3.2.3 Piezoelectric Ultrasonic Generator

A piezoelectric ultrasonic generator is a device that uses the piezoelectric effect of piezoelectric materials to generate ultrasound. Its working principle is to apply an electric field to a piezoelectric crystal, causing it to produce mechanical vibrations, thereby generating ultrasound.

7.3.2.4 Magnetostrictive Ultrasonic Generator

A magnetostrictive ultrasonic generator is a device that uses the magnetostrictive effect to generate ultrasound. The magnetostrictive effect refers to the phenomenon where certain ferromagnetic materials change in size under the influence of a magnetic field. This change can be used to generate ultrasound, widely used in non-destructive testing, cleaning, welding, and other fields.

7.3.2.5 Electric Ultrasonic Generator

An electric ultrasonic generator is a device that uses an electric transducer to produce ultrasonic waves. Its working principle involves the transducer generating mechanical vibrations under the influence of an alternating electric field, thereby producing ultrasonic waves. It can efficiently convert electrical energy into mechanical vibrations, allowing precise control of the frequency and amplitude of the ultrasonic waves by adjusting the power supply and control circuits. Compared to other types of ultrasonic generators, the electric generator has a simpler structure.

7.3.2.6 Frequency-Adjustable Ultrasonic Generator

A frequency-adjustable ultrasonic generator is a device that can adjust the output ultrasonic frequency according to application needs. This flexibility allows it to adapt to various application scenarios, such as industrial cleaning, welding, and medical imaging.

7.3.2.7 Power-Adjustable Ultrasonic Generator

A power-adjustable ultrasonic generator is a device that can adjust the output power according to application needs. It achieves power adjustment by changing the input voltage, duty cycle, or using direct power regulation functions. Due to its flexibility and adaptability, the power-adjustable ultrasonic generator is widely used in industrial, medical, and environmental fields. With technological advancements, the performance and control precision of these devices will further improve to meet broader application demands.

7.3.3 Ultrasonic Transducer

An ultrasonic transducer can convert any other form of energy into ultrasonic frequency. It is a device that converts electrical signals into ultrasonic waves and vice versa, with minimal power loss. It can convert the excitation signal provided by the excitation device into mechanical vibrations, thereby emitting ultrasonic waves into the medium.

The structure of an ultrasonic transducer typically includes piezoelectric ceramics, an acoustic matching layer, and a damping layer. Piezoelectric ceramics are the core of the transducer, capable of converting electrical energy into mechanical energy (or vice versa). The acoustic matching layer is used to optimize the transmission of sound waves from the transducer to the medium (such as air or water), while the damping layer helps reduce vibrations and improve signal stability. The main function of the horn is to enhance and concentrate energy. When the end surface of the horn vibrates, these vibrations are amplified and transmitted into the fluid, producing specific effects. Chen et al. [20, 21] have elaborated on ultrahigh frequency and fiber optic ultrasonic transducers, which have achieved wide bandwidth and high sound pressure ultrasonic excitation, holding significant importance in the field of ultrasonic processing.

Ultrasonic transducers can be classified according to different standards, and here are some common classification methods:

7.3.3.1 Vibration Mode

Divided into shear vibration transducers, torsional vibration transducers, longitudinal vibration transducers, and bending vibration transducers. Currently, multiple vibration composite mode transducers are often used in processes to meet the needs of new power ultrasound technologies, such as longitudinal-torsional composite mode and longitudinal-bending composite mode.

7.3.3.2 Operating State

Divided into receiving and transmitting transducers. Receiving transducers are devices that can convert ultrasonic waves into electrical signals, where certain materials (such as quartz, ceramics, etc.) generate charges under mechanical stress. When sound waves act on the transducer, they cause mechanical deformation of the transducer material, leading to changes in the internal electric field. Transmitting ultrasonic transducers are those that convert electrical signals into ultrasonic mechanical vibration signals.

7.3.3.3 Material Type

The material type is divided into magnetostrictive transducers, piezoelectric ceramic transducers, and single-crystal transducers. Magnetostrictive transducers are made from ferromagnetic materials and ceramic materials with magnetostrictive effects, and piezoelectric ceramic transducers are designed to work stably and reliably in high-pressure extractors by placing ceramic crystals in the middle of metal cover plates, using a sandwich transducer structure [22, 23]. The working principle of single-crystal transducers is based on the piezoelectric effect, where single-crystal materials deform under electric field excitation, generating sound waves. Single-crystal materials can effectively improve the sensitivity and bandwidth performance of ultrasonic transducers.

7.3.3.4 Working Principle

Divided into piezoelectric, electrostatic, and electromagnetic transducers. Piezoelectric transducers operate based on the piezoelectric effect principle. Electrostatic transducers can output highly directional ultrasonic arrays, facilitating the nonlinear self-demodulation of ultrasonic modulated signals in the air. Electromagnetic acoustic transducers are a new type of device that uses electrodynamic methods to generate ultrasonic waves in conductive metals. Ultrasonic transducers are widely used in industrial, agricultural, medical, and military fields, and different application scenarios and requirements determine the use of specific types of transducers.

7.3.4 *Ultrasonic Processing System*

Ultrasonic technology is applied in various processing procedures, such as ultrasonic emulsification, dispersion, and coagulation; ultrasonic degassing, defoaming, and fractionation; ultrasonic cutting, homogenization, and nanonization; ultrasonic extraction and separation; ultrasonic crystallization and anti-fouling; ultrasonic freezing, thawing, and drying; and ultrasonic cleaning, sterilization, etc. Although each processing system is different, the basic principle is consistent, mainly consisting of two parts: the ultrasonic vibration system and the processing system.

7.3.4.1 Ultrasonic Vibration System

The ultrasonic vibration system is the core part of ultrasonic processing equipment, including the ultrasonic generator, ultrasonic transducer, horn, and tool head. This system can convert electrical energy into mechanical vibration, thereby achieving the processing of the workpiece. The horn is a key component in the ultrasonic vibration system, and its main function is to amplify the smaller amplitude generated by the transducer to the required larger amplitude and concentrate the ultrasonic energy on a smaller area to meet different application needs. With the continuous development of ultrasonic technology, the design and application of horns are also continuously advancing. The design of new horns, such as the cosine half-cycle cylindrical composite horn, aims to achieve more efficient amplitude amplification and energy transfer. Meanwhile, the application of modern simulation design methods such as finite element analysis further optimizes the performance of horns, enabling them to better meet the diverse needs of industrial applications. The design and application of horns are an important research direction in the field of ultrasonic technology, and their performance directly affects the working effect and efficiency of ultrasonic equipment. With the advancement of computational methods and material science, future horns will be more efficient, compact, and durable. In ultrasonic processing technology, the tool head is the part that directly contacts the material being processed, transferring the vibration energy to the material to achieve the purpose of processing. When designing the tool head, it is necessary to consider the service life of the tool head and its stability in harsh environments, ensuring effective transmission of ultrasonic vibrations and maintaining precision requirements during the processing.

7.3.4.2 Processing System

The processing system includes processes such as the processing medium, tool selection, and control of processing operations.

Processing Medium: Liquid media (such as water or oil) are usually used to help transfer energy and improve cutting effects. Additives that aid in cleaning and cooling may also be added to the liquid.

Processing Tools: The tool head is usually connected to the vibration system, enhancing cutting ability through ultrasonic vibration, suitable for processing hard and brittle materials.

Processing Control: The system is usually equipped with a control unit to adjust vibration frequency, amplitude, and processing parameters to adapt to different materials and processing requirements.

Ultrasonic processing is widely used in fields such as micromachining, precision cutting, cleaning, and deburring, especially in industries with high precision requirements such as semiconductors, optoelectronics, and medical devices. The combination of these systems enables ultrasonic processing to achieve efficient and high-precision processing effects, particularly in handling hard and brittle materials, reducing mechanical stress and thermal effects brought by traditional processing methods.

7.4 Application of Acoustic Technology in Food and Agricultural Product Detection and Processing

Acoustic detection technology has broad prospects and practical value in the detection of food and agricultural products. Acoustic detection technology uses the propagation characteristics of sound waves to evaluate the physical and chemical properties of food, thereby achieving the detection and monitoring of food quality and safety. Acoustic detection technology can be used in the following aspects:

1. Evaluating the internal structure and density of food. For example, ultrasound can detect internal cavities, damage, and density changes in fruits, thereby assessing the maturity and quality of fruits, helping consumers and the supply chain better manage inventory and transportation [24].
2. Real-time monitoring of food quality changes [25]. For example, by detecting the attenuation and propagation speed of sound waves, changes in the pore structure during the fermentation process of bread can be monitored, thereby assessing the degree of fermentation and baking quality of the bread.
3. Detecting foreign objects in food, such as metal, glass, or other materials that should not be present [8]. By analyzing abnormal situations in sound wave reflection and attenuation, foreign objects can be quickly identified and located, improving food safety.
4. Detecting moisture, muscle quality, fat content, and the proportion of connective tissue in meat [26].
5. Detecting the integrity of food packaging. For example, by detecting the propagation of sound waves in packaging materials, it can be identified whether there is

air leakage, damage, or poor sealing in the packaging, thereby ensuring the freshness and safety of the food [27].

7.4.1 Application in Fruit Quality Detection

7.4.1.1 Application in Watermelon Maturity Detection

Maturity is an important indicator of watermelon quality. Generally speaking, watermelon maturity can be divided into immature, mature, and overripe. Traditional methods for determining watermelon maturity mainly involve manual picking and experience-based judgment, which are time-consuming, inefficient, and often result in misjudgment. Chen et al. developed a device using an acoustic detection method to measure watermelon maturity [28]. The application of acoustic detection technology in assessing watermelon maturity primarily relies on analyzing the characteristics of sound waves emitted by the watermelon to evaluate its internal state. There are differences in internal structure and density between mature and immature watermelons, which in turn affect the way sound waves propagate within them.

Ungho et al. [29] used a self-made acoustic detection platform (as shown in Fig. 7.9) to perform tapping excitation on watermelon samples and measure sound velocity, while also using a portable refractometer to determine the sugar content in the middle of the watermelon. They conducted artificial neural network analysis and correlation coefficient evaluation using the sound velocity values and the sugar content values of the watermelon samples. By measuring sound velocity and considering the relationship between sound velocity and watermelon sugar content, they evaluated the maturity of the watermelon. The correlation coefficients between the sound velocity (center, top, and bottom) and sugar content of 56 watermelon samples were evaluated using a backpropagation network model (BPANN), with 39 samples (70%) used for training, 11 samples (20%) for validation, and 6 samples (10%) for testing, resulting in a high correlation coefficient [29, 30].

7.4.1.2 Application in Apple Hardness Detection

The acoustic characteristics of fruit change with the alteration of the internal structure of the fruit, and the natural frequency of the fruit also varies with increasing

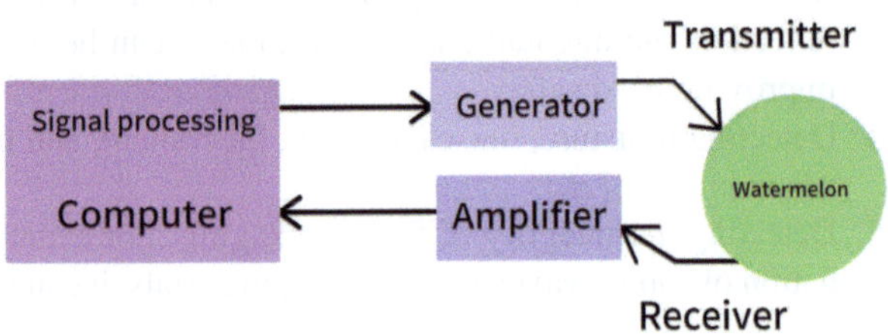

Fig. 7.9 Schematic diagram of watermelon ripeness detection

maturity [31]. Therefore, acoustic detection technology is often used to measure the hardness of fruit. Researchers found that by suspending an apple from the bottom and using a vibration rod to touch the apple and then using an audio generator to produce vibrations, a waveform detector can record the frequency distribution of sound waves from 20 to 2400 Hz. The experimental results show that, in most cases, apples exhibit four distinct resonance frequencies, corresponding to $f_n = 1, f_n = 2$, $f_n = 3$ and $f_n = 4$. Additionally, a correlation was observed between the second resonance frequency (f_n^2m) and the hardness of the apple. Here, (f_n^2m)represents the resonant frequency of the audio, and m is the mass of the apple. As storage time increases, apples of better quality tend to exhibit higher values of f_n^2m, while softer or lower-quality apples usually have lower f_n^2m values.

7.4.1.3 Application in Durian Ripeness Detection

Durian ripeness has a significant impact on durian export and transportation, with experienced experts responsible for detecting durians of different ripeness levels. However, this manual detection method involves high costs in terms of time and personnel, resulting in subjective and inaccurate outcomes. Chang [32] and others designed a method to detect durian ripeness using transmission spectroscopy and acoustic characteristics. First, spectral information is recorded and collected in transmission mode using a fiber optic spectrometer on the conveyor belt of the spectral detection system (as shown in Fig. 7.10a). Next, the durian is placed on an acoustic detection platform (as shown in Fig. 7.10b), where a motor-driven steel ball strikes the equatorial region of the durian for acoustic detection. Two symmetrical microphones capture and record the sound signals generated by the durian impact. Spectral and acoustic detection is performed on durians stored for different days (durian ripeness is indicated by different storage days). The impact of durian ripeness on its spectral and acoustic characteristics is studied through correlation analysis.

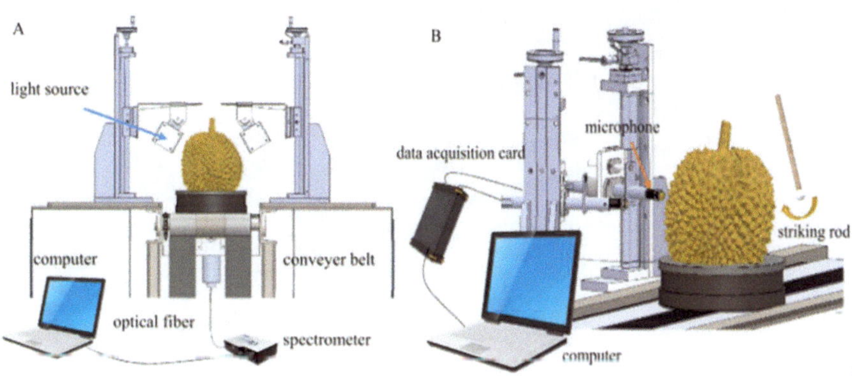

Fig. 7.10 Schematic of spectral detection system (**a**) and acoustic detection system (**b**)

7.4.2 Application in Agricultural Product Detection

7.4.2.1 Application in Poultry Eggshell Crack Detection

Poultry egg crack detection is of great significance in the transportation and sale of poultry eggs. Cracks in eggshells not only affect appearance but may also lead to egg contamination. In general, these cracks are difficult to detect in time with the naked eye, which may lead to the consumption of contaminated eggs, posing certain health risks. Additionally, cracked eggs may affect surrounding intact eggs, causing significant economic losses. Therefore, the screening of poultry eggshell cracks is urgently needed [33].

Lin [34] and others developed a novel egg crack detection technology. The egg crack detection system they designed is shown in Fig. 7.11, relying on the comprehensive analysis of multiple vibration sensors. Vibration signals are collected by tapping different parts of the eggshell. The signal characteristics and related data captured by each sensor are thoroughly analyzed. The effectiveness of the selected features is evaluated using an index called the Discrimination Index (DI), and these features are then fed into a linear discriminant analysis model. The response signal characteristics of intact and damaged eggs are used as input data, and classification is performed using linear discriminant analysis (LDA) to obtain classification results.

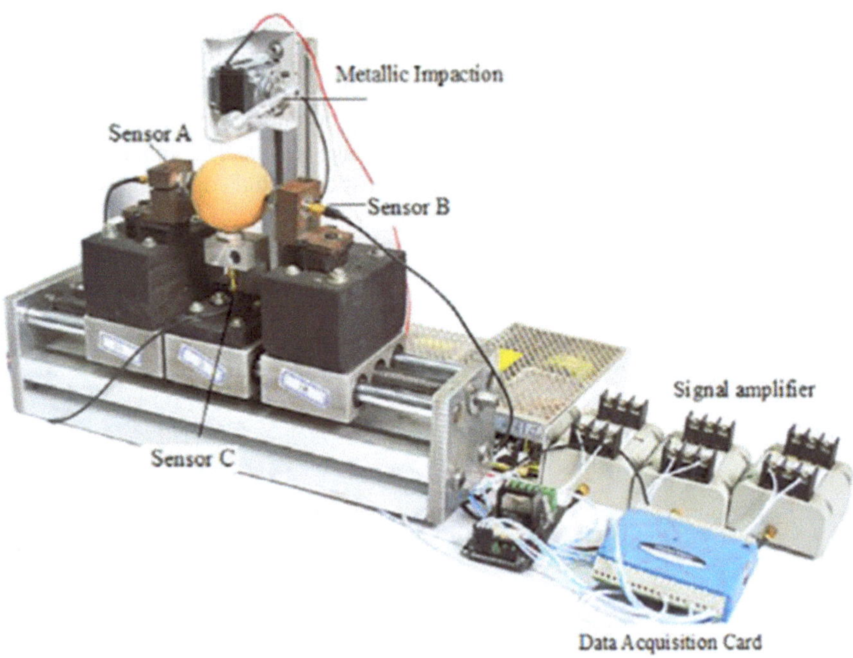

Fig. 7.11 Eggshell crack detection system

7.4.2.2 Application in Internal Damage Detection of Purple Sweet Potatoes

Fauzi [35] and his team developed a non-destructive method to predict the internal damage of purple sweet potatoes caused by weevils by studying the ultrasonic characteristics of purple sweet potatoes. They analyzed 60 healthy purple sweet potatoes and 45 weevil-infected purple sweet potatoes. These samples were divided into two groups: two-thirds for the first group, used to establish the prediction model, and one-third for the second group, used for model validation. The study used a frequency of 50 used ultrasound measurement equipment at kHz frequencies to detect the ultrasonic properties of purple sweet potatoes. In addition, Mo, zero moment power, is considered as the amount energy transmitted in the medium where the waves transmitted and can be calculated based on the area under power spectral density. The study found that the average ultrasonic speed of infected purple sweet potatoes was 247.32 m per second, with an Mo value of 19.45, while healthy purple sweet potatoes had an ultrasonic speed of 235.15 meters per second and an Mo value of 18.72. Additionally, the average attenuation coefficient for infected purple sweet potatoes was 40.06 decibels per meter, whereas for healthy ones, it was 42.69 decibels per meter. Through discriminant analysis, researchers established a linear discriminant model that effectively distinguished between healthy and infected purple sweet potatoes, achieving an accuracy of 85.71%.

7.4.3 Detection of Air Leaks in Vacuum Packaging

Before sealing food, vacuum packaging is required to protect and extend the shelf life of the food. In a vacuum environment, if there is an air leak in the packaging of sealed food and ingredients, it may lead to product spoilage, causing it to lose its original function. Although the proportion of defective packaging is relatively small in actual production, all packaging must undergo strict inspection before shipment to promptly identify and discard these noncompliant packages, which increases production costs. Manufacturers, to ensure product quality, usually implement strict quality control procedures, but this often comes with high costs. Therefore, developing an economical and efficient automated quality control technology to detect the integrity of vacuum packaging is particularly necessary. Such technology can help reduce costs while maintaining or improving product quality.

Hasan Koruk et al. [36] evaluated an acoustic method for detecting air leaks in vacuum-sealed instant dry yeast packaging. In an acoustic laboratory, the sound pressure generated by air in defective packaging was captured using low-noise microphones, allowing the detection of packaging integrity. By analyzing changes in sound pressure levels within specific frequency bands, defective packaging can be identified. A mathematical model was also established to predict the pressure change over time inside yeast packaging with certain defect sizes, which can determine the size of the defect causing the leak and the time required for the internal pressure to reach a threshold.

7.4.4 Application in Detecting Foreign Matter in Food

The presence of foreign matter in packaged food is a serious food safety issue that directly threatens consumer health. Therefore, foreign matter detection in packaged food is essential during production. Current conventional detection methods include sensory detection, which is not highly accurate and is time-consuming and labor-intensive; physical property detection, which is effective but can lead to food waste and is usually limited to sample testing, making it difficult to ensure overall quality; and metal detection, which can find metal foreign objects but is limited to metals, presenting obvious limitations.

Milk, with its high nutritional content, is considered a healthy beverage and is widely consumed worldwide. However, incidents of milk contamination occur from time to time. Khairi et al. [37] designed a method using an ultrasonic sensing system to detect foreign matter in milk. By using the two ultrasonic parameters of amplitude and acoustic impedance, they mathematically modeled several different artificially added foreign objects (steel, rubber, air) to determine the voltage amplitude detected by the ultrasonic receiver when the ultrasound is obstructed by foreign matter during propagation. The difference in waveform between milk samples obstructed and unobstructed by foreign objects using ultrasound can effectively distinguish the foreign objects. The experimental results indicate a good correlation between the received voltage and the size of the foreign matter, with a correlation coefficient greater than 0.95.

7.4.5 Application in Food Processing

7.4.5.1 Application in Meat

Marination processing with societal development, the World Health Organization has set a goal to reduce global dietary sodium intake by 30% by 2025. The focus on dietary sodium intake has led to increased interest in methods to reduce sodium intake, especially in meat processing. Traditional meat marination often requires a large amount of salt to extend shelf life and maintain the moisture and quality of the meat, prompting researchers to explore methods that can reduce sodium salt usage while ensuring the quality of meat processing. Inguglia et al. [26] conducted a controlled experiment, marinating meat treated with ultrasound for 1 h and untreated meat for 4 h. After marination, the excess brine was removed, and the meat was vacuum-packed. Subsequently, meat quality analysis, total viable count (TVC), and lactic acid bacteria count (LAB) were performed, and the inoculated microorganisms were counted. Principal component analysis was used for data analysis. The experimental results showed that ultrasound treatment can effectively reduce the salting time of meat without affecting its tenderness, internal tissue, and structure.

7.4.5.2 Application in Industrial Cleaning

Ultrasonic cleaning technology is very popular in the cleaning and disinfection processes of the food and beverage industry. This technology uses specific equipment to generate ultrasonic waves, which can penetrate the cleaning liquid and help remove grease and dirt attached to food, especially effective in the degreasing process. The working principle of an ultrasonic cleaner is to use a piezoelectric transducer, which can convert low-frequency AC signals into high-frequency sound waves. These transducers are usually installed at the bottom of the cleaning container or directly immersed in the cleaning liquid. When the transducer operates, it emits high-energy sound waves into the cleaning liquid, creating a compression effect in the liquid, leading to the formation of millions of tiny bubbles, a phenomenon known as "cavitation." These bubbles rapidly expand and explode in the cleaning liquid under high temperature and pressure, releasing tremendous energy (as shown in Fig. 7.12). This instantaneous energy release can effectively remove dirt from the surface of products soaked in the cleaning liquid. The entire process is very fast and can clean all products being washed in a very short time [38].

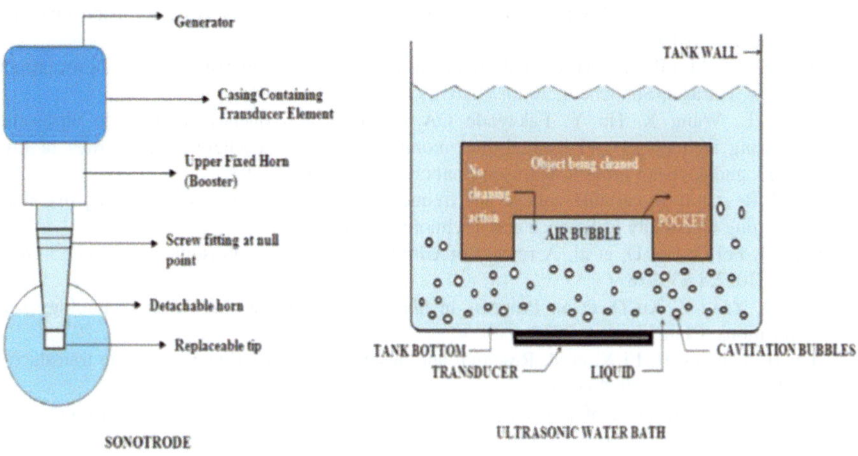

Fig. 7.12 Ultrasonic cleaning device

References

1. Chen R, Wen X, Shucai W. Detection and fuzzy recognition of egg damage using knocking sound signals. J Agric Eng. 2004;4:130–3.
2. Gonghuan D, Zhemin Z, Xiufen G. Fundamentals of acoustics. 3rd ed. Nanjing: Nanjing University Press; 2012.
3. Hughes LF. The fundamentals of sound and its measurement. J Am Assoc Lab Anim Sci (JAALAS). 2007;46:14.
4. Ze H. Fundamentals of recording acoustics [M]. Beijing: Communication University of China Press; 2021. p. 13–31.
5. Tianming S, Guofu Z, et al. Ultrasonic testing [M], vol. 1–2. Beijing: China Petrochemical Press; 2012. p. 22–3.
6. Jiang S, Zhao JW. Food physics. Beijing: Chemical Industry Press; 2015.
7. Bühling B, Maack S, Strangfeld C. Fluidic ultrasound generation for non-destructive testing. Adv Mater. 2024;36:e2311724.
8. Mohd Khairi MT, Ibrahim S, Md Yunus MA, et al. Finite element simulation for detecting the foreign body based on ultrasonic sensor. J Food Process Eng. 2017;41:e12595.
9. Bartlett B, Lei J. Choosing the right microphone by understanding design principles (Part 1), vol. 11. Performing Arts Technology; 2017. p. 14–9.
10. Rong C, Xiu L, et al. Electret condenser microphone—a newly developed standard test microphone. Chinese Science Bulletin (Chinese Version); 2019.
11. Luo Z, Lu B, Huang J, et al. Sound source direction-of-arrival estimation method for microphone array based on ultra-weak fiber Bragg grating distributed acoustic sensor. Opt Express. 2023;31:31419.
12. Aiadi KE, Rehouma F, Bouanane R. Theoretical analysis of the membrane parameters of the fiber optic microphone. J Mater Sci Mater Electron. 2006;17:293.
13. Wei H, Gong Z, Wu W, et al. Broadband fiber-optic acoustic sensors. J Lightwave Technol. 2022;40:4033.
14. Yan LT, Zhang QJ. Ultrasonic energy field technology and application basics. Beijing: Chemical Industry Press; 2023.
15. Yiqin N. Active noise reduction methods in digital audio signal processing. Satellite TV Broadband Multimed. 2024;11:126–8.
16. Carone D, Harston GWJ, Garrard J, et al. ICA-based denoising for ASL perfusion imaging. NeuroImage. 2019;17:363.
17. Bhargava N, Mor RS, Kumar K, et al. Advances in application of ultrasound in food processing: a review. Ultrason Sonochem. 2020;70:105293.
18. Zhang L, Wang X, Hu Y, Fakayode OA, et al. Dual-frequency multi-angle ultrasonic processing technology and its real-time monitoring on physicochemical properties of raw soymilk and soybean protein. Ultrason Sonochem. 2021;80:105803.
19. Cao J. Research on variable voltage and frequency generator of ultrasonic cleaning machine. Chongqing University of Science and Technology; 2023.
20. Chen J, Fei C, Lin D, et al. A review of UltraHigh frequency ultrasonic transducers. Front Mater. 2022;8:733358.
21. Sun Q, Yang L, Xu D, et al. Progress in fiber optic ultrasonic transducer technology and applications. Chin J Lasers. 2022;
22. Yin H, Yanping H, Li X, et al. Research and design of sandwich-type ultrasonic transducer. Mech Eng. 2024;09:40–43+46.
23. Lin S. The cornerstone of ultrasonic technology—principle and design of ultrasonic transducers. Physics. 2009;38(3):141148.
24. Dias C, Brandão TRS, Cassoni C, et al. Firmness prediction of 'Rocha' pear based on non-destructive methods and shelf-life using PLS-regression modeling. Postharvest Biol Technol. 2023;205:112493.

25. Sun L, Feng S, Chen C, et al. Identification of eggshell crack for hen egg and duck egg using correlation analysis based on acoustic resonance method. J Food Process Eng. 2020;43 https://doi.org/10.1111/jfpe.13430.
26. Inguglia ES, Granato D, Kerry JP, et al. Ultrasound for meat processing: effects of salt reduction and storage on meat quality parameters. Appl Sci. 2020;11:117.
27. Ozguler A, Morris SA, O'Brien WD. Ultrasonic monitoring of the seal quality in flexible food packages. Polym Eng Sci. 2001;41:830.
28. Chen X, Yuan P, Deng X. Watermelon ripeness detection by wavelet multiresolution decomposition of acoustic impulse response signals. Postharvest Biol Technol. 2018;142:135.
29. Ungho C, Hyonchol K, Jin H, et al. Maturity assessment of watermelon by acoustic method. Sci Hortic. 2021; https://doi.org/10.1016/j.scienta.2021.110735.
30. Abbaszadeh R, Moosavian A, Rajabipour A, et al. An intelligent procedure for watermelon ripeness detection based on vibration signals. J Food Sci Technol. 2013;52:1075.
31. Vasighi-Shojae H, Gholami-Parashkouhi M, Mohammadzamani D, et al. Ultrasonic based determination of apple quality as a nondestructive technology, vol. 21. Sens Bio-Sens Res; 2018. p. 22.
32. An C, Yu Y, Qu M, et al. Transmittance spectra and acoustic properties of durians with different ripening: an exploration of application for complex-structured and large-sized fruit. Postharvest Biol Technol. 2024;217:113103.
33. Sun L, X-K B, Lin H, et al. On-line detection of eggshell crack based on acoustic resonance analysis. J Food Eng. 2013;116:240.
34. Lin H, Xu PT, Sun L, et al. Identification of eggshell crack using multiple vibration sensors and correlative information analysis. J Food Process Eng. 2018;41
35. Fauzi RM. Determination of the internal damage of purple sweet potatoes non-destructively using ultrasonic wave characteristics. IOP Conf Ser Earth Environ Sci. 2018;147:012002.
36. Koruk H, Sanliturk KY. Detection of air leakage into vacuum packages using acoustic measurements and estimation of defect size. Mech Syst Signal Process. 2019;114:528–38.
37. Mohd Khairi MT, Ibrahim S, Md Yunus MA, et al. Detection of foreign objects in milk using an ultrasonic system. Indones J Electr Eng Comput Sci. 2019;15:1241.
38. Mason TJ. Ultrasonic cleaning: an historical perspective. Ultrason Sonochem. 2016;29:519–23.

Chapter 8
Electromagnetic Properties in Foods

Yanyu Li and Leiqing Pan

8.1 Introduction

In daily life and basic physics, we learn that matter has an electrical nature, with charge being a fundamental property of its basic particles. Matter consists of protons, neutrons, and electrons, with protons and electrons carrying charge. Atoms are organized so that they share electrons, resulting in negative (anions) and positive (cations) charges. The movement of charge in conductive media generates electric current. According to Faraday's terminology from 1833, there are two types of conductors: first-type conductors, where electrons conduct current, and second-type conductors, or electrolytes, where ions carry current. Electrolytic conduction, driven by ions from dissolved salts, is common in nature.

The electrical properties of materials can be classified as active or passive. Active properties involve an energy source within the material, generating electromotive force or potential differences, such as bioelectric potentials in biological systems or strain-induced potentials in piezoelectric crystals. In contrast, passive properties influence electromagnetic field distribution and how materials behave in these fields, determined by their chemical composition and molecular structure. Both active and passive properties depend on the material and environmental factors.

Further categorizing electrical properties is complex. Key parameters like permittivity, permeability, and conductivity describe how electromagnetic waves interact with matter. DC values of these parameters are specific cases of broader AC

Y. Li
Academy of National Food and Strategic Reserves Administration, Beijing, Beijing, China
e-mail: lyy@ags.ac.cn

L. Pan
College of Food Science and Technology, Nanjing Agricultural University, Nanjing, Jiangsu, China
e-mail: pan_leiqing@njau.edu.cn

H. Lin et al., *Advanced Technologies for Physical Properties Measurement of Food and its Processing*, https://doi.org/10.1007/978-981-96-2347-1_8

characteristics. Surface properties are crucial in electrostatics, but volume properties also affect charge distribution. Dielectric breakdown is important for electrical insulation materials. Permanent polar dielectrics, or electrets, resemble permanent magnets, though their occurrence in agricultural products is still being studied. Piezoelectric effects may also be present in agricultural products, though their significance is unclear. Nuclear magnetic resonance and electron spin resonance are additional electrical properties. At the high-energy end of the radio frequency spectrum, particularly in the submillimeter wave and far-infrared regions, the distinction between electrical and optical properties becomes ambiguous.

Food's primary role is to provide nutrition and sensory experiences. Historically, electrical properties of food have received little attention, with electromagnetic technology mainly used for power and external heating sources. However, advances in electromagnetic processing and detection technologies from electromechanical, medical, and communication fields are increasingly being applied in the food industry, leading to a greater focus on the electromagnetic properties of food.

8.1.1 Basic Concept

The Poisson, Laplace, and Maxwell equations are fundamental laws of electricity that apply to describing electrostatic fields in various conductors and dielectrics at both the microscopic and macroscopic levels.

One useful equation linking volumetric charge to potential distribution is the Poisson equation:

$$\nabla^2\varphi = -\frac{\sigma}{\varepsilon} \tag{8.1}$$

where ε is the dielectric constant of the medium and ∇^2 is the Laplace operator. The Poisson equation is a partial differential equation that can be solved when the functional dependence of $\sigma(x, y, z)$ and the appropriate boundary conditions are known. The operator ∇^2 is not tied to any specific coordinate system, and we will use the Poisson equation in various coordinate systems. The main conclusion of Eq. (8.1) is that the potential variation in any direction within a conductor is proportional to the volumetric charge density.

Special electrostatic cases: In a class of electrostatic problems involving conductors, all charges are either located on the surface of the conductor or are present as point charges. In such cases, σ is zero at most points in space, and the Poisson equation simplifies into a more straightforward form of the Laplace equation:

$$\nabla^2\varphi = 0 \tag{8.2}$$

Electric and magnetic fields are interconnected. The motion of charges (current) generates magnetic fields, and changes in magnetic fields produce currents. For stationary charge and current distributions, the independent form of Maxwell's equations describes the behavior of electric and magnetic fields. A key conclusion from Maxwell's equations is that stationary charges create only an electric field, which can be studied separately from the magnetic field. The potential of the electrostatic field is determined by the volumetric charge density, as shown in the Poisson equation (Eq. 8.1).

8.1.2 Importance of Studying the Electromagnetic Properties of Food

As agricultural technologies evolve, and new methods for utilizing electric energy are developed, understanding the electrical properties of agricultural materials has become increasingly important. Even when focused only on agricultural products, the topic of electrical properties is broad. The use of electromagnetic waves to analyze materials and measure moisture content is beneficial for quality control in industries such as forestry, construction, agriculture, and food quality evaluation (e.g., meat, fruits, and coffee) [1, 2]. For example, non-destructive quality assessments of fruits and vegetables based on their components have been successfully applied [3, 4].

The electrical properties of agricultural materials have been a topic of research for years and have applications in various sectors of human activity. Non-destructive testing methods that assess the quality and properties of agricultural products are useful for producers, processors, and traders. There is a close correlation between dielectric constant and moisture content, especially in materials with high moisture content. Understanding the dielectric properties of partially frozen materials is crucial for determining the heating rate and uniformity during microwave thawing. Electrical properties are important when processing food with pulsed electric fields, ohmic heating, inductive heating, radio frequency (RF), and microwave heating, and these properties can also be useful in detecting food processing conditions and quality.

Given the increasing demands for food quality and safety, developing fast and effective non-destructive testing methods for food quality assessment is necessary. The dielectric properties of grains and oilseeds are essential for understanding their electrical behavior and developing indirect non-destructive methods to determine their physical properties. These properties are also critical for designing equipment for RF or microwave dielectric heating and for potential agricultural applications, such as drying grains, improving seed germination, and controlling insects in stored grains using high-frequency and microwave fields.

Dielectric spectroscopy, a simple, fast, and non-destructive measurement technique, provides information about how materials respond to electromagnetic fields, making it a convenient method for evaluating food quality, especially moisture content. This technique has been extensively researched and applied for analyzing and monitoring the quality of biological materials. It allows for rapid, non-destructive measurements, making it suitable for online monitoring. The application of dielectric properties in sensing agricultural materials and food quality has been well-documented. This chapter aims to review the fundamental concepts of electrical properties, various measurement methods, their roles in the agricultural food sector, and the potential use of thermal imaging (TI) in food processing quality and safety assessments. The current research status in this field is summarized, along with recent developments.

Food has three main electromagnetic properties that can be applied to food analysis, detection, and processing:

1. Dielectric Properties: Useful for heating and detection. Applications include dielectric moisture meters, high-frequency waves, microwaves, infrared heating, and drying.
2. Conductive Properties: Can be used for heating and detection, such as with conductive moisture meters and ohmic heating.
3. Charged Properties: These are utilized in electrostatic field processing, including cleaning, separation, smoking, powder spraying, molding, and mold prevention. Charged properties and electrokinetic phenomena are also applied in food analysis and separation processes, such as electroosmosis, electrodialysis, electrophoresis, and electrolocation.

The separation, analysis, preservation, and processing of food often rely on differences in the properties of its components. For example, distillation separates food components based on boiling point differences, while sedimentation, centrifugation, and separation rely on differences in density. Electromagnetic properties of food and raw materials are often utilized for analysis, separation, and processing in ways that offer advantages over other methods.

1. Rapid Response and Convenience: Electromagnetic properties allow for fast, easy, and typically non-destructive detection, making them a key tool for automated food production and control.
2. Efficient Separation and Processing: The differences in electromagnetic properties between different foods or components enable efficient, low-pollution processing that is easy to control, promoting the production of high-quality products.
3. Preserving Food with Low Energy Consumption: External electric fields can be applied to intervene in electrochemical reactions and bioelectric fields within food, preserving it in a manner that is energy-efficient, safe, and hygienic.

8.2 Electromagnetic Properties of Food

8.2.1 Introduction

Food is made up of common chemical substances and exhibits certain electromagnetic properties. However, food differs from the typical objects studied in physics as it is usually a mixture of different components, each with unique electromagnetic properties. These properties form the basis for both electromagnetic processing and testing of food. This section introduces relevant electrical knowledge, focusing on the main components of food such as water, starch, proteins, and electrolytes. It also explores the relationship between the physical and geometric properties of food and its electrical characteristics.

8.2.2 Conductivity Properties

In terms of conductivity, materials can generally be classified into two categories: conductors and insulators. When a conductor is connected between two other conductors with unequal electric potentials, a certain amount of charge flows through the conductor until equilibrium is established. For insulators under the same conditions, little to no charge movement is observed. However, the distinction between conductors and insulators is largely one of degree.

For example, metals can establish equilibrium very quickly, within around 10–10 to 10–9 seconds, whereas typical insulators such as glass or ceramics take much longer, potentially days or even months.

Conductors are divided into two categories: first-class conductors like metals, which conduct electricity through free electrons and second-class conductors like salts, acids, and bases dissolved in water (electrolytes). These substances don't have free electrons but do have positive and negative ions that can move freely under an external electric field, allowing conduction.

Insulators, on the other hand, have electrons in their molecules or atoms that are strongly attracted to the nucleus and cannot easily escape. Therefore, under normal conditions, insulators do not conduct electricity. However, insulators do have a very small number of free electrons, which gives them a weak conductivity under normal conditions. Under certain circumstances, such as the application of a strong electric field, insulators can become conductors.

While metals are not typically found in food, they are crucial materials in food processing equipment, and their electrical properties can affect food processing. Some foods contain ions that can conduct electricity through ion movement, making food generally a mixture of electrolyte solutions and dielectric (insulating) components. In a direct current (DC) electric field, electrolytes conduct electricity while dielectrics do not. In an alternating current (AC) electric field, dielectrics can also exhibit some conductivity, and this conductivity is frequency-dependent.

The Electrical Conductivity Properties of Water: Most foods contain water, and water's properties significantly influence a food's electrical properties. Pure water does not conduct electricity, but the water in food usually contains ions, giving it some conductivity in DC and low-frequency AC fields.

The Electrical Conductivity Properties of Small-Molecule Electrolytes: These are substances like citric acid, monosodium glutamate (MSG), and sodium chloride (table salt), which dissociate in water and behave similarly to typical electrolytes in terms of conductivity.

The Electrical Conductivity Properties of Ionic Polymers: These are large-molecule compounds found in food, such as polysaccharides and proteins. Examples of ionic carbohydrates include pectins, xanthan gum, sodium alginate, and arabic gum. Proteins, composed of amino acids linked by peptide bonds, also contain groups like carboxyl and amino groups, making them ionic polymers. In different pH environments, proteins can carry different charges; they are positively charged in acidic conditions and negatively charged in alkaline conditions. At their isoelectric point, proteins carry no overall charge. Ionic polymers in solution can conduct electricity, either through their own movement or by facilitating movement of the surrounding liquid.

Non-Electrolyte Components: Foods often contain non-electrolyte substances like starch, sucrose (sugar), and fats, which have very weak conductivity.

8.2.3 Electrical Resistance and Resistivity

Resistance is a physical quantity that represents a conductor's opposition to electric current. The higher the resistance, the greater the opposition to current flow. The unit of resistance is the ohm, denoted by the symbol Ω.

$$R=\rho\frac{L}{A} \tag{8.3}$$

where R is the resistance (Ω), ρ is the resistivity (Ω·m), L is the conductor length (m), and A is the cross-sectional area of the conductor (m^2).

Resistivity is a physical measure of a material's resistance characteristics. It depends on the type of material and is also influenced by external factors such as temperature, pressure, and surrounding magnetic fields. The formula for calculating resistivity is:

$$\rho=\frac{m}{e^2Nt} \tag{8.4}$$

where m is the electron mass (kg), e is the charge (C), N is the number of electrons per unit volume, and t is the average free flight time between electron collisions (s).

Because food materials generally contain moisture, the resistance and resistivity of food materials are related to their moisture content and temperature. Generally, as temperature increases, the resistance and resistivity of food materials also increase. Resistivity is inversely proportional to the number of free electrons (charges) in a material. Since food materials contain far fewer free electrons (charges) than metal materials, the resistivity of food materials is significantly higher than that of typical metal conductors. Food is often heterogeneous, and some food items contain a certain amount of air. As a result, the resistance and resistivity of food depend not only on moisture content and temperature but also on the orientation of the sample during measurement.

At higher temperatures, when the temperature variation is not significant, resistivity can be expressed with the following empirical formula:

$$p_t = p_o(1 + a_t) \tag{8.5}$$

where p_t is the resistivity at temperature t°C, p_o is the resistivity at o°C, and a_t is the temperature coefficient of resistance.

8.2.4 Electrical Conductance and Conductivity

Conductance is a physical quantity that describes the electrical conductivity of a conductor, representing how easily a particular conductor allows current to pass through it. Numerically, conductance is the reciprocal of resistance and is denoted by G. The unit of conductance is the Siemens, abbreviated as S. The lower a conductor's resistance, the higher its conductance. For a purely resistive circuit, the relationship between conductance and resistance is:

$$G = \frac{1}{R} \tag{8.6}$$

This leads to Ohm's conductance law equation:

$$G = \frac{I}{U} \tag{8.7}$$

where I is the current (in amperes) and U is the voltage (in volts).

Conductivity, also known as electrical conductivity, is a physical concept that describes the ease of charge flow within a substance. In a medium, conductivity (σ) multiplied by the electric field strength E gives the conduction current density J. For isotropic media, conductivity is a scalar; for anisotropic media, it is a tensor. In ecology, conductivity is expressed numerically to represent a solution's ability to conduct current. The Greek letter σ denotes conductivity, with a standard unit of Siemens per meter (S/m), the reciprocal of resistivity ρ:

$$\sigma = \frac{1}{\rho} = \frac{L}{RA} \tag{8.8}$$

The resistance and conductance of agricultural products depend on factors such as the specific material and characteristics (e.g., size, shape). However, the resistivity and conductivity of agricultural materials are independent of these specifics; they only reflect the material's conductive properties and are influenced by factors such as material characteristics, temperature, and moisture content.

Conductivity has potential applications in fluid pasteurization, making it important to understand the effective conductivity or overall resistance of liquid-particle mixtures. Conductivity measurements help determine various characteristics of agricultural materials and food, such as frost sensitivity, cold resistance, freeze tolerance, moisture content, seed germination stress, pasteurization, and other qualities for grains, seeds, meat, sugar, milk, wood, soil, fruits, vegetables, and infected foods (e.g., moisture content, fruit ripeness, egg freshness, insect control in seeds, radio frequency heating). The conductivity of various food groups, as measured with a commercial electrical conductivity meter (YSI model 30, YSI Incorporated, Yellow Springs, OH), is presented in Table 8.1.

Table 8.1 Dielectric constants of common food components

Component	Dielectric constant	Dielectric loss coefficient tanδ	Dielectric loss factor $\in$ tan δ
Water (30 °C, 3000 MHz)	77	0.15	11.5
Water (25 °C)			12.3
Water (55 °C)			4.62
Water (85 °C)			3.1
Ice (−12 °C, 3000 MHz)	3.2	0.00095	0.003
Carbohydrates	3–5		
Protein	4–6		
Air	1		
Fat	2–5		
Potato (2450 MHz)	4.5	0.2	0.9
Peas (2450 MHz)	2.5	0.2	0.5
Spinach (2450 MHz)	13	0.5	6.5
Apple (water content 86%, 23 °C, 2500 MHz)	43	0.21	9.03
Carrot (water content 86%, 23 °C, 2500 MHz	57	0.32	18.24
Cucumber (water content 96%, 23 °C, 2500 MHz	63.5	0.20	12.7
Peach (water content 89–91%, 23 °C, 2500 MHz)	51–62	0.31	15.8–19.2
Potato (water content 80%, 23 °C, 2500 MHz)	52–54	0.32–0.37	16.6–20
Watermelon (water content 91%, 23 ° C, 2500 MHz	59	0.26	15.34
Wheat flour (8% water content)	2.6	0.03	0.078
Wood (hard)	3		0.09
Wood (soft)	5	0.065	0.325
Polystyrene	2.6–30	0.0002–0.0004	0.00052
Polyvinyl chloride	3–5	0.025–0.05	0.075–0.25
Polyethylene	2.3	0.0005	0.012
Paper			0.16
Glass			0.05
Ceramics			0.085

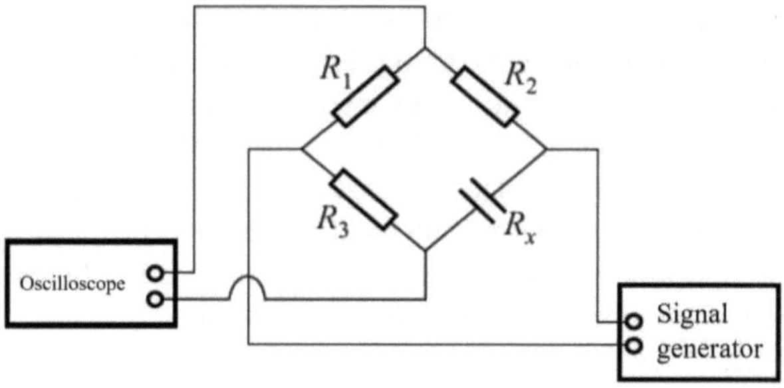

Fig. 8.1 Schematic diagram of the conductivity measurement circuit

Food generally belongs to nonhomogeneous dispersion systems, composed of electrolytes and dielectric components. In a direct current electric field, conductivity is primarily achieved through ion conduction. In an alternating current electric field, the effect of ionic conductivity on medium loss can be represented by the following equation:

$$\varepsilon_l'' = \varepsilon_c'' - \sigma/(2\pi f \varepsilon_0) \tag{8.9}$$

where ε_l'' is the dielectric loss caused by dipole moment polarization, ε_c'' is the measured value of the dielectric constant, σ is conductivity, and f is the frequency of the electric field used for measuring conductivity.

The AC bridge is a classic method for measuring conductivity. In electrolyte solutions, cations and anions migrate toward the electrodes under the action of the electric field. It is evident that the electrolyte solution cannot satisfy pure resistance conditions, and the resistance box (resistance ratio arm) is also difficult to regard as pure resistance. Therefore, under alternating current conditions, the balancing condition of the bridge should consider the impedance of each resistance (including resistance, capacitance, and inductance). When the frequency of the alternating current is not too high, the influence of the resistance box's capacitance and inductance is minimal; the commonly used power supply frequency is 1–4 kHz. Within this frequency range, accurate results can still be obtained without employing strict shielding measures. A schematic diagram of the conductivity measurement circuit is shown in Fig. 8.1.

8.2.5 *Dielectric Properties*

A dielectric is essentially an insulator. Its main characteristic is that its electrons are tightly bound to the atoms they belong to, and they cannot move freely. Therefore, under normal electric fields, dielectrics do not conduct electricity. However, when subjected to an electric field, the positive and negative charges inside the dielectric material shift slightly, forming electric dipoles. These dipoles create a secondary electric field that overlaps with the original electric field, thereby altering the overall field.

The dielectric properties of food materials measure how food interacts with electromagnetic fields. Permittivity is one of the dielectric properties that determines how electromagnetic waves interact with a substance. It also defines the density of charges under an applied electric field. Permittivity in solids, liquids, and gases depends on the dielectric constant (ε'), which relates to the material's capacity to store electrical energy, and the dielectric loss factor (ε''), which describes the energy loss in food when exposed to alternating electric fields. This loss is associated with dielectric relaxation and ionic conduction.

8.2.5.1 Permittivity

To better understand the concept of the dielectric constant, it's useful to first discuss capacitance. Capacitance refers to the ability of a system to store electric charge (or electrical energy). A capacitor is a device used for storing charge and can be made up of two metal plates separated by an insulating material, called the dielectric. The two plates are referred to as electrodes, and the material between them is the dielectric.

A capacitor has infinite resistance to direct current (DC), meaning it blocks DC, but its behavior toward alternating current (AC) depends on the frequency of the AC. For a given capacitance, the capacitor's resistance to AC (known as capacitive reactance) varies with the frequency of the current. Capacitors work by charging and discharging, and the higher the frequency of the AC, the lower the reactance or resistance to the AC.

When a capacitor is filled with a uniform dielectric material, its capacitance increases. If the capacitance in a vacuum is denoted as C_0 and the capacitance when filled with a dielectric material is C, the ratio is given by:

$$\frac{C}{C_0} = \in_r' \geq 1 \tag{8.10}$$

where $\in_r'$ is the relative dielectric constant (also called relative permittivity). It is a dimensionless number that depends solely on the dielectric properties of the material. This value explains how food interacts with electric fields, and it influences how electromagnetic waves interact with materials and defines the charge density under an applied electric field.

For a parallel-plate capacitor with a dielectric material, the capacitance becomes:

$$C = \in_r' C_o \tag{8.11}$$

$$C = \frac{\in_o \in_r' S}{d} \tag{8.12}$$

where $\in = \in_o \in_r'$ is the dielectric permittivity (or dielectric constant), $\in_o$ is the permittivity of free space (vacuum), S is the area of the electrodes, and d is the distance between the electrodes.

The dielectric constant in food can be influenced by its chemical composition, physical structure, frequency, and temperature, with moisture content being the most critical factor.

8.2.5.2 Polarization of Dielectrics

When two charges are separated by a certain distance, they form a dipole, which can be represented by a vector called the dipole moment. Molecules with a permanent dipole moment are known as polar molecules. Even nonpolar molecules can acquire a dipole moment in an electric field due to distortions in their electron distribution and nuclear positions. Polarization refers to the average dipole moment per unit volume. The relationship between the dielectric constant and polarization is given by:

$$P = (\varepsilon - 1)\varepsilon_0 E \tag{8.13}$$

where P = polarization, $\varepsilon - 1$ = electric susceptibility, ε_0 = *absolute permittivity of a vacuum*, E = electric field.

Thus, the relative dielectric constant (ε), or electric susceptibility ($\varepsilon - 1$), measures the ease with which a dielectric material becomes polarized under an external field. Polarization can occur in four ways: ionic polarization, orientation polarization, atomic polarization, and electronic polarization. Electronic polarization occurs when electrons within an atom shift relative to the nucleus and is present in all materials. Atomic polarization happens when atoms move within a crystal or molecule. Both types of polarization cause the permittivity of most dry solids to be low, typically $\varepsilon < 10$. In such cases, materials are nearly lossless at microwave frequencies. Atomic polarization has a lower resonance frequency than electronic polarization due to the larger mass of the atoms involved.

In a microwave or high-frequency field, dipoles attempt to follow the rapidly changing field, leading to energy loss through random thermal motion in water, resulting in temperature increase. This is known as orientation polarization. In the case of hydrated ions, they try to move in the direction of the electric field, transferring energy through this movement.

Various effects can lead to polarization, ranging from surface charge accumulation at the interface of materials with different electrical properties (interface polarization) to dipole orientation and other mechanisms [5]. Dielectric materials can be classified into gaseous, solid, and liquid dielectrics. These can further be divided into nonpolar and polar molecular dielectrics.

In nonpolar molecular dielectrics, such as carbon dioxide, there is no dipole moment when no external field is present because the centers of positive and negative charges overlap. However, when exposed to an electric field, the charge centers shift, turning the molecules into dipoles, with the degree of displacement proportional to the field strength. This process is called displacement polarization. Once the external field is removed, the charge centers return to their original positions, and the dielectric material loses its polarization. This is also called electronic displacement polarization, and it is only effective in high-frequency fields, which is why it is referred to as optical polarization.

In polar molecular dielectrics, such as water, hydrogen chloride, or ammonia, molecules have a dipole moment even in the absence of an electric field. In such materials, the dipole moments are randomly oriented due to thermal motion, resulting in no net polarization. However, when an external electric field is applied, the dipole moments align with the field, a process known as orientation polarization. Additionally, in polar dielectrics, the electrons within molecules can shift relative to the nucleus, leading to electronic displacement polarization, though this is usually much smaller than orientation polarization.

For solid dielectrics, besides the above mechanisms, ionic displacement polarization can occur. For example, in an ionic crystal like sodium chloride (NaCl), sodium ions move in the direction of the electric field while chloride ions move in the opposite direction, causing the crystal to exhibit electrical properties.

8.2.5.3 Dielectric Loss

Most foods have weak electrical conductivity and are classified as dielectrics. In a static electric field, dielectric materials store most of the energy with minimal energy loss through polarization friction. However, in an alternating electric field, the bound charges in the dielectric continuously realign with the changing field, causing collisions and friction between molecules, converting electrical energy into heat. This energy loss is known as dielectric loss or dielectric dissipation.

The energy absorbed by the dielectric in an AC electric field can be represented as:

$$E = UI \cos\varnothing = E(E/x_0)\cos\varnothing \tag{8.14}$$

where E is the absorbed energy; U is the effective voltage $U = 0.707U_{\max}$, in volts; I is the effective current $I = 0.707I_{\max}$, in amperes; $\cos\varnothing$ is the power factor of the alternating current; x_0 is the capacitive reactance; $x_0 = 1/(2\pi fC)$, in ohms; and f is the frequency of the electric field, in hertz.

The heat energy released per unit volume of dielectric per unit time is given by:

$$Q = 55.6 \times 10^{-14} f E^2 \in \tan\delta \tag{8.15}$$

where Q is the power consumed by the electric field (i.e., heat energy released per unit time per unit volume), in $W/cm^3 \cdot s$; f is the electric field frequency, in hertz; E is the electric field strength, in V/cm; $\in$ is the dielectric constant; $\tan\delta$ is the tangent of the dielectric loss angle of the medium; and $\in \tan\delta$ is the loss tangent, representing the ratio of dielectric loss to stored energy. It is not a constant and is related to both the electric field frequency and the structure of the material. The common dielectric loss coefficient and dielectric loss factor for food media are shown in Table 8.1.

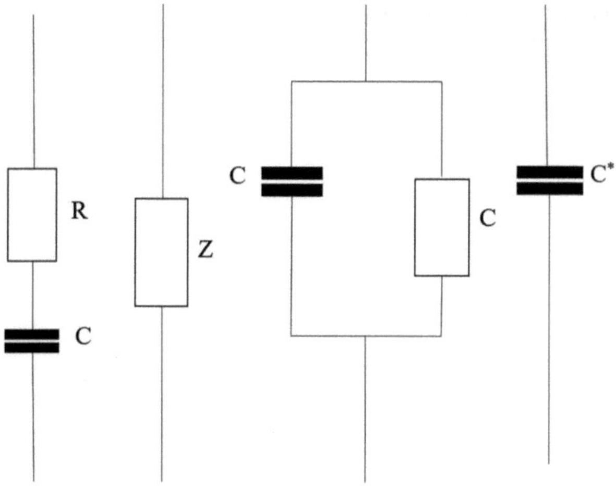

Fig. 8.2 Impedance and complex capacitance

8.2.5.4 Impedance and Complex Capacitance

Most foods are neither perfect conductors nor perfect insulators but lie somewhere between the two. In an alternating electric field, foods experience energy loss due to their resistance and capacitance. This energy loss can be divided into two parts: conductive loss due to the material's resistance, which is converted into heat, and dielectric loss due to the continuous polarization of the dielectric, which also generates heat. Thus, food can be modeled as a complex equivalent circuit of resistors and capacitors arranged either in series or parallel (as shown in Fig. 8.2).

In practice, food can generally be represented well by a parallel circuit of resistors and capacitors. When an AC voltage (U) is applied across the parallel circuit, current flows through both the resistor and capacitor. The current through the resistor (I_R) is in phase with the applied voltage and is responsible for heat dissipation. The current through the capacitor (I_C), on the other hand, is out of phase by 90° and is stored as electrical energy.

The total current I is the vector sum of the currents through the resistor and capacitor, i.e.,

$$I = I_R + I_C \tag{8.16}$$

The angle ∅ in Fig. 8.3 is called the phase angle, cos∅ is the power factor, δ is the loss angle, and tan δ is the loss tangent, which represents the ratio of energy lost to energy stored,

$$\tan\delta = I_R / I_C = \in''_r / \in'_r \tag{8.17}$$

There are three key dielectric properties: relative dielectric constant ($\in'_r$), relative dielectric loss factor ($\in''_r$), and dielectric loss tangent (tanδ). The relationship between these quantities can be expressed as:

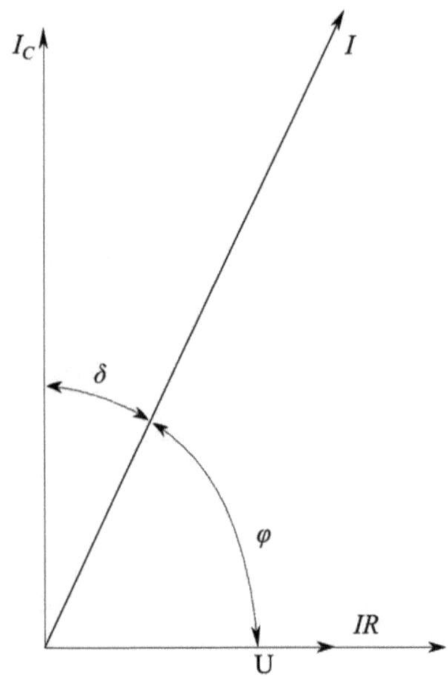

Fig. 8.3 Dissipation at the cellular level

$$\in^0{}_r = \in'_r - j\in''_r = \left|\in^0{}_r\right|e^{-j\delta} \quad (8.18)$$

and

$$\tan\delta = \in''_r / \in'_r \quad (8.19)$$

where $\in^0{}_r$ is the complex relative dielectric constant, $\in'_r$ is the relative dielectric constant, $\in''_r$ is the relative dielectric loss factor, **tan(δ)** is the dielectric loss tangent, δ represents the dielectric loss angle, and j is the imaginary unit $\sqrt{-1}$.

8.2.5.5 Dielectric Relaxation Characteristics

When a dielectric material is exposed to (or removed from) an external electric field, it transitions from an instantaneous polarized state to a new equilibrium. The time it takes for this polarization to stabilize is called the relaxation time, and it is closely related to the polarization mechanisms. Dielectric relaxation is one of the factors contributing to dielectric loss.

In an alternating electric field, polar molecules in agricultural products or food experience an applied AC voltage, causing dipoles to align with the changing field. As the frequency of the electric field increases, the dipoles can no longer keep up, leading to a time lag in their orientation. At this point, the value of I_C decreases, and the real part of the complex dielectric constant also decreases. This frequency-dependent decrease is called dissipation, while a frequency-dependent increase is called absorption. Together, these changes are referred to as relaxation.

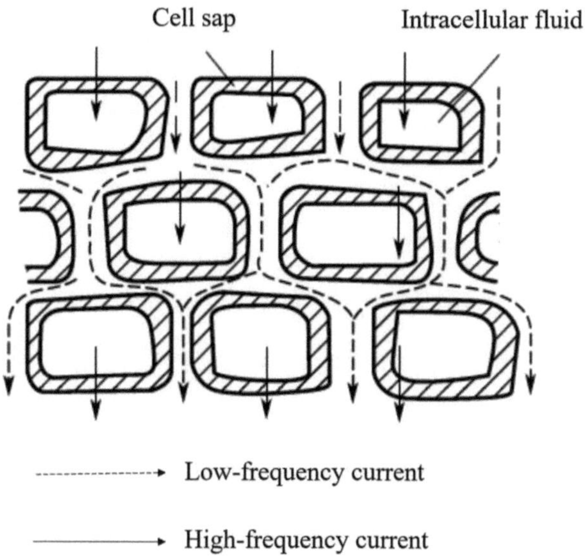

Fig. 8.4 Electric field frequency distribution in cell tissues

Apart from water dipoles, other factors also contribute to relaxation phenomena in agricultural products and food, particularly the cell structure. For example, as shown in Fig. 8.4, cell membranes (walls) have high resistance and capacitance. At low frequencies, the current only flows through the extracellular fluid, resulting in high resistance. At high frequencies, the capacitance between cell membranes becomes large, reducing the impedance, and current flows through the intracellular fluid, causing the resistance to decrease. This resistance change due to structural inhomogeneity is called structural dissipation (β-dissipation).

In biological tissues, there are also α-dissipation and γ-dissipation (Fig. 8.5). α-dissipation occurs at low frequencies due to changes in the cell membrane, while γ-dissipation occurs at high frequencies. β-dissipation lies between these two. These frequency-dependent resistance changes must be considered in food testing and ohmic heating processes.

Figure 8.6 illustrates the various factors that cause relaxation phenomena and the frequency ranges they affect. In actual measurements, the observed effect is a combination of all these factors. These combined effects can be modeled as a complex equivalent circuit made of multiple series and parallel components. Understanding the measurement frequency is crucial for accurately determining relaxation phenomena in electrical properties.

The imaginary part of the complex dielectric constant ($\in_r''$), unlike the real part ($\in$), first increases with frequency, then decreases as the frequency continues to rise, creating a peak in the middle. On a complex plane made up of the x-axis and y-axis, this behavior forms a semicircular curve as shown in Fig. 8.7a.

In water, which contains permanent dipoles, this change is evident and can be described by the Debye equation:

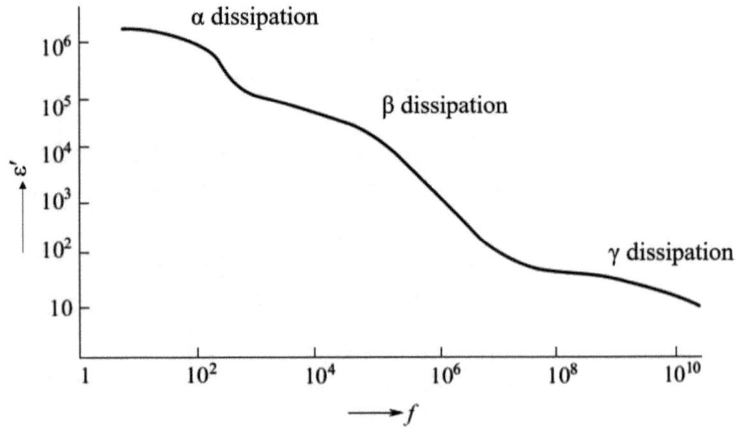

Fig. 8.5 Typical relationship between dissipation and frequency

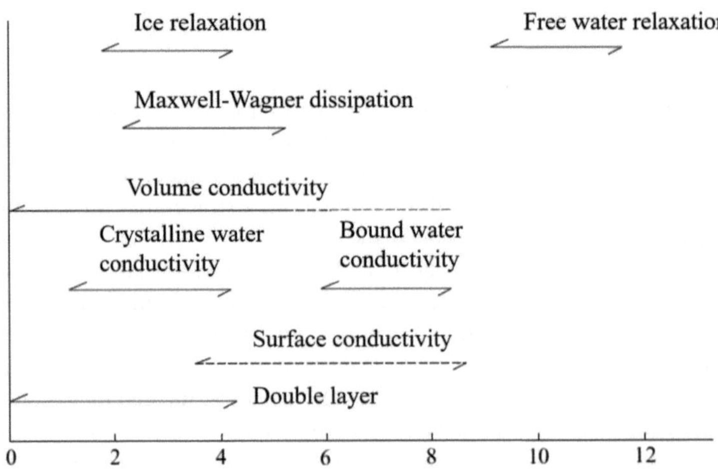

Fig. 8.6 Main factors influencing relaxation phenomena and their frequency ranges

$$\in = \in_{\infty} + (\in_0 - \in_{\infty})/(1 + iw\tau) \tag{8.20}$$

where $\in_{\infty}$ is the dielectric constant at very high frequencies (approaching infinity), $\in_0$ is the dielectric constant at very low frequencies (approaching zero), and τ is the relaxation time, which represents the time delay in dipole orientation.

τ has units of time and indicates how long it takes for dipoles to align with the electric field. The reciprocal of τ, $1/\tau$, represents the frequency (in radians) corresponding to the peak of the semicircle in Fig. 8.7a. The relaxation frequency (f) is given by:

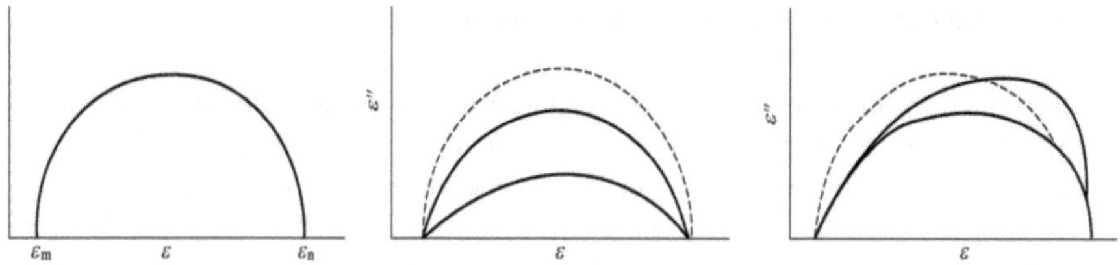

(a) Debye Semicircle (b) Cole-Cole Circular Arc (c) Cole-Davidson Oblique Arc

Fig. 8.7 Schematic diagram of different types of electric fields

$$f = 1/(2\pi\tau) \tag{8.21}$$

In pure water, only dipoles exist, so there is a single relaxation structure with a single relaxation time. However, in real foods, various relaxation phenomena occur, as shown in Fig. 8.6, many of which behave differently from water. For example, in Fig. 8.7b, the shape on the complex plane becomes a circular arc with its center below the x-axis. This arc can be described using the Cole-Cole empirical equation:

$$\in = \in_{\infty} + (\in_0 - \in_{\infty})/\left[1 + (iw\tau)^{\beta}\right] \tag{8.22}$$

where β is an experimental coefficient (with values between 0 and 1).

In this relaxation structure, there is a distribution of relaxation times, and β represents the degree of this distribution. Research shows that biomaterials with cell structures, particularly in the low-frequency range, exhibit structural dissipation that often follows the Cole-Cole arc rule.

Similarly, for the slanted arc shown in Fig. 8.7c, the Cole-Davidson equation is used:

$$\in = \in_{\infty} + (\in_0 - \in_{\infty})/(1 + iw\tau)^{\alpha} \tag{8.23}$$

where α is another experimental coefficient (with values between 0 and 1).

When $\alpha = 1$, this equation becomes equivalent to the Debye model (Eq. 8.19). The smaller α is (less than 1), the more the arc bends outward.

Additionally, there is the Maxwell-Wagner model, which describes relaxation phenomena that occur at two-layer interfaces, commonly seen in emulsions. Like the Debye model, it occurs at much lower frequency ranges.

The dielectric constant of biological materials is also affected by temperature. In polar materials, the relaxation time decreases as temperature increases. Experiments using Eq. 8.19 show that in the dissipation or dielectric loss region, the dielectric constant increases with rising temperature. In conditions where there is no dielectric loss, the dielectric constant tends to decrease with temperature. Additionally, the dielectric constant may undergo abrupt changes when a phase transition occurs in the material.

8.2.5.6 Factors Affecting Dielectric Properties

Various factors influence the dielectric properties of food materials, such as the frequency of the applied alternating electric field, moisture content, bulk density, temperature [2, 6, 7], the ionic properties of food materials, concentration (density), structure, and composition [2, 8–11]. The primary state variables affecting dielectric performance at a given frequency are temperature and moisture content. Table 8.2 summarizes the effects of parameters such as moisture content, temperature, bulk density, and frequency on the dielectric properties of agricultural food materials. The following section reviews the relationships between these factors, compositions, and the dielectric performance of food and food materials.

The Effects of Salt, Fat, and Other Components

At higher moisture levels and lower fat content, the dielectric loss factor slightly increases with rising temperature, while an increase in fat/lipid content reduces the dielectric properties of various foods [12]. The dielectric loss factor rises with increased ash content in oils and decreases as the fat content in oils is reduced [13]. The influence of water and salt or ash content largely depends on how these components are bound or limited in movement by other food constituents [14]. Most common foods have a loss factor below 25, indicating a penetration depth of 0.6–1.0 cm. However, literature data mostly pertains to pure foods and individual food components. For complex or formulated foods, their dielectric properties must be measured or estimated [15, 16].

The addition of salt lowers the dielectric constant (ε') and increases the loss factor (ε'') of various foods [17], with particularly significant effects at 450 and 900 MHz [18]. At the higher microwave frequency of 2800 MHz, only the dielectric loss in salted foods rises with temperature, while at lower microwave frequencies, the loss factor generally increases [19, 20].

The Effects of Moisture Content, Temperature, and Frequency

For dry granular solids, no temperature dependence of dielectric performance has been observed; however, at higher moisture content above 9.4 GHz, dielectric performance sharply increases [21]. The dielectric constant and loss factor of various foods increase with the increase in moisture content [19]. The rate of change in the dielectric constant and loss factor with temperature depends on the free water and bound water content of the food material [22]. In general, the dielectric constant of fruits and vegetables tends to increase with rising temperature, while the loss factor may either increase or decrease depending on the operating frequency [22]. At lower frequencies, the dielectric constant increases with temperature; however, at higher frequencies, it decreases with rising temperature. Within the frequency range of 0.01 to 1.8 GHz, the temperature dependence of the dielectric constant is minimal [23].

Table 8.2 Effects of the parameters such as moisture content, temperature, bulk density, and frequency on the dielectric properties of food and food products

Food/product source	Range of parameters	Comments
Food materials, Salty foods	2800 MHz freq.	At constant temperatures, dielectric constant (ε') and loss factor (ε") of food materials increase with decreasing frequency. The penetration depth (PD) decreased with increasing frequency, temperature, and moisture content. At higher microwave frequencies, only salty foods show an increase in ε" with temperature, but at lower microwave frequencies, there is a general increase in ε". Both ε' & ε" of various foods increased with increasing moisture content
Moist foods	3 GHz freq.	Loss factor (ε") increases with falling frequency
Fruits and vegetables	Freq. (1–1800 MHz), temp. (20–60 °C)	Generally, ε' increases with temperature, whereas ε" increased with increasing temperature and ε' increased with temperature at lower frequencies but decreased with temperature at the higher frequencies
Mango	Freq. (1–1800 MHz), temp. (20–60 °C)	Dielectric properties values decreased with increasing frequency, but this reduction was larger for ε" than for the dielectric constant. Loss factor increased but ε' decreased with increasing temperature. Both values decreased with storage time
Grape juice and wine	Freq. 200 MHz and 3 GHz	Both ε' and ε" of grape juice are greater than wine at 200 MHz. At 3 GHz, ε' of grape juice is slightly less than of wine, whereas ε", at 3 GHz, is practically the same for grape juice and wine
Apples	915 MHz or 2.45 GHz: moisture content range of 4% to 87.5% at 22 and 60 °C	An increase in temperature at low moisture contents resulted in increased dielectric properties. The penetration depth increased as moisture was removed from the sample
Garlic	Freq 2450 MHz, m. c. (6–185%, db) temp. (35–75 °C)	Both ε'& ε" directly proportional to the moisture content. Both ε' and ε" decreased with increase in temperature at high moisture levels and a reverse trend in lower moisture

(continued)

Table 8.2 (continued)

Food/product source	Range of parameters	Comments
		range. The relationship of e' as well as ε" was found to be linear with temperature. PD of microwaves increased with decrease in moisture content, and the temperature had the in significant effect on it
Legume flours (chick-pea green pea, lentil and soybean flour)	Freq. (10–1800 MHz), temp. (20–90 °C) and m.c. (8–21 g/ 100 g)/water activities (0.3–0.9)	Both ε' and ε" of the four legume flours decreased with increasing frequency but increased with increasing temperature and moisture content. At low frequencies and high temperature and moisture content, negative linear correlations for both ε' and ε"
Common bean, nuts	Freq (75 kHz–5 MHz), moisture content range from 13.4% to 15.5% w.b	The value of relative permittivity at each moisture content decreased regularly with increasing frequency. The frequency dependence of ε" and loss tangent was less regular than that of permittivity. Relative permittivity increased with moisture content at every frequency, Both ε' and ε" of nuts increased regularly with moisture content at all frequencies and decreased as the frequency increased
Flaxseed	Freq (50 kHz–10 MHz), m.c. (5.92–22.18%, db)	Both ε' and ε" increased with an increase in moisture content and bulk density and with a decrease in frequency
Safflower	Freq. (50 kHz–10 MHz); m.c. (5.33–16.48%, d b)	The ε' increased with an increase in the moisture content and bulk density, whereas it decreased with an increase in the frequency.
Fresh eggs	Freq.(10–1800 MHz) at 24 °C temperature	The ε' increases with temperature, whereas ε" increased with increasing temperature and ε' increased with temperature at lower frequencies, but decreased with temperature at the higher frequencies
Ground beef	Freq. 915 MHz	Loss factor independent of moisture content <20% and >45% but linear increase with moisture content between 20% and 45%
Beef and beef products	2800 MHz freq.	Both ε' and ε" increases with decreasing frequency at constant temperature; however, ε' decreases and ε" increases with increasing temperature at constant frequency

(continued)

Table 8.2 (continued)

Food/product source	Range of parameters	Comments
Beef burgers	2430 MHz freq.	Adding 15% water had no effect on ε' or ε" as most of 15% water lost during deep frying
Aqueous and beef	27.12, 2450 MHz freq.	25% water slightly increased both ε' and ε" in beef blends at both frequencies
Tuna fish, catfish, and shrimp, fish meal	2450 and 915 MHz freq.	A sharp increase in dielectric properties was observed around the freezing point. Both ε' and ε" increased with increased water content at constant temperature, ε' and ε" of lean tuna were larger than those of fatty tuna. PD as the temperatures below the freezing point increased rapidly as the temperature decreased. ε' of marinated catfish and shrimp, generally decreased with increasing temperature, whereas ε" increased with temperature. Both ε' and ε" of fish meal increased nonlinearly with moisture content and also increased with temperature in a relatively linear manner
Salmon fillets	Freq. (27–1800 MHz), temp. (20–120 °C)	At RF frequencies of 27 and 40 MHz, ε' decreased with increasing temperature. But at microwave frequencies (e.g., 915, 1800 MHz), an opposite trend was observed. ε" increased with increasing temperature over the tested frequency range
Macaroni and cheese	Freq. (915–1800 MHz), temp. (20–121 °C) temperature	ε' increased as temperature increased and ε" increased sharply with increase in temperature
Cheeses	Freq. (0.3–3 GHz); temp (55–75 ° C)	Both ε' and ε" decreases with increasing frequency. Dielectric constants generally decreased with increasing temperature, within this frequency range
Sugar/starch H20	1000, 3000 MHz freq.	ε' increased with moisture content while ε" changed with the addition of sugars and starch
Pure honeys and water added honey	Freq. (10–4500 MHz) at 25 °C temp., m.c. (18–42.6%)	ε’ decreased monotonically with increasing frequency and increased with increasing water content. Dielectric relaxation was evident in ε".. The critical frequency and the maximum ε" increased with increasing water content

(continued)

Table 8.2 (continued)

Food/product source	Range of parameters	Comments
Edible oils and fatty acids	Freq. (100 Hz–1 MHz), temp. (20–45 °C), m.c.(0.02–0.31%)	Both ε' and ε" of oils decreased with increasing temperature, and ε' increased with increasing moisture content
Starch		ε" decreases rapidly with increasing temperature, ε' increases slightly with decreasing frequency
Powder potato starch, locust bean gum	Freq. 2.45 GHz; 0–20% m.c, wb	The dielectric properties increase regularly with moisture content
Salt solutions	3000, 10,000	ε' of the salt solutions decreased as temperature increased. For salt contents greater than 2%, the predicted ε" increased with temperature, which is in accord with the behavior of aqueous salt solutions
Soy protein isolate	200–2500 MHz concentration (5, 10, and 15 g/100 g water), temperature (20–90 °C)	ε' decreased with temperature (except at 90 °C) and frequency while increased with concentration. ε" increased with frequency and concentration; however, temperature showed mixed effect. PD decreased with an increase in frequency, concentration, temperature, and pH
Potato flour-water dispersions (slurry)	500–2500 MHz; concentration (10–25% w/w); temperature (20–75 °C)	ε' decreased with temperature and frequency while increased with concentration. ε" increased with frequency and concentration; however, temperature showed mixed effect. PD decreased with an increase in frequency and non-systematic with temperature. Addition of salt substantially reduced Penetration depth of potato slurry

The dielectric constant and loss factor of flaxseed increase with increasing moisture content, bulk density, and decreasing frequency [24]. The dielectric constant of fatty acids increases with the number of double bonds or the length of the molecular chain. The dielectric constant and dielectric loss of oils decrease with increasing temperature, while the dielectric constant increases with increasing moisture content [25]. The loss factor increases with temperature and transitions from the solid phase to the liquid phase [26].

The dielectric constant and loss factor of nuts consistently increase with increasing moisture content at all frequencies but decrease with increasing frequency. At low moisture content, there is little temperature dependence; however, at high moisture levels, both the dielectric constant and loss factor rapidly increase with rising temperature [27]. For salty foods at lower microwave frequencies, the dielectric constant sharply decreases with increasing temperature. For moist foods, the loss factor increases with decreasing frequency [18, 19]. At constant temperature, both the dielectric constant and loss factor increase with decreasing frequency [28]. The penetration depth decreases with increasing frequency, temperature, and moisture content; at 27 MHz, the penetration depth is sufficient to develop large-scale industrial RF (radio frequency) processing.

Research [18] observed that for most tested foods, in most cases, dielectric performance sharply increases with rising temperature during the transition from −10 °C to 0 °C (thawing). Below the freezing point, the dielectric constant of codfish slightly increases from a value below 10 to between 10 and 20 and then suddenly increases to between 60 and 90 upon thawing, depending on the frequency [19].

The dielectric performance of fish meal relates to temperature and moisture content [29]. Both the dielectric constant and loss factor increase nonlinearly with moisture content and relatively linearly with temperature [29]. The dielectric constant of pickled catfish and shrimp generally decreases with rising temperature, while the loss factor increases with increasing temperature [30]. For tuna products, a sharp increase in dielectric performance was observed near the freezing point. Under isothermal conditions, the dielectric constant and loss factor increase with increasing moisture content; the dielectric constant and loss factor of lean tuna are greater than those of fatty tuna. When the temperature is below the freezing point, the penetration depth rapidly increases as the temperature decreases [31]. For beef products, at constant temperature, both the dielectric constant and loss factor increase with decreasing frequency; however, at constant frequency, the dielectric constant decreases with increasing temperature, while the loss factor increases with rising temperature [32].

The Physical Structure of Materials

Dielectric performance varies with many physical properties, including bulk density, particle size, and uniformity. Bulk density has been identified as a primary source of variation in the dielectric constant and loss factor [33]. When elucidating the function of determining grain moisture content, it is essential to consider the density dependence of the material's dielectric performance, a relationship that can also be used to control the continuous online processing of grains [34]. Venkatesh et al. (1998) [35] obtained measurements of various sieve fractions of crushed granular powders. One of the issues derived from the particle size effect is ensuring that the moisture content and composition of various particle size fractions are identical and that no changes occur during the grinding process due to heating. If there is a pure particle size effect, it seems very small compared to the influences of bulk density and other factors.

8.2.5.7 Methods for Measuring Dielectric Properties

The electrical properties of food in an electric field are related to the frequency of the electric field. Currently, the frequency range used in practice is between 0-10^{13} Hz. In such a wide frequency band, different methods are needed to measure the electrical properties of food. These methods and their suitable electric field frequency bands are shown in Fig. 8.8. From the figure, it can be seen that there can be various methods for measuring electrical properties corresponding to each frequency band. When choosing a measurement method, considerations should include the shape and properties of the food material, sample size, and anisotropy.

Determination of Dielectric Constant Under DC Conditions

The static dielectric constant is determined by the impulse current method. The principle involves charging the container being tested with static electricity and accurately reaching a certain static voltage, followed by discharge measurement using an impulse current meter to record the swing of the ammeter. Based on the instrument's parameter values, the capacitance can be determined. If the charge, voltage, and capacitance are known, the static dielectric constant can be calculated. This method is most suitable for materials with low conductivity.

Determination of Dielectric Constant Under AC Conditions

(a) Bridge Circuit Method: The bridge method is the main method for measuring the dielectric constant and the tangent of the loss of the material at low frequencies. The principle of this measurement mainly utilizes various forms of the

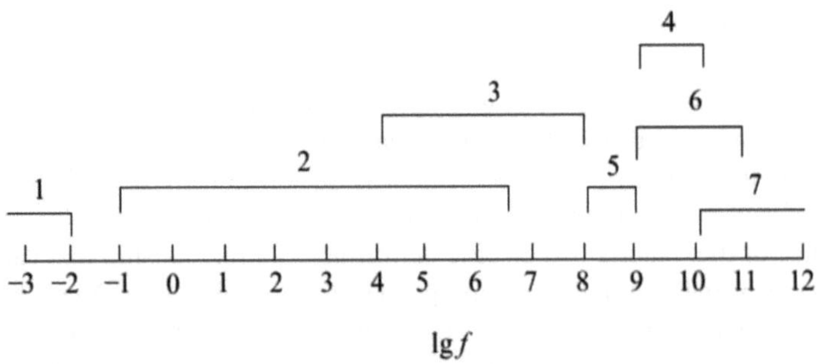

Fig. 8.8 Frequency domain of the electric field and methods for measuring electrical properties. (1) Methods include impact current meter; (2) electric bridge circuit; (3) resonance method; (4) constant wave method; (5) coaxial resonator method; (6) cavity resonator method; (7) waveguide and optical methods

Wheatstone bridge circuit. The measurement is usually conducted at electromagnetic wave frequencies of 1–10 MHz because, in this case, the electrodes do not produce polarization phenomena.
The specific measurement method involves using the tested sample as one arm of the bridge, while the impedances of the other three bridge arms are known. The bridge is adjusted to achieve balance. Based on the balance condition, the parallel equivalent capacitance and resistance of the sample are calculated, thereby determining the sample's relative dielectric constant and loss tangent. However, this method can lead to significant errors due to parasitic (or residual) capacitance and inductance. For measuring the dielectric constant of materials with low loss at frequencies below 500 kHz, the highly accurate Schelling circuit is widely used. The main errors come from the inductance and residual capacitance of the standard bridge components or parasitic capacitance generated between the components of the bridge itself or with the ground. Therefore, strict shielding of each component is required. Each branch of the bridge must have a shielding film, and wires must be grounded. Building on the development of transformer bridges, an impedance measurement method has been successfully developed. This method can avoid issues that are difficult to mitigate with other AC bridges.

(b) Microwave Dielectric Complex Permittivity Measurement Method: In a high-frequency electromagnetic field, the dielectric constant is a composite quantity. By determining the relationship between the geometric dimensions of the sample holder of the measuring device, sample dimensions, and the material's electrical parameters, the relative dielectric constant and loss tangent value can be calculated. There are two major types of microwave dielectric performance parameter measurement: transmission characteristic method and resonance method, as shown in Fig. 8.9.

The resonance method for detecting dielectric constant is realized through an adjustable frequency oscillator exciting an RLC resonant circuit (Fig. 8.10). When voltage U is applied to the loop, adjusting C allows the circuit to reach resonance (where current is maximum at a certain frequency):

$$I_{\max} = \frac{U}{R} \tag{8.24}$$

At this time, Q_1 and C_1 are recorded; the sample's parallel plate capacitor is connected, the circuit is adjusted to achieve resonance, and Q_2, C_2, and ε_r are recorded. Then, the relative dielectric constant and dissipation tangent can be calculated using the following formulas:

$$\varepsilon_r' = \frac{C_s d}{\varepsilon_0 A} \tag{8.25}$$

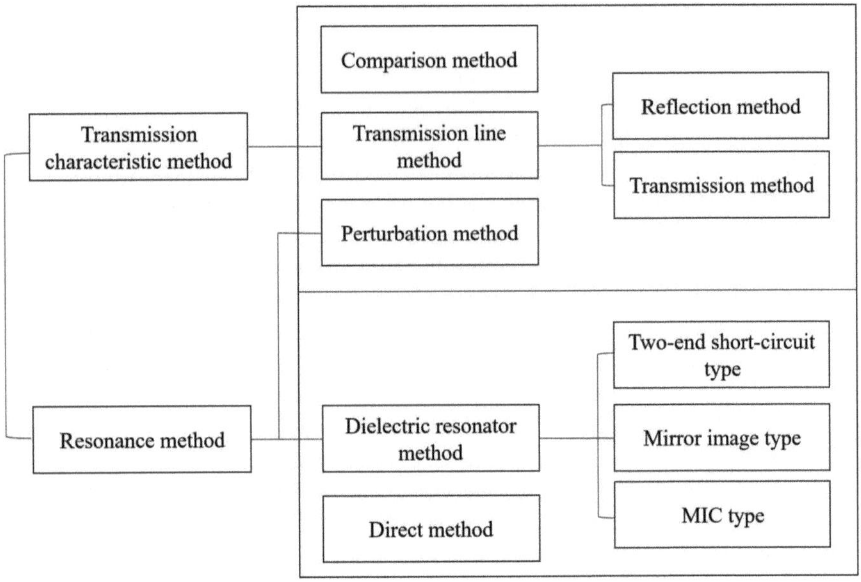

Fig. 8.9 Schematic diagram of AC measurement

Fig. 8.10 Resonance method for detecting dielectric constants (Q meter method)

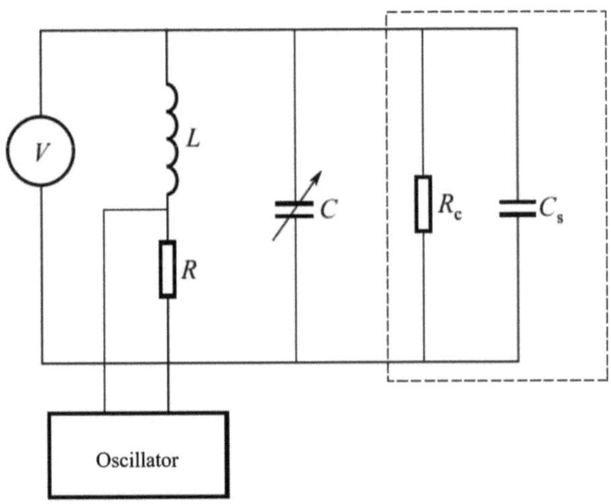

$$\tan\delta = \frac{C_1}{C_1 - C_2} \times \left(\frac{1}{Q_2} - \frac{1}{Q_1}\right) \tag{8.26}$$

where C_s is the capacitance of the capacitor, $C_s = C_2 - C_1$, C_1 is the capacitance before adding the material, C_2 is the capacitance after adding the material, A is the plate area of the capacitor (m^2), and d is the distance between the plate electrodes (m). Q_1 is the Q meter reading when the circuit resonates before adding the material; Q_2 is the Q meter reading when the circuit resonates after adding the material.

The resonance method for detecting dielectric constant is simple and easy to perform, but it is challenging to accurately detect the dielectric constant at various resonant frequencies. Both of the above methods have shortcomings, such as the inability of the material to fill the plates completely, a nonproportional relationship between the dielectric constant and the capacitance between plates, and complex calculations.

Microwave methods for measuring dielectric constant can be divided into time-domain detection and frequency-domain detection. The time-domain detection method infers the dielectric constant by detecting the reflection coefficient, transforming the time-domain response into frequency-domain response using Fourier transformation. The frequency-domain method detects steady-state effects of the measured signal using continuous periodic electromagnetic waves as the probe source within the frequency domain. Specific methods can be further categorized into waveguide methods, resonant cavity methods, and free space methods, among others.

8.2.6 Charging Characteristics

Charging characteristics include electrostatic charging properties, charge density, and charge distribution. Electrostatic charging mainly involves contact electrostatics and ionization gas charging.

Whenever two different objects or the same object in different states undergo a contact-separation process, charge transfer occurs, resulting in electrostatic charging. However, some charging processes are extremely weak, and the static charges generated in some processes are neutralized or transferred, not exhibiting electrostatic charging phenomena macroscopically. Contact charging can be roughly divided into solid charging, liquid charging, and powder charging.

Contact charging of solids can be categorized into three types: contact charging between metals, contact charging between metals and insulators, and contact charging between insulating materials. Initially, electrostatic charging was referred to as triboelectric charging. In reality, friction is not a necessary condition for electrostatic charging; simple contact-separation processes can cause objects to become charged. However, friction can enhance the effect of contact charging. The friction process involves a continuous series of contact and separation events between different contact points on the surfaces of two objects. For metallic conductors, only the final moment of separation has an effect on electrostatic charging, while for insulators, the entire friction process is related to electrostatic charging.

In addition to triboelectric charging, solids can also experience other charging methods such as peeling charging, breaking charging, electrolytic charging, piezoelectric charging, thermoelectric charging, induction charging, adsorption charging, and spraying charging. The main factors affecting solid electrostatic charging include: ① the nature of the material: This encompasses the chemical composition within the material, the chemical composition of the material's surface (contamination, oxidation, adsorption), molecular structure, orientation, crystallinity, and the strain state of the material; and the shape, size, and spatial position of the tested object. ② Environmental factors: These include ambient temperature, humidity, the composition of the gas medium surrounding the object, atmospheric pressure, and wind speed. ③ Charging history of the object: The object's prior exposure to electrification events.

Liquids can exhibit electrostatic phenomena during processing operations such as transportation, spraying, mixing, stirring, filtering, blending, and coating of liquid objects. Liquid charging primarily involves streaming charging, sedimentation charging of particles in the liquid, liquid spraying charging, impact charging of liquids, and splashing charging of liquids.

Powders are solids in a special state. Compared to bulk solid materials, powders possess two main characteristics: dispersibility and suspension. The primary mechanism for powder charging is the contact-separation charging generated during friction and separation between rapidly flowing or agitated powders and pipes, walls, and conveyor belts, as well as the mutual friction, collision, and separation of the powder particles themselves, along with processes like solid particle breakage and fragmentation.

In industry, in addition to the aforementioned contact charging methods, charged particles can also be generated through ionized gases. This is a primary method for electrostatic processing in the food industry. The commonly used gas ionization methods include the activated ionization method and the self-ignition ionization method.

The activated ionization method utilizes ionizing agents (such as X-rays, shortwave radiation, ultraviolet radiation, and high temperatures) between electrodes for ionization. When the external ionizing agent is removed, ionization stops, and the resulting oppositely charged ions will recombine.

Self-excitation ionization involves reaching a certain voltage within a circuit, accelerating charged particles in an electric field, and colliding them with neutral gas molecules to produce ionization. This gas collision ionization can continue in the presence of an external excitation source (ionizing agent). This conductive behavior is also referred to as self-sustaining conductivity in gases.

Self-excitation ionization often utilizes discharge in nonuniform electric fields. As the voltage increases, gas ionization occurs at the maximum electric field strength, entering a stable corona discharge state. When the electric field is very uneven, local gas ionization discharge may cause the entire gap electric field strength to disappear.

Ions produced in the corona discharge region move within the electric field at a speed proportional to the electric field strength, expressed as:

$$v = \mu E \tag{8.27}$$

where v is the ion drift velocity, and μ is the ion mobility (the drift velocity in an electric field strength of 100 V/m).

The mobility of positive and negative ions in the electric field is not equal. Therefore, the mobility of negative ions is 1.87×10^{-2} m^2/(V·s), while that of positive ions is 1.35×10^{-2} $m^2/(V \cdot s)$. This means that the drift velocities of positive and negative ions are:

$$\begin{cases} V_+ = \mu + E \\ V_- = \mu - E \end{cases}, \tag{8.28}$$

The ionization occurring in gases always produces both positive and negative ions. When positive and negative ions meet, they recombine to form neutral molecules. Under the influence of the external electric field, ions migrate to the electrodes, neutralizing the opposite charges present there.

When the ionization system reaches equilibrium, the number of generated ions is determined by the following equation:

$$n = \mu n_0^2 \tag{8.29}$$

where μ is the recombination coefficient and n_0 is the number of positive (or negative) ions.

Thus, the number of like-sign ions per unit volume is:

$$n_0 = \left(\frac{n}{\varphi}\right)^{1/2} \tag{8.30}$$

When ionizing air in an actual electric field, negative corona discharge is used. The two discharge poles consist of a cylindrical electrode and a coaxial rod-shaped core electrode. The minimum voltage U required to generate corona discharge is given by:

$$U = 31\delta\left(1 + \frac{0.308}{\sqrt{r\delta}}\right) r \ln(R/r) \tag{8.31}$$

where R is the outer cylindrical electrode radius, r is the rod electrode radius, δ is 0.392P/(273 + t), P is the gas pressure, and t is the gas temperature.

8.2.7 Bioelectricity

The potential difference generated within a living organism due to its own energy is referred to as bioelectricity. Generally, bioelectricity can be divided into three categories: (1) damage potential, which refers to the potential difference between intact and damaged areas of biological tissue; (2) membrane potential, which refers to the potential difference inside and outside the biological tissue or cell membrane; and (3) action potential, which refers to the potential changes generated when an organism is stimulated and excited.

The phenomenon of bioelectricity is a fundamental attribute of life activities. Bioelectricity is produced during all life processes within the organism. The so-called bioelectric phenomenon refers to the potential changes and current conduction generated within the organism and its relationship with life phenomena and functions. In the study of the electrical properties of food, the active electrical properties of food are generally referred to as bioelectricity. Because bioelectricity is related to life and is often relatively weak, it is rarely used in food processing, but mainly in food preservation and detection. For example, during seed germination, there exists a potential difference between the embryo region and other parts. This germination current is an important indicator for detecting germination potential. After germination, the cotyledons are positively charged, while the roots are negatively charged. The more active the cell division and the more vigorous the growth in a given area, the higher the potential. Based on these potential changes, the germination status of food materials can be determined.

Since the life activities of organisms are accompanied by the generation of bioelectricity, it is possible to influence the life activities of organisms through external electromagnetic fields to achieve the purpose of modifying food texture and preservation.

8.2.8 Electromagnetic Radiation Characteristics

8.2.8.1 Electromagnetic Spectrum

Not only radio waves are electromagnetic waves; light, X-rays, and γ-rays are also electromagnetic waves. The only difference between them lies in their frequency or wavelength. The frequency of light waves is much higher than that of radio waves, and the wavelength of light waves is much shorter than that of radio waves; X-rays and γ-rays have even higher frequencies and shorter wavelengths. To gain a comprehensive understanding of various electromagnetic waves, these waves are arranged according to their wavelengths or frequencies, forming the electromagnetic spectrum.

Due to the sharp decline in radiation intensity with decreasing frequency, low-frequency electromagnetic waves with wavelengths around 10^5 m have very weak intensity and are usually unnoticed by people. In practice, radio waves used begin at wavelengths of about several kilometers (with frequencies of hundreds of kilohertz). Waves with wavelengths of 3000–50 m (frequencies of 100 kHz–6 MHz) are classified as medium waves, those with wavelengths of 50–10 m (frequencies of 6–30 MHz) are classified as short waves, and those with wavelengths of 10 m–1 cm (frequencies of 30–30,000 MHz) or even below 1 mm (frequencies of 3×10^5 MHz) are classified as ultrashort waves (or microwaves). Sometimes, based on the order of magnitude of the wavelengths, terms like meter waves, decimeter waves, centimeter waves, and millimeter waves also appear. Medium and short waves are used for radio broadcasting and communication, while microwaves are used for television and radio positioning technology (radar).

The wavelength range of visible light is very narrow, ranging from 7600 to 4000 Å (where Ångströms, Å, is commonly used as a unit of length in spectroscopy, with 1 Å = 10^{-8} cm). Extending from visible light in both directions, wavelengths longer than visible light are referred to as infrared rays, with wavelengths approximately from 7600 Å up to several tenths of a millimeter. The thermal effect of infrared rays is particularly significant; wavelengths shorter than visible light are called ultraviolet rays, with wavelengths ranging from 50 to 4000 Å, which have notable chemical effects and fluorescence effects. Both infrared and ultraviolet rays are invisible to the human eye and can only be detected using special instruments. Whether visible light, infrared, or ultraviolet, they are all excited by microscopic objects such as atoms or molecules. In recent years, on one hand, due to the development of ultrashort wave radio technology, the range of radio waves has continuously advanced toward shorter wavelengths; on the other hand, due to the development of infrared technology, the range of infrared rays has continuously expanded toward longer wavelengths. Currently, the boundary between ultrashort waves and infrared rays no longer exists, and their ranges overlap.

X-rays, which are emitted by inner-layer electrons in atoms, have a wavelength range of approximately 10^2 to 10^{-2} Å. With the advancement of X-ray technology, its wavelength range has also continuously expanded in both directions. Currently, in the long wavelength band, it overlaps with ultraviolet rays, while in the short wavelength band, it has entered the domain of γ-rays. The wavelength of radioactive γ-rays ranges from about 1 Å to infinitely short wavelengths.

The various bands mentioned above in the electromagnetic spectrum (see Fig. 8.11) are primarily categorized based on the different methods used to generate and detect them. With the advancement of science and technology, each band has broken through boundaries and overlaps with adjacent bands. Currently, aside from the extremely short wavelengths (below 10^{-4} to 10^{-5} Å), there are no longer any unknown gaps in the electromagnetic spectrum.

8.2.8.2 Ionizing Radiation

Ionizing radiation refers to all types of radiation that can cause ionization in matter. There are many kinds, including high-speed charged particles such as α particles, β particles, and protons, as well as uncharged particles like neutrons, X-rays, and γ-rays.

α rays are a type of charged particle flow; due to their charge, they easily cause ionization wherever they travel. α rays have strong ionizing capabilities, which can be both utilized and harmful, particularly in terms of their ability to damage tissues within the human body. Because they have a relatively large mass, their penetrating ability is poor, with a range of only a few centimeters in air, easily blocked by a sheet of paper or healthy skin.

β rays are also a type of high-speed charged particle flow, but their ionizing ability is much less than that of α rays, while their penetrating ability is greater. However, compared to X-rays and γ-rays, β rays have a shorter range and can be easily absorbed by materials such as aluminum foil or acrylic glass.

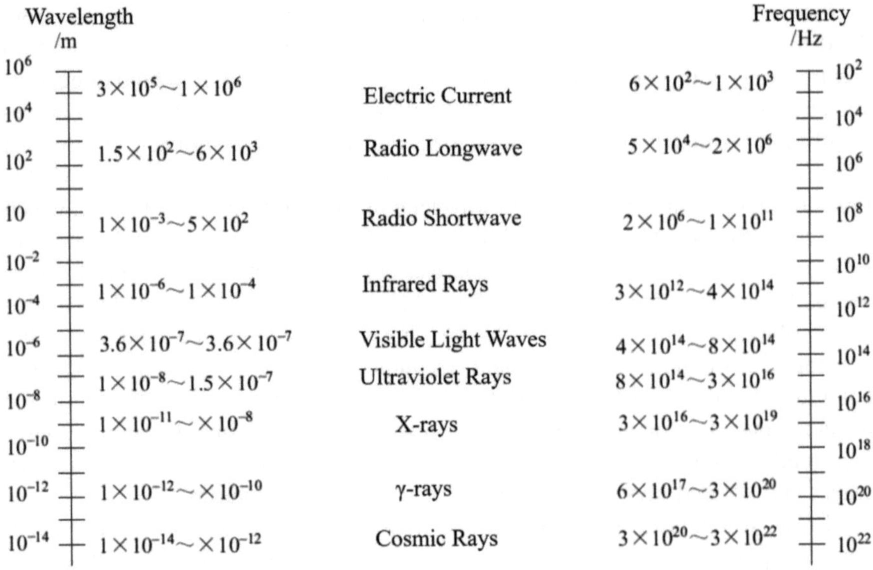

Fig. 8.11 Electromagnetic spectrum

X-rays and γ-rays have roughly similar properties; they are uncharged, short-wavelength electromagnetic waves. Both have extremely strong penetrating power, requiring special attention to protection against accidental exposure. The absorption effects of various ionizing radiations are shown in Table 8.3.

Radiation doses from radiation rays on agricultural products and foods vary, resulting in different effects. The generation of free radicals and molecular damage is the material basis of ionizing radiation.

1. Biological Effects: The damage to biological molecules is the material basis for all radiation biological effects, and the damage to biological molecules is closely related to the generation of free radicals. Biological effects include sterilization and insecticidal effects; abnormal growth and development of fruit trees; inhibition of rooting and sprouting in potatoes, onions, garlic, sweet potatoes, etc.; prevention of mushrooms from opening; delayed ripening of bananas and tomatoes; and promotion of ripening in peaches and persimmons.
2. Chemical Effects: The chemical effects of radiation are also based on the generation and transformation of free radicals. These chemical effects include increasing the rehydration performance of dried foods, improving the performance of wheat flour in bread processing, enhancing the quality of wine, causing denaturation of proteins and starches, and increasing the decomposition ability of various enzymes in fermented feed.

The process of ionizing radiation acting on biological organisms and foods can be described as Fig. 8.12

Table 8.3 Absorption effects of various ionizing radiations

Radiation type and wavelength	Effects of radiation on materials
Long waves (radio waves and microwaves)	Water vapor and gases are excited into rotational states; if the field strength is strong, liquids and solids undergo dielectric and resistive heating
Far-infrared	Water vapor is excited into rotational states, liquids experience hydrogen bond vibration, and solids undergo lattice vibrations
Infrared (2.5×10^{-6}–2.0×10^{-4} m)	At shorter wavelengths, water vapor and gases are excited into fundamental vibrational-rotational states (not pure rotational), and liquids and solids undergo fundamental molecular vibrations
Infrared (1.0×10^{-6}–2.0×10^{-4} m)	Absorption weakens
Near-infrared (1.0×10^{-7}–1.0×10^{-6} m)	Neutrons have enough energy to excite electronic oscillations, break chemical bonds in less stable organic molecules (e.g., changes in dyes, leading to chemical reactions), cause fluorescence, and eject electrons from surfaces through photoelectric emission
Visible light (3.8×10^{-7}–7.6×10^{-6} m)	Activates the visible colors in the human eye's retina and reduces carbon dioxide in plant chloroplasts (in this region, water vapor is nearly transparent, allowing visible light radiation to penetrate Earth's atmosphere)
Ultraviolet	If neutrons are absorbed, it can have lethal effects
Below 1.0–10.0×10^{-8} m (including X-rays and gamma rays)	If neutrons are absorbed, ionization occurs

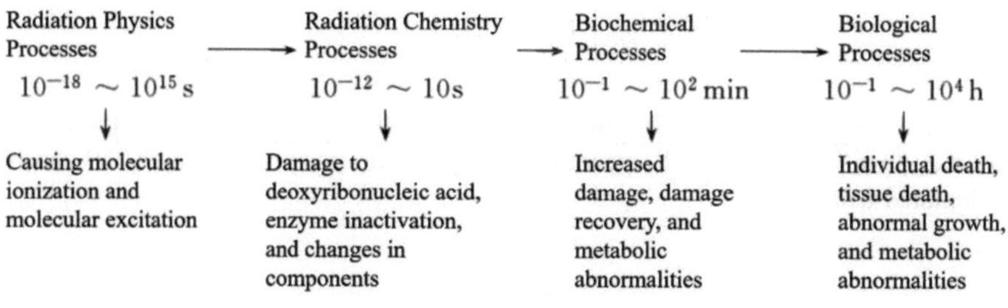

Fig. 8.12 Process of ionizing radiation acting on biological organisms and foods

8.2.9 *Microwave Characteristics*

Food materials are composed of polar and nonpolar molecules. Under the influence of a high-frequency microwave electromagnetic field, polar molecules transition from their original thermal motion state to an arrangement that varies according to the direction of the electromagnetic field, generating frictional heat. In this microscopic process, the energy of the alternating electromagnetic field is converted into thermal energy within the medium, resulting in a macroscopic increase in temperature. Thus, microwave heating occurs as the material itself dissipates the energy of the electromagnetic field as heat. For metallic materials, the electromagnetic field cannot penetrate the interior and is instead reflected, so metallic materials cannot absorb microwaves. Water is the best medium for absorbing microwaves; therefore, any substance containing water will inevitably absorb microwaves. Some materials composed of nonpolar molecules can also absorb microwaves to varying degrees. Glass and ceramics transmit microwaves with very little absorption. Hence, glass and ceramic containers can be used in household microwaves to defrost, heat, and cook moist foods. In food processing plants, microwaves are sometimes used to dry or cook food. High-energy microwaves are commonly generated by vacuum magnetrons. Low-energy microwaves can be generated by solid devices or matrix vacuum tubes. The power absorbed by materials in the microwave field can be expressed by Eq. (8.15).

Microwaves can penetrate the heated object, but attenuation occurs due to absorption. If the incident wave power absorbed by the food is P_0, then the power P at a distance d from the food surface is given by:

$$p = p_0 e^{-2\alpha d} \tag{8.32}$$

where α is the attenuation coefficient.

Decareau and Pestersan provided a calculation formula for the attenuation coefficient α, which is a function of the material's dielectric properties and the free space wavelength λ_0:

$$\alpha = \frac{2\pi}{\lambda_0}\left[\frac{\epsilon'}{2}\left(1 + \frac{\varepsilon''^2}{\varepsilon'^2}\right)^{0.5} - 1\right]^{0.5} \tag{8.33}$$

When P decreases to P_0 $1/e$, the distance from the incoming wave to the heated food surface is defined as the penetration depth. The penetration depth DP can be calculated using α:

$$D_P = \frac{1}{2\alpha} \tag{8.34}$$

8.2.10 Electromagnetism and Nuclear Magnetic Resonance

An electric field is a special substance that exists in the space around charges and changing magnetic fields. Although it is not composed of molecules or atoms, it possesses objective properties such as force and energy typical of ordinary matter. The fundamental property of an electric field is its strong action on charges placed within it, referred to as electric field force. The magnitude and direction of the electric field force depend on the charge amount and type (positive or negative) of the charge placed in the electric field, as well as the position of the charge within the field. Electric fields are mainly classified into two types: one is the static electric field, which is produced by stationary charges in the surrounding space; the static electric field is an active, irrotational field, with charges as the source. The other is the induced electric field (or vortex electric field), which is generated by a time-varying magnetic field in the surrounding space. The induced electric field is a passive, rotational field with closed electric field lines that have no starting or ending points. As shown in Fig. 8.13, the electric fields formed by two charges differ based on their charge signs and magnitudes. The electric field strength (or field intensity) at a certain point in the electric field describes the strength and direction of the electric field at that point. Electric field strength is a vector whose direction is that of the electric field force acting on a positive charge at that point. The definition of electric field strength is as follows:

$$E = \frac{F}{q} \tag{8.35}$$

where E is the electric field strength and F is the electric field force on charge q at that point.

A magnetic field is composed of moving tiny particles (mainly electrons and protons), but under existing conditions, it is invisible and intangible, representing a special form of matter. The magnetic field is not composed of atoms or molecules, yet it objectively exists and possesses particle radiation characteristics. The strength of a magnetic field is determined by the magnetic flux density within the field, measured in teslas (Tesla). The stronger the magnetic field, the denser the magnetic lines. Figure 8.14 shows the schematic of magnetic fields formed by different types

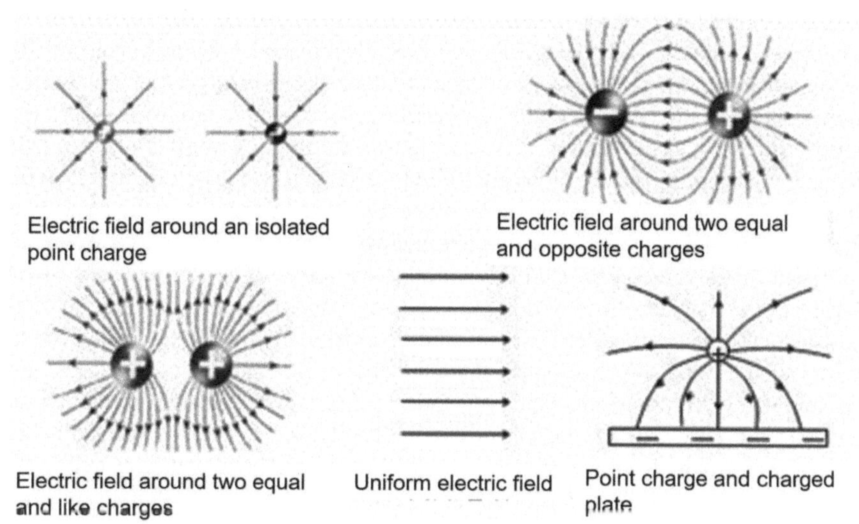

Fig. 8.13 Schematic diagram of different types of electric fields

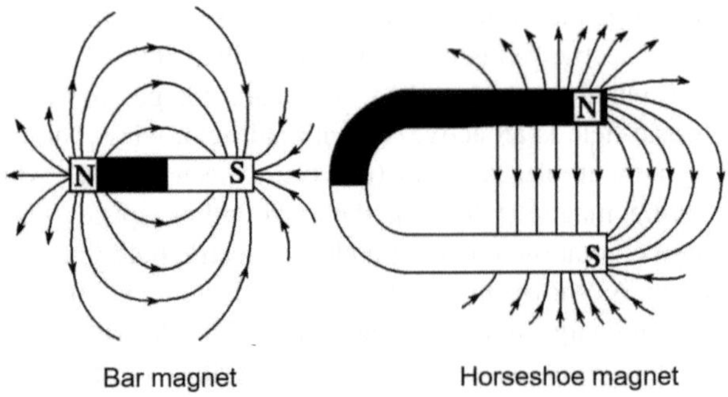

Fig. 8.14 Schematic diagram of different types of magnetic fields

of magnets. Electromagnetic interaction is one of the four fundamental interactions in nature, describing the interaction between charges and magnetic fields. According to Maxwell's electromagnetic field theory, a changing electric field produces a magnetic field, and conversely, a changing magnetic field produces an electric field. This alternating changing electric and magnetic field constitutes the electromagnetic field.

8.2.10.1 Principle of Electromagnetic Heating

When current passes through a conductor, a magnetic field is generated around the conductor, which interacts with the free electrons (or other charge carriers) inside the conductor, leading to changes in the motion state of the electrons. According to the Lorentz force law, when the free electrons inside the conductor are acted upon by the magnetic field, they will exhibit directional movement and acceleration. These accelerated electrons move at high speeds and, in a chaotic manner within the conductor, colliding with the atoms or other electrons of the conductor. During these collisions, part of the kinetic energy of the electrons is converted into thermal energy. This occurs because collisions between electrons and atoms or other electrons result in energy exchange, releasing some kinetic energy as heat. High-frequency current (typically between 20 and 40 kHz) produces a rapidly changing alternating magnetic field through induction coils. When conductors (such as copper or aluminum) are placed within this alternating magnetic field, eddy currents are generated inside the conductor. Eddy currents are loops of current induced by the change in the magnetic field within the conductor, and these currents flow through the conductor, generating heat. The magnitude of the eddy currents is influenced by factors such as the geometric shape of the conductor, its electrical conductivity, and the frequency of the magnetic field. Since low-frequency magnetic fields can penetrate the metal surface, eddy currents primarily occur near the surface of the metal. To reduce eddy current generation and slow down the temperature rise inside the conductor, methods such as outer insulation or longitudinal slicing are often employed for support or isolation. Additionally, magnetic materials exhibit hysteresis in alternating magnetic fields, meaning that magnetic domains within the material require energy to reorient to a new magnetization direction when the magnetic field changes. The energy consumed in this process is released in the form of heat, generating additional thermal energy. Hysteresis loss is related to the energy consumption of the magnetic domain walls and the rate of movement of these walls; as the energy consumption or movement rate increases, hysteresis loss also correspondingly increases. The eddy currents generated within the heated object by the alternating magnetic field produced by high-frequency current result in the heating effect. Since electromagnetic heating directly acts on the interior of the heated object, it boasts advantages such as high thermal conversion efficiency and rapid heating speed.

8.2.10.2 Principle of Nuclear Magnetic Resonance Detection

Nuclear magnetic resonance (NMR) refers to the process in which nonzero magnetic moment atomic nuclei experience Zeeman splitting and absorb electromagnetic waves of specific frequencies in a constant magnetic field and an alternating magnetic field. Each atomic nucleus with a nonzero magnetic moment continuously spins around its axis, creating a microscopic spin magnetic field. The disordered state of protons results in the overall zero magnetism of the material. When the material is placed in a constant magnetic field B0, all protons align directionally, simultaneously spinning while performing circular motion around the external magnetic field. This phenomenon is known as Larmor precession, and the frequency of rotation of protons around the external magnetic field is referred to as the Larmor precession frequency (angular frequency). The precession frequency is influenced by the inherent properties of the material and the strength of the external magnetic field, expressed as:

$$w = r\,\mathrm{B0} \tag{8.36}$$

where w is the precession frequency, B0 is the strength of the external magnetic field, and r is the gyromagnetic ratio.

In the external magnetic field B0, protons exist in two states: one with the magnetic field's north and south poles aligned parallel to B0 (low energy state) and the other aligned antiparallel to B0 (high energy state). When protons reach thermal equilibrium, they form a macroscopic magnetization vector M0 aligned with the direction of B0. For convenience, the direction of B0 is generally set as the z-axis, with M0 also aligned along this axis. The rotation of protons is rapid and difficult to detect. By applying a 90° radio frequency magnetic field in the z-axis direction, M0 rotates from the z-axis to the y-axis, with phase coherence. When the radio frequency pulse magnetic field is turned off, the protons in the high-energy state become unstable and release energy in a non-radiative form, returning to the thermal equilibrium state or losing phase coherence, which constitutes the relaxation process. The relaxation process consists of two parts: longitudinal relaxation (spin-lattice relaxation) and transverse relaxation (spin-spin relaxation). The lattice refers to the external environment or particles surrounding the protons; lattice motion generates a magnetic field. When the frequency of the lattice magnetic field matches that of the processing proton magnetic field, energy is transferred from the protons to the lattice, causing the processing protons to return to a steady state. This energy exchange between the processing protons and their surroundings, releasing absorbed electromagnetic energy and returning to thermal equilibrium, is known as longitudinal relaxation. Longitudinal relaxation occurs when low-energy protons absorb energy to transition to a high-energy state and then release energy from the high-energy state back to the steady state, represented by the transition of Mz from 0 to M0. The process of Mxy decaying from its maximum value M0 to 0 represents transverse relaxation, indicating the interaction between adjacent spin protons and the transfer of energy, gradually losing phase coherence without energy gain or loss.

The time for the longitudinal relaxation process is represented as T1, while the time for the transverse relaxation process is denoted as T2. T1 and T2 are measured using nuclear magnetic pulse sequences. The main testing sequences include free induction decay (FID), inversion recovery (IR), and Carr-Purcell-Meiboom-Gill (CPMG) sequences. The FID process, as shown in Fig. 8.15, is often used to set system parameters for equipment due to its simplicity. The detection of T1 generally employs the inversion recovery (IR) sequence. Since it is challenging to detect magnetization along the z-axis, a 180° pulse is first applied to convert M0 in the z-axis to −M0. After a relaxation time of $\tau 0$, a 90° pulse is applied to begin collecting decay signals. After a collection time of TR (ensuring complete relaxation), a 180° pulse is applied again, and after a relaxation time of $\tau 1$ (where $\tau 1 > \tau 0$), another 90° pulse is applied to collect the decay signal. This process is repeated until the z-axis signal returns to M0. T2 describes the decay of Mxy; theoretically, the FID sequence can be used for detection. However, it is found that the T2 relaxation time is significantly shorter than the theoretical value because the spin relaxation is influenced by magnetic field inhomogeneity and diffusion, leading to accelerated decay. Therefore, the CPMG sequence resolves the issue of amplitude error in relaxation echoes caused by magnetic field inhomogeneity

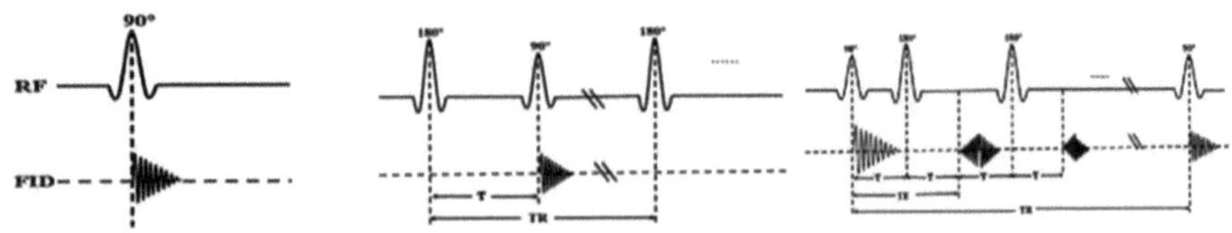

Fig. 8.15 Free induction decay signal

8.3 Application of Electromagnetic Properties in Food Analysis and Detection

8.3.1 Application of Conductive Properties in Detection

The electrical properties of food that can be used for detection mainly include resistive and capacitive characteristics. The principle of measuring alternating current (AC) is shown in Fig. 8.16. First, an alternating signal is input into the AC terminal by a signal generator, and the frequency of the input signal is measured with a frequency counter. Then, a millivoltmeter is used to simultaneously measure the input voltage U_{AC} at the input terminal and the voltage drop U_{BC} across the standard resistance R_{H}. A phase meter is used to measure the phase difference θ between U_{AB} and U_{BC}. From electrical and vector knowledge, we have:

$$\vec{U}_{\mathrm{AC}} = \vec{U}_{\mathrm{AB}} + \vec{U}_{\mathrm{BC}}, \vec{I}_H = \vec{I}_R + \vec{I}_C \tag{8.37}$$

$$R = \frac{R_H\sqrt{U^2{}_{\mathrm{AC}} + U^2{}_{\mathrm{BC}} - 2U_{\mathrm{AC}}U_{\mathrm{BC}}\cos\theta}}{U_{\mathrm{BC}}\cos(\theta+\delta)} \tag{8.38}$$

$$C = \frac{U_{\mathrm{BC}}\sin(\theta+\delta)}{2\pi f R_H\sqrt{U^2{}_{AC} + U^2{}_{\mathrm{BC}} - 2U_{\mathrm{AC}}U_{\mathrm{BC}}\cos\theta}} \tag{8.39}$$

$$\delta = \arcsin\frac{U_{\mathrm{BC}}\sin\theta}{\sqrt{U^2{}_{\mathrm{AC}} + U^2{}_{\mathrm{BC}} - 2U_{\mathrm{AC}}U_{\mathrm{BC}}\cos\theta}} \tag{8.40}$$

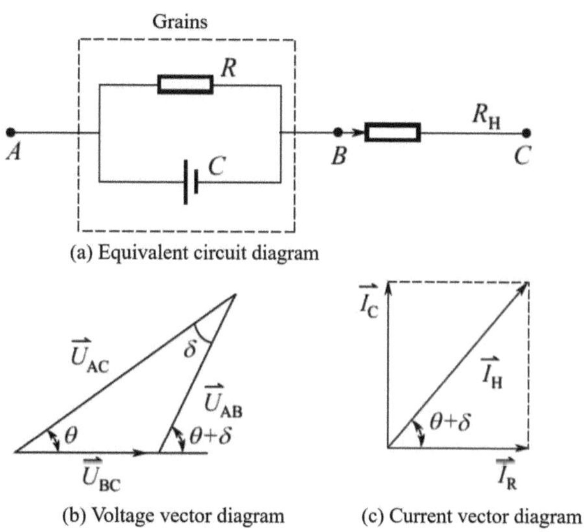

Fig. 8.16 Schematic diagram of AC measurement principles

where R is the resistance of the grain in ohms (Ω), C is the capacitance of the grain in farads (F), δ is the dielectric loss angle of the grain, and f is the testing frequency in hertz (Hz). After conversion, the above equations become:

$$\rho = R\frac{S}{L} \tag{8.41}$$

$$\varepsilon = \frac{CL}{\varepsilon_0 S} \tag{8.42}$$

$$\tan\delta = \frac{1}{2\pi\varepsilon_0 f \varepsilon\rho} \tag{8.43}$$

where ρ is the resistivity in ohms (Ω), ε is the dielectric constant in farads per meter (F/m), tanδ is the tangent of the dielectric loss angle, S is the cross-sectional area of the measuring container in square meters (m^2), and L is the distance between the electrodes in meters (m).

The resistivity of grains is primarily related to their moisture content and testing frequency. The relationship between the resistivity of grains and their moisture content is as follows:

$$\ln\rho = -a_\omega + C_\omega \tag{8.44}$$

where a_ω and C_ω are experimental constants. From this equation, it can be observed that the resistance of grains decreases exponentially with increasing moisture content. This is because grains, as living organisms, have protoplasm in a gel state when the moisture content is very low, causing slow ion movement within the cells, resulting in extremely weak physiological activity. At this point, the cell resistance is high, and the macro resistivity of the grains is also high. As the moisture content of the grains increases, the cell water potential rises sharply. Under the influence of water, the metabolism of grains accelerates, including enzyme activation and resynthesis, leading to a rapid increase in ion movement within the cells, thereby reducing cell resistance and consequently causing the macro resistivity of the grains to decrease rapidly.

Conductivity plays a fundamental role in ohmic heating, where electricity is transformed into thermal energy as an alternating current (a.c.) flows through food. Understanding the effective conductivity or overall resistance of liquid-particle mixtures is crucial due to its potential use in fluid pasteurization. Additionally, liquid-particle mixtures can be pasteurized using pulsed electric field technology, where products with low conductivity are processed more effectively and energy-efficiently. Therefore, conductivity can be used for acidity studies and for monitoring the process of increasing acidity, such as during fermentation. The crystallization process (e.g., in sugar solutions) can also be monitored through conductivity measurements, as conductivity is inversely related to viscosity, which is closely related to supersaturation.

8.3.2 Application of Dielectric Properties in Detection

Understanding the relationship between frequency and dielectric properties is essential for identifying the optimal frequency range needed to achieve desired dielectric characteristics for specific applications. This knowledge supports the development of heating processes and classification techniques using electromagnetic energy. It also aids in selecting appropriate packaging materials, cookware, and designing microwave and radio frequency heating equipment.

Moisture-dependent dielectric properties can inform the creation of online moisture meters for drying processes and other food industry operations. Additionally, these properties help analyze the behavior of food during radio frequency and microwave processing, affecting the interaction between food and high-frequency electromagnetic energy. As such, they play a crucial role in designing processing equipment and microwave-prepared foods, impacting energy penetration depth and the physical properties of biological materials.

Research into selective dielectric heating for controlling stored grain pests has also been conducted [7]. Nelson emphasized the importance of dielectric performance data for improving seed germination rates, while Engelder and Buffler [36] noted that dielectric measurements could determine density and moisture activity. Several applications of these principles are discussed in the literature.

8.3.2.1 Water Quality Detection

Microwave technology has been employed to study and assess water quality. This technology can also be used to evaluate oil-water mixtures in food applications.

8.3.2.2 Dielectric Behavior of Foods and Food Materials

The dielectric constant or dielectric properties of food materials can be correlated with the mass of the substance; for example, high or low frequencies can be used to measure the mass of fish and meat [37].

8.3.2.3 Grains

In agriculture, dielectric properties of grains are essential for determining moisture content, which is affected by factors such as moisture levels, detection frequency, grain type, and bulk density. Figure 8.17 illustrates the relationship between the dielectric loss tangent of wheat, moisture content, and frequency, expressed by the equation:

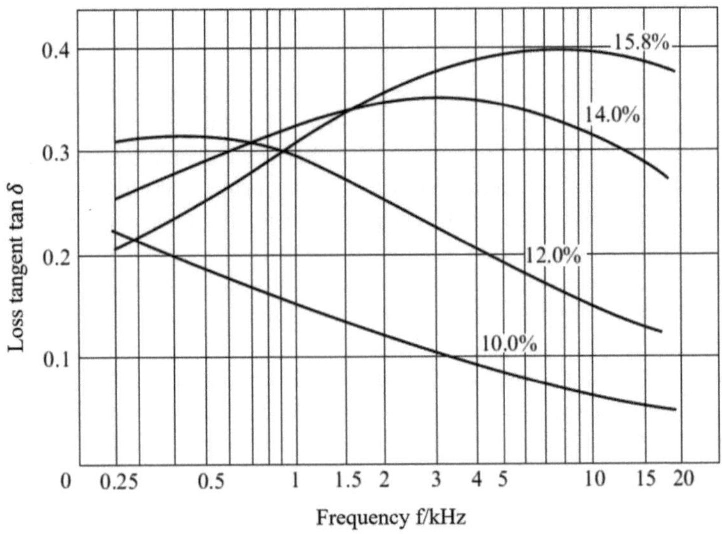

Fig. 8.17 Relationship between the loss tangent of wheat at 24 °C and moisture content and frequency

$$\varepsilon = 1 + \exp(a_\varepsilon \omega + b_\varepsilon) \tag{8.45}$$

where a_ε and b_ε are experimental constants.

These dielectric properties are vital for designing applications in dielectric heating at radio and microwave frequencies, with potential uses in grain drying, seed treatment for improved germination, and pest control in stored grains [38]. Holaday (1964) [39] utilized DC resistance and capacitance measurements to assess moisture distribution in corn kernels, successfully detecting heat damage in artificially dried corn. Dexter (1965) [40] differentiated live and dead corn seeds by measuring current between electrodes connected to a 6 V DC power source on soaked kernels.

Quantitative data on dielectric properties are crucial for predicting material behavior under high-frequency fields during dielectric heating. For example, the dielectric constant and loss factor of baking dough at 27 MHz are significant for the baking industry using RF and microwave frequencies [41]. Many grains and seeds exhibit dispersion in the audio frequency range, particularly at high moisture levels, where the dielectric constant decreases rapidly with rising frequency, and loss factor curves show peaks. At higher frequencies, wheat with 10.6% moisture demonstrates a dielectric constant drop from nearly 8 at 250 Hz to about 2.5 at 12 GHz, with dispersion zones and a loss peak between 10 and 100 MHz [42].

Dielectric constant serves as a key indicator for monitoring frying oil quality in commercial applications, as well as in oil processing, storage, and food production [2]. Dielectric spectroscopy can even detect olive oil adulterated with less than 5% of other oils [43].

8.3.2.4 Fruits and Vegetables

Microwave dielectric properties of fresh fruits and vegetables are valuable for assessing non-destructive quality factors like peach ripeness and cold damage in sweet potatoes. Vegetables have high absolute dielectric constants due to their moisture content; for example, cooked peas and mashed potatoes range from 54 to 65, while cooked carrots range from 61 to 76 at 3 to 6 °C and a frequency of 2.8 GHz [19]. Dried vegetables exhibit lower dielectric constants due to reduced moisture.

The dielectric properties of mangoes were measured using open coaxial probes at frequencies (1–1800 MHz) and temperatures (20–60 °C), with values decreasing at higher frequencies, especially in the loss factor. This method helps identify the optimal ripeness stage for dielectric heating during postharvest pest control. Similarly, the dielectric properties and sweetness of freshly harvested melons were analyzed from 10 MHz to 1.8 GHz [44]. For apples, dielectric performance varies with ripeness from 300 to 900 MHz, showing significant declines with aging [45]. Over a 10-week refrigeration period at 4 °C, the dielectric performance of apples remained relatively constant [46].

Dielectric properties of four apple varieties measured from 30 to 3000 MHz showed no significant variation based on variety, flesh cross-section, or ripeness. Below 2450 MHz, insects have much higher dielectric constants than apples, suggesting cod larvae in apple hosts may not survive adverse thermal conditions at higher frequencies [47]. The temperature dependence of dielectric characteristics in apple juice closely resembles that of pure water, with relaxation frequencies shifting to higher values as temperature increases.

8.3.2.5 Livestock and Fishery Products

Dielectric technology provides a simple, rapid, and accurate method for assessing quality parameters in fish, seafood, and meat, with advantages for evaluating meat freshness, such as being non-destructive, quick, reliable, and practical. Bengtsson et al. (1963) [48] conducted extensive measurements of meat and fish dielectric properties within the 10 to 200 MHz range, noting a sharp rise in dielectric constant after thawing frozen materials. At lower frequencies, the loss tangent increased significantly upon thawing, though there was little temperature impact at 100 and 200 MHz. For chopped samples, dielectric properties fell between whole meat samples with fibers parallel and perpendicular to the electric field. Van Dyke et al. (1969) [49] measured reconstituted ground beef at 915 MHz, observing minimal effects on the loss factor with moisture content below 20%, while levels between 20% and 45% caused a significant increase. The loss factor stabilized at 70% moisture, increased with ash, and decreased with fat. Evaluating meat quality via dielectric properties shows significant potential [50]. Bengtsson and Risman (1971) [19] found no major differences in dielectric properties between ground and whole meat, while To et al. (1974) [32] found turkey's loss factor at 300, 915, and 2450 MHz higher than beef's, mainly due to moisture. The dielectric constant and loss factor of cooked beef were about 60% of raw beef's values [28]. Mudgett (1985) [51] reported that dielectric properties of meats and vegetables vary greatly above and below freezing. Fresh eggs showed dielectric constants and loss factors for egg whites higher than yolks across the 10–1800 MHz range, with storage-related changes not aligning closely with traditional quality indicators [52].

8.3.2.6 Dairy Products

Dielectric properties of dairy products are relatively understudied. Nunes et al. (2006) [53] examined the complex dielectric constant of milk at room temperature (17–20 °C) over a frequency range of 1–20 GHz, suggesting this could help estimate ion compounds, fat, carbohydrates, and protein content in milk. Dielectric measurement techniques have been widely applied in analyzing cheese components and maturity [54]. Dielectric methods are also used to detect mastitis in milk [1]. Butter's dielectric properties show little seasonal variation, while at 2.45 GHz, butter's dielectric constant and loss factor are temperature-dependent [55]. In the 0.3–3 GHz range, cheese's dielectric constant decreases gradually with frequency, and for cheeses with high and medium moisture/fat ratios, the loss factor generally increases with temperature [54]. Salted and unsalted butter exhibit significant differences in dielectric spectra due to temperature and moisture content [17]. Accurate prediction of milk's dielectric loss factor based on ash content requires adjustments for bound and unbound interactions of milk salts [56].

8.3.2.7 Stored Grain Pests

Dielectric Properties of Storage Pests

Research on controlling stored grain pests through selective dielectric heating measured the dielectric properties of the rice weevil and confused flour beetles [57]. Results indicated that effective insect control requires differential heating, particularly in the 5 to 100 MHz range for rice weevils [42]. The dielectric constant for rice weevils was estimated at about 36 at 10.2 GHz and 24 °C. Hamid et al. (1968) [58] reported dielectric constants for confused flour beetles, while Shackelford (1970) [59] measured the dielectric constant for the exoskeletons of tomato hornworm moths.

Thermal Mortality Kinetics Model of Pests

Pest mortality rates are crucial for assessing heat treatment effectiveness. The survival rate S is calculated as:

$$S = \frac{N}{N_0} \times 100 \tag{8.46}$$

where N_0 is the number of pests before heat treatment and N is the number of surviving pests after heat treatment.

A basic kinetics model estimates the thermal mortality time of various pests, such as codling moth, navel orange worm [60], Indian meal moth [61], and rice weevil [62]. The relationship between survival rate S and heating time at a constant temperature is described by the equation:

$$\frac{d(N/N_0)}{d_t} = -k(N/N_0)^n \tag{8.47}$$

where t represents the heating time (minutes) at a specific temperature, k is the thermal mortality rate constant (1/min), and n is the reaction order of the kinetics model. Equation (8.47) can be transformed into the following:

$$\begin{aligned} \ln(N/N_0) &= -kt + c \quad (n = 1) \\ (N/N_0)^{1-n} &= -kt + c \quad (n \neq 1) \end{aligned} \tag{8.48}$$

Once data on pest survival rates and heat retention times are collected, linear regression using Eq. (8.48) with varying n values (0, 0.5, 1, 1.5, 2) can determine the optimal reaction order based on the highest average R^2 across temperatures. This allows calculation of constants k and c and prediction of times needed for 95%, 99%, 99%, 99%, and 68% pest mortality at different temperatures.

Plotting temperature against heat retention time for 100% mortality yields a thermal death time (TDT) curve, where the negative slope indicates the z-value, reflecting pest sensitivity to temperature. Higher z-values signify greater heat resistance. The activation energy (Ea, J/mol) required for pest mortality can be calculated from reaction rates at two temperatures using the Arrhenius equation:

$$E_a = \frac{2.303RT_{\min}T_{\max}}{z} \tag{8.49}$$

where R is the gas constant (8.314 J/mol/K); and $T_{\min}$ and $T_{\max}$ are the minimum and maximum absolute temperatures (K) within the tested range.

From Eq. (8.49), we know that the mortality rate of pests is proportional to the rate constant k, which is strongly temperature-dependent. Therefore, the activation energy required for pest thermal death can also be determined by calculating the reaction rates k and k_{ref} at two absolute temperatures T and T_{ref} and substituting them into another form of the Arrhenius equation to obtain E_a:

$$k = k_{\text{ref}} e^{-\frac{E_a}{R}\left(\frac{1}{T} - \frac{1}{T_{\text{ref}}}\right)} \tag{8.50}$$

Table 8.4 Determination coefficients of different kinetic models for the thermal mortality rate of various storage pests

Species	Temperature(°C)	N_0	$n = 0$	$n = 0.5$	$n = 1.0$	$n = 1.5$	$n = 2.0$
C. pusillus	46	150	0.98	0.99	0.92	0.79	0.66
	48	150	0.97	0.92	0.85	0.76	0.68
	50	150	0.99	0.96	0.89	0.80	0.71
	52	150	0.96	0.96	0.83	0.72	0.65
	Average	150	0.97	0.96	0.87	0.87	0.67
R. dominica	48	150	0.98	0.96	0.88	0.77	0.67
	50	150	0.98	0.93	0.83	0.73	0.64
	52	150	0.96	0.91	0.83	0.74	0.66
	54	150	0.97	0.88	0.98	0.97	0.95
	Average	150	0.97	0.92	0.88	0.80	0.73
T. castaneum	46	150	0.98	0.91	0.78	0.65	0.58
	48	150	0.98	0.95	0.82	0.67	0.57
	50	150	0.93	0.87	0.80	0.84	0.65
	52	150	0.96	0.99	0.99	0.72	0.75
	Average	150	0.96	0.93	0.72	0.75	0.64

Note: N_0 represents the number of pests before each experiment

where T represents the absolute temperature (K) and k_{ref} is the thermal mortality rate constant at the reference temperature T_{ref}(K). By plotting the relationship between k and temperature T on a semilogarithmic graph, the activation energy can be calculated from the slope. According to the Arrhenius theory, the activation energy represents the extent to which temperature affects the reaction rate k. A higher activation energy indicates that k increases significantly with rising temperature, meaning that the reaction rate k is more sensitive to temperature changes.

Table 8.4 shows coefficients of determination for different kinetic models applied to *Cryptolestes pusillus*, *Rhyzopertha dominica*, and *Tribolium castaneum*, confirming a zero-order kinetics model as the best fit. Table 8.5 lists constants for this model, highlighting that heat retention time decreases significantly with increasing temperature. For instance, complete mortality of *R. dominica* at temperatures of 48 °C, 50 °C, 52 °C, and 54 °C required 75.2, 28.7, 11.7, and 5.0 minutes, respectively. *R. dominica* exhibits the highest heat resistance, followed by *C. pusillus* and *T. castaneum*.

8.3.3 Applications of Electrophoresis in Separation and Analysis

Impurities, contaminants, and microorganisms in food often carry electric charges. When their solutions or dispersions are placed in an electric field, these charged substances will move. The technique that separates charged particles based on their

Table 8.5 Constants for the zero-order kinetics model and time required for 100% mortality at different temperatures

Species	Temperature(°C)	$(N/N_0)^{1-0} = -kt + c$		Exposure time for 100% mortality
		k	c	
C. pusillus	46	0.01	0.99	153.3
	48	0.02	1.17	58.5
	50	0.10	1.16	11.1
	52	0.74	1.03	1.4
R. dominica	48	0.01	0.98	75.2
	50	0.04	1.03	28.7
	52	0.08	0.98	11.7
	54	0.20	0.99	5.0
T. castaneum	46	0.02	1.18	62.4
	48	0.08	1.03	12.9
	50	0.40	1.03	2.6
	52	1.04	0.98	0.9

differing migration speeds in an electric field is called electrophoresis. Under specific conditions, the distance (i.e., mobility) that charged particles move per unit time under a unit electric field strength is constant, representing a characteristic constant of the physical and chemical properties of the charged particle. Different charged particles, due to differing charges or, even with the same charge, differing mass-to-charge ratios, will migrate and separate in the same electric field over a certain period due to their different travel distances. The separation distance is proportional to the voltage of the applied electric field and the electrophoresis time.

Electrophoresis has been increasingly widely used in various fields, including analytical chemistry, biochemistry, clinical chemistry, toxicology, pharmacology, immunology, microbiology, and food chemistry. In the 1960s and 1970s, with the introduction of media such as filter paper and polyacrylamide gel, electrophoresis technology rapidly developed. The diverse forms of electrophoresis have made its applications very broad. In addition to the separation and analysis of small molecules, electrophoresis is primarily used for the study of proteins, nucleic acids, enzymes, and even viruses and cells. Due to its simple equipment, ease of operation, and characteristics of high resolution and selectivity, electrophoresis has become a commonly used technique in medical and food safety inspections.

8.3.4 *Application of Nuclear Magnetic Resonance (NMR) in Detection*

Nuclear magnetic resonance (NMR) occurs when atomic nuclei's spin angular momentum processes in an external magnetic field. Each type of nucleus has a unique nuclear spin quantum number, with only those having a spin of 1/2, such as

^{1}H, ^{13}C, and ^{19}F, producing detectable NMR signals. When spinning, charged atomic nuclei create a magnetic moment that aligns with their spin. If misaligned with an external magnetic field, this moment processes, akin to a wobbling top, at a frequency determined by the magnetic field strength and the nucleus's properties.

The energy of this precession depends on the magnetic moment, the magnetic field, and the angle between them. Transitions between energy levels occur when nuclei absorb energy from an external radio frequency (RF) field, which matches their precession frequency. This principle underlies NMR spectroscopy, a key method for analyzing molecular structures, particularly for ^{1}H and ^{13}C nuclei.

In isolated nuclei, sensitivity to specific RF frequencies is consistent. However, within molecular structures, variations in electron cloud distribution create different effective magnetic field strengths for different nuclei, leading to chemical shifts in NMR signals. These shifts provide insight into the chemical environment surrounding the nuclei.

NMR also reveals coupling constants, which indicate interactions between neighboring nuclei and result in peak splitting in the spectra. Analyzing these peaks helps determine atomic connectivity in molecules. Signal intensity is crucial as well; peaks correspond to nuclei in similar environments, with peak area reflecting their quantity—particularly important for ^{1}H NMR, though less so for ^{13}C.

Historically, ^{1}H was favored for NMR due to its abundance and strong signals. Advances in Fourier transform techniques now allow simultaneous RF emissions, improving the detection of weaker signals, including those from ^{13}C. Two-dimensional NMR has further enhanced molecular analysis, enabling spatial structure resolution of small proteins.

In food science, NMR is used to analyze components like fats, carbohydrates, and moisture. However, high costs and limited basic research investment in the low-margin food industry have restricted its application. Despite this, large companies are collaborating to advance NMR research, increasing its potential in food analysis. Low-field NMR is particularly useful for monitoring moisture migration in food, as demonstrated by Li et al., who found that storage temperature affects the migration rate of water in passion fruits, linking it to skin structure. Additionally, low-field NMR is valuable in characterizing moisture during food drying processes.

8.4 Application of Electromagnetic Characteristics in Food Processing

8.4.1 Electrostatic Separation

Electrostatic separation relies on the distinct paths materials take under the combined influence of electrostatic forces, centrifugal forces, and gravity within an electric field. While centrifugal force and gravity enhance the separation, the key factor is the electric field force, which depends on the electric field strength and the materials'

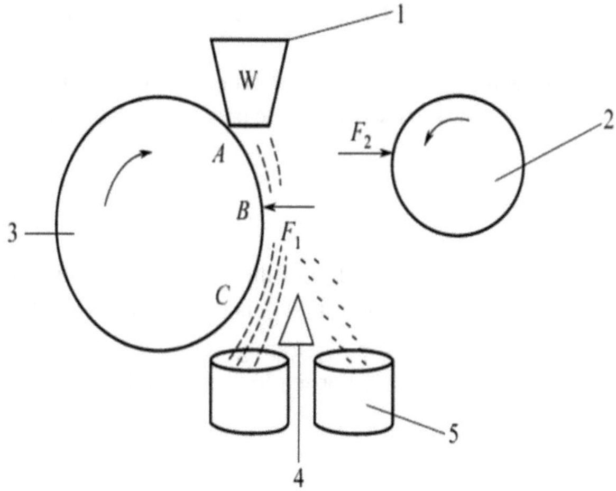

Fig. 8.18 Induced polarized charge. (1) Hopper; (2) negative electrode; (3) positive electrode; (4) separation plate; (5) container

charging properties. These properties are influenced by conductivity, dielectric characteristics, geometry, and chemical composition. Even if materials start with the same charge, differences in their ability to dissipate electrostatic energy can lead to varying charge levels in the electric field, allowing for separation based on these differences.

Electrostatic separation is a type of electro-powered grading, utilizing machines like polarization, dielectric, corona, and friction-type separators. Polarization-type separators use a static electric field, while dielectric separators use an alternating field. Corona-type separators charge materials via corona discharge, followed by static field separation, and friction-type separators combine friction charging with static field separation.

Below is an example of how an electrostatic separator works, using tea leaf separation as an example. On the left of Fig. 8.18 is a large-diameter metal cylinder, which serves as the positively charged electrode and is grounded. On the right is a negatively charged metal cylinder, which has a smaller diameter and thus a greater curvature. The higher curvature results in a stronger electric field, so the electric field strength at the negative electrode is greater than that at the positive electrode, creating a nonuniform electric field.

For example, in tea leaf separation, a positively charged metal cylinder serves as the electrode (Fig. 8.18), while a negatively charged, smaller-diameter cylinder creates a stronger electric field. As the tea mixture moves through the electric field, it becomes charged through induction and polarization. The charged mixture is influenced by the varying electric fields, with tea stems—more conductive and containing more free electrons—experiencing a greater force toward the negative electrode compared to tea leaves.

To quantify the forces acting on the tea stems and leaves, we consider both electrical and mechanical forces:

1. *Coulomb Force*: F1 = qE (where q is the charge on the tea leaves and E is the electric field strength).
2. *Nonuniform Electric Field Force*: The force due to the nonuniform field attracts the polarized tea stems and leaves to areas of greater strength:

$$F_e = \alpha V E \frac{de}{dx} \quad (8.51)$$

where α is the polarizability of the tea leaves, V is the volume of the tea particle, E is the electric field strength, and $\frac{d_e}{d_x}$ is the electric field gradient.

The conditions for separating tea stems and leaves are (1) For tea stems: $F_1 + F_g < F_2 + F_v$; (2) For tea leaves: $F_1 + F_g + F_v > F_2$

8.4.2 Electrostatic Smoking

There are several methods of electrostatic smoking, and the principle is quite simple. As shown in Fig. 8.19a, to maintain stable self-sustaining ionization, an industrial nonuniform electric field is generated using wire electrodes and plate electrodes. A nonuniform electric field, which is independent of the product size, exists between the corona electrode (active electrode) and the positive plate. Near the active electrode, where the electric field strength is greatest, corona discharge occurs, causing the smoke components delivered from below to become ionized. Since the mobility of negative ions is greater than that of positive ions, the corona electrode uses the negative electrode. The ions formed in the corona region are adsorbed by smoke particles, charging the smoke particles. The charged smoke particles move directionally in the electric field and collide with and deposit onto the surface of the meat products.

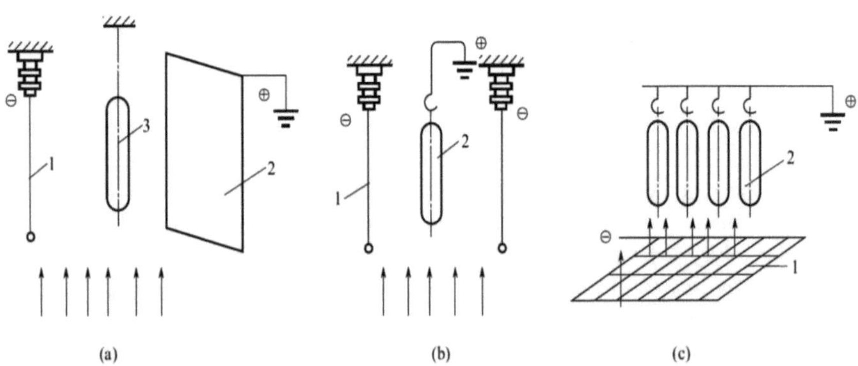

Fig. 8.19 Principles and major methods of electrostatic smoking. (1) Corona electrode; (2) passive electrode; (3) product

In the electrostatic smoking method shown in Fig. 8.19b, the product itself becomes the passive electrode, with the corona electrodes placed on both sides. This setup makes it difficult to ensure a stable nonuniform electric field. Consequently, sharp protruding parts of the product may accumulate excessive smoke substances, forming black and white shells and causing reverse corona discharge.

In Fig. 8.19c, the smoke is first ionized in an ionization grid and then drifts toward the product for deposition. The disadvantage of this method is that products close to the ionization grid may be overly smoked. The main feature of electrostatic smoking is its high efficiency. Under moderate smoke density conditions, the smoking process is very rapid (2–5 minutes). However, the drawback is that it does not provide the usual drying effect of traditional smoking, so additional microwave or far-infrared treatment is required after electrostatic smoking.

8.4.3 Electrostatic Molding and Powder Spraying Devices

Electrostatic fields are primarily used in the processing of bulk food materials, and in addition to separation and smoking, they can also be applied to molding and powder spraying operations. The principles of these operations are similar to those of separation and smoking. To achieve better results in molding and powder spraying, specialized devices are needed. A typical electrostatic molding device is shown in Fig. 8.20. The device consists of a set of hoppers, feeders, raw material meters (for dosing salt, sugar, yeast powder, yeast liquid, etc.), electrostatic sprayers, and metering and electrostatic spraying equipment for vegetable oil. All electrostatic sprayers are connected to the negative terminal of a high-voltage power supply. First, the vegetable oil droplets become charged particles, which are attracted to the surface of a heated drum in the electric field, forming an oil layer. Then, charged particles of flour and other raw material droplets are cross-mixed in the space between the electrodes and sprayed onto the drum surface to form a dough sheet of a certain thickness. The drum serves not only as the positive electrode of the electric field but also as a heating body, using electric heating wires. As the drum rotates, the materials sprayed onto it are shaped, heated, expanded, and dried, and eventually cut into finished products.

Similar devices are also used for powdering operations on the surfaces of fish, meat, and other products, such as breading.

8.4.4 High-Voltage Electrostatic Field Processing

Applying a high-voltage electrostatic field to fruits and vegetables can enhance preservation through three main mechanisms:

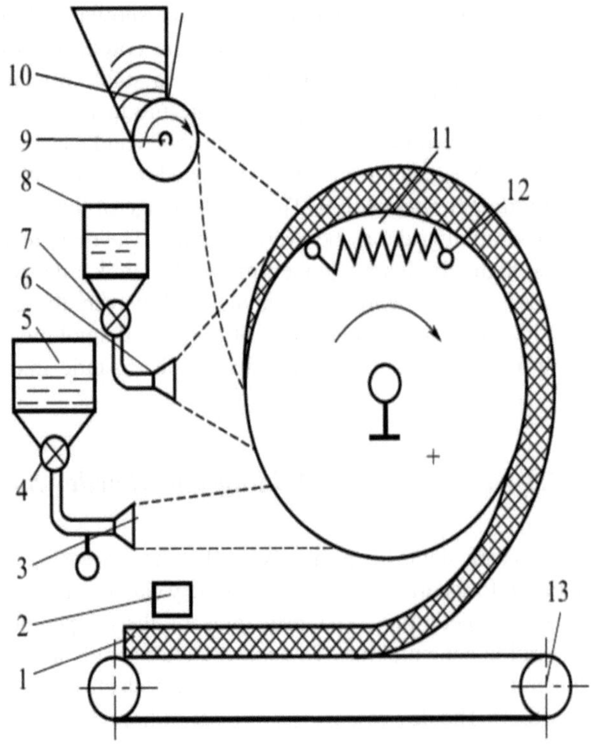

Fig. 8.20 Electrostatic forming device for electrostatically banded products. (1) Product; (2) cutter; (3, 6, 9) electrostatic sprayers for vegetable oil, salt, sugar, fermentation powder, or yeast liquid; (4, 7) automatic metering device; (5) container; (8) material tank; (10) refill hopper; (11) electric heater; (12, 13) conveyor belt

8.4.4.1 Alteration of Cell Membrane Potential

The electric field modifies the transmembrane potential of cells, influencing physiological metabolism. Ion movement across the membrane requires carriers and is driven by both chemical and electrical gradients. When the electric field aligns with the membrane potential, it increases, promoting bioelectric currents and enhancing metabolic processes. Conversely, an opposing field can decrease membrane potential and slow metabolism, aiding preservation.

8.4.4.2 Impact on the Intrinsic Electric Field of Fruits and Vegetables

Fruits and vegetables have inherent electric fields that can be disturbed by external fields, leading to physiological changes. Postharvest, fruit skins are typically positively charged while cores are negatively charged. High-voltage fields can intensify these charge states.

8.4.4.3 Influence on Water Structure and Enzyme-Water Binding

Water molecules cluster due to dipole interactions, which can release energy as heat. Applying an external resonant energy field can alter water structure, potentially creating "activated water." This affects enzyme activity, crucial for fruit and vegetable metabolism. Changes in water structure can inhibit enzyme activity, thereby aiding preservation. Additionally, high-voltage fields can generate negative air ions and ozone, which can inhibit metabolism and possess antibacterial properties.

Advantages of this preservation method include (1) cost-effective and energy-efficient (the small electric current (0–10μA) minimizes energy use), (2) flexible operation (the high voltage can be easily adjusted), and (3) no chemical residues (the process avoids secondary pollution).

Disadvantages include (1) safety risks (high voltage can be dangerous) and (2) humidity control (high humidity can cause electrical breakdown and short circuits).

8.4.5 Electrostatics in Controlling Agricultural Pests

The effectiveness of insecticides in pest control depends on the biological traits of the pests and the properties of the insecticides. All living organisms possess some charge, and insects acquire static electricity through movement. Understanding the charge characteristics of both pests and insecticides can enhance pest control strategies. By using insecticides with an opposite charge to that of the pests, more of the insecticide can adhere to the pests, improving efficiency and reducing pesticide use and environmental impact. This method is applied in agriculture, such as with electrostatic sprayers that enhance pesticide deposition on crops.

8.4.5.1 Common Electrostatic Properties of Stored Grain Insects

Insects gain charge from contact with insulating materials, with the charge level varying by interaction. As shown in Table 8.6, stored grain insects like *Rhyzopertha dominica*, *Sitophilus oryzae*, *Sitophilus zeamais*, *Oryzaephilus surinamensis*, and *Cryptolestes ferrugineus* often carry positive charges, while *Tribolium castaneum* and *T. confusum* has a negative charge.

8.4.5.2 Charge Properties of Different Grain Types

Table 8.7a shows that wheat, corn, and rice exhibit negative charges on glass and filter paper at 24.5 °C and 28.5% relative humidity, with higher charges on glass. Lower moisture content in wheat correlates with greater charge accumulation (Table 8.7b).

Table 8.6 Common electrostatic properties of stored grain insects at 24.5 ± 0.5 °C and 28.5 ± 1%

Species	Quality ($\times 10^{-3}$ kg)/50 insects	Electric charges ($\times 10^{-9}$C)/50 insects		Charge mass ratio ($\times 10^{-6}$C/kg)	
		Filter paper	Glass	Filter paper	Glass
R. dominica	0.0525	0.109 ± 0.012	0.020 ± 0.003	2.076 ± 0.229	0.381 ± 0.006
S. oryzae	0.0741	0.197 ± 0.003	0.180 ± 0.023	2.659 ± 0.040	2.429 ± 0.031
S. zeamais	0.0924	0.196 ± 0.005	0.429 ± 0.04	2.118 ± 0.054	4.643 ± 0.043
O. surinamensis	0.0351	0.046 ± 0.003	0.022 ± 0.002	1.311 ± 0.085	0.627 ± 0.057
C. ferrugineus	0.0082	0.027 ± 0.004	0.005 ± 0	3.333 ± 0.488	0.650 ± 0
T. castaneum	0.0903	−0.039 ± 0.005	−0.035 ± 0.003	0.436 ± 0.055	0.387 ± 0.003
T. confusum	0.0998	−0.041 ± 0.005	−0.039 ± 0.004	0.414 ± 0.050	0.391 ± 0.004

Table 8.7a Electrostatic properties of different grain types at 24.5 ± 0.5 °C and 28.5 ± 1%

Species	Moisture content (%)	Quality (g)	Electric charges ($\times 10^{-9}$C)/30 kernels		Charge mass ratio ($\times 10^{-6}$C/kg)	
			Filter paper	Glass	Filter paper	Glass
Wheat	10.2	1.2263	−0.317 ± 0.010	−0.808 ± 0.031	0.259 ± 0.008	0.659 ± 0.025
Corn	10.2	10.9688	−2.420 ± 0.885	−4.300 ± 0.331	0.221 ± 0.081	0.392 ± 0.030
Paddy rice	10.2	1.1562	−0.062 ± 0.005	−0.493 ± 0.027	0.065 ± 0.005	0.426 ± 0.023

Table 8.7b Electrostatic properties of wheat with different moisture contents at 24.5 ± 0.5 °C and 28.5 ± 1%

Moisture content (%)	Quality (g)	Electric charges ($\times 10^{-9}$C)/30 kernels	Charge mass ratio ($\times 10^{-6}$C/kg)
10.0	1.2777	−0.307 ± 0.005	0.240 ± 0.005
12.1	1.2687	−0.269 ± 0.008	0.212 ± 0.006
12.7	1.3860	−0.233 ± 0.006	0.168 ± 0.005
14.2	1.3884	−0.184 ± 0.007	0.133 ± 0.005
15.6	1.3787	−0.160 ± 0.010	0.116 ± 0.007

8.4.5.3 Comparison of Electrostatic Efficacy

When an insect flies through the air or walks on an insulated material, it is confronted with electrical currents, and its body will be electrostatically charged with "static" electricity. As an insect carrying an electrical charge approaches the powder which carried the opposite charge, an electric field is formed between the insect and the powder. This electric field grows in strength as the gap between them narrows, and the forces of attraction temporarily created between the insect and the powder initiate the detachment of airborne powder and its deposition on the body of the insect. The electrostatic force can also function as a short-term sticking factor, especially if the powder is deposited on a few hairs of the insect, enabling it to remain on the receptive surface long enough for an insecticidal effect.

In Table 8.8, the food grade inert powder shows a negative charge of −0.061 nC on filter paper, while *Oryzaephilus surinamensis* has an average positive charge of 0.040 nC. After mixing, the charge is 0.006 nC. On glass, the powder carries −0.167 nC, and the insect has 0.006 nC, resulting in a mixed charge of −0.174 nC. For *Sitophilus granarius*, the powder on filter paper has a charge of −0.056 nC, while the insect's positive charge is 0.067 nC, giving a mixed charge of −0.038 nC. On glass, the powder's charge is −0.131 nC, and the insect's charge is 0.017 nC, resulting in a mixed charge of −0.056 nC. These lower mixed charges suggest some charge neutralization occurs.

Table 8.8 Electrostatic properties of food grade inert dust and insects on each of two dielectrics at 24.5 ± 0.5 °C and 28.5 ± 1%

	Electric charges ($\times 10^{-9}$C)			
	O. surinamensis		*R. dominica*	
Sample	Filter paper	*Glass*	Filter paper	Glass
Food grade inert dust	−0.061	−0.167	−0.056	−0.131
Insect	0.040	0.006	0.067	0.017
Mixture	0.006	−0.174	−0.038	−0.056

Note: Each charge test uses 50 insects, and the amount of food grade inert dust is calculated at 2 g/m^2.

Table 8.9 Mortality of two insect species exposed to two different surfaces treated with food grade inert dust at 24.5 ± 0.5 °C and 28.5 ± 1%

	Mortality(%)			
	O. surinamensis		*R. dominica*	
Time(h)	Filter paper	Glass	Filter paper	Glass
1	0.4	0.8	3.2	2.4
2	0.8	0.8	4.4	3.6
3	0.8	3.2	20	11.6
4	10.8	7.2	66.8	44
5	46.4	18.8	88.4	78.8

After measuring the charges, average mortality rates of *O. surinamensis* and *R. dominica* were recorded hourly over 5 hours on filter paper and glass slides at 25 °C and 23% RH (Table 8.9). For the first 3 hours, mortality rates for *O. surinamensis* were similar on both substrates. By the fourth hour, the mortality rate on filter paper rose significantly, reaching 46.4% by the fifth hour, compared to 18.8% on glass. *R. dominica* also showed lower mortality on glass. The higher charge accumulation on filter paper suggests that both species adsorbed more of the inert powder, leading to increased mortality. *R. dominica* demonstrated stronger charge accumulation, indicating greater efficacy of the inert powder.

8.4.6 Radio Frequency Control of Agricultural Product Pests

Radio frequency (RF) refers to electromagnetic waves between 10–300 MHz. The US Federal Communications Commission (FCC) designates three RF frequencies—13.56, 27.12, and 40.68 MHz—for industrial, scientific, and medical uses. RF heating operates like a parallel plate capacitor, with agricultural products placed between two electrodes. This results in two key phenomena: (1) space charge polarization, where ions move due to the electric field, and (2) the rotation of polar water molecules, which align with the alternating field. The combination generates heat through friction, raising the product's temperature uniformly both on the surface and internally. Unlike traditional heating methods that rely on conduction and

Fig. 8.21 Temperature distribution at the top layer of the sample after heating 6 kg of rice in plastic baskets and PP woven bags to 54 °C using radio frequency

convection, RF energy penetrates materials deeply, offering consistent heating without contact, even through packaging materials. RF technology is widely applied in postharvest grain storage and processing for pest control, sterilization, enzyme inactivation, drying, and baking.

8.4.6.1 Heating Uniformity of Rice in Different Packaging Forms

In experiments, samples were heated to 54 °C using a 27.12 MHz, 15 kW RF system. Two kilograms of rice in a woven PP bag were placed in a plastic basket (350 mm × 260 mm × 90 mm, with a sample thickness of approximately 80 mm), with RF power turned off upon reaching the target temperature. Figure 8.21 shows the temperature distribution of a 6 kg sample heated to 54 °C, illustrating the "edge effect," where temperatures at the edges were higher due to the denser electric field.

8.4.6.2 Determination of Optimal Coupling Thickness and Electrode Gap

Figure 8.22a illustrates temperature changes over time for one, two, and three layers of rice during RF heating with electrode gaps of 115, 180, and 245 mm. The time to heat the samples from 30 °C to 54 °C was 1.4, 2.9, and 7.1 minutes, respectively, resulting in heating rates of 16.8, 8.3, and 3.4 °C/min. As the number of layers increases, heating time rises and heating rate declines. Figure 8.22b presents temperature changes for two layers of rice at various electrode gaps. At gaps of 180, 190, and 200 mm, the heating rates were 8.3, 6.6, and 4.3 °C/min, respectively, indicating that heating rate decreases with increasing gap. Consequently, an electrode gap of 190 mm, with a heating rate of 6.6 °C/min and conveyor belt speed of 12.38 m/h, was chosen for scale-up testing.

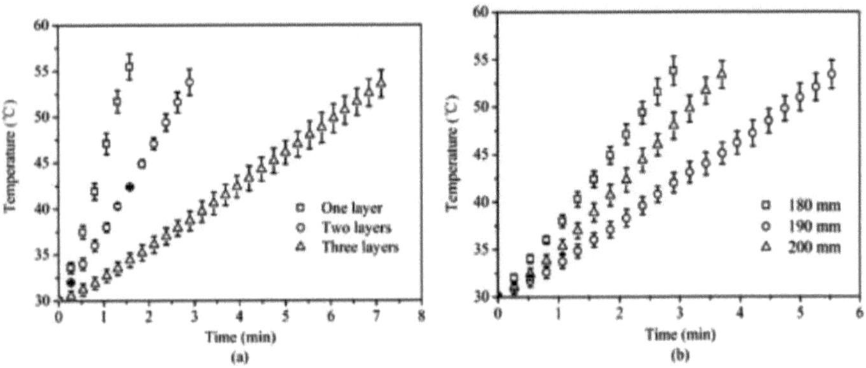

Fig. 8.22 The variation curve of the central temperature of rice during radio frequency heating with sample thickness, plate spacing, and time

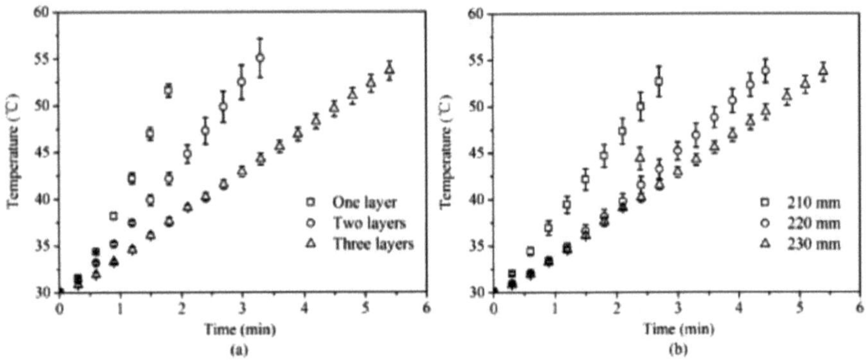

Fig. 8.23 The variation curve of the central temperature of brown rice during radio frequency heating with sample thickness, plate spacing, and time

Figure 8.23a depicts temperature changes over time for one, two, and three layers of brown rice during RF heating at electrode gaps of 110, 170, and 230 mm. The time to heat from 30 °C to 54 °C was 1.9, 3.2, and 5.3 minutes, with corresponding heating rates of 12.3, 7.6, and 4.5 °C/min. Figure 8.23b illustrates temperature changes for three layers of brown rice at gaps of 210, 220, and 230 mm, yielding heating rates of 8.5, 5.5, and 4.5 °C/min, respectively. As electrode gaps increase, heating rates decrease, but uniformity improves. Thus, an electrode gap of 220 mm, with a heating rate of 5.5 °C/min and conveyor belt speed of 10.31 m/h, was chosen for further studies.

Figure 8.24a displays temperature changes over time for two, three, and four layers of paddy rice during RF heating at electrode gaps of 140, 185, and 230 mm. Heating from 30 °C to 54 °C took 0.8, 3.3, and 10.2 minutes, resulting in heating rates of 29.0, 7.2, and 3.4 °C/min. In Fig. 8.24b, for three layers at gaps of 185, 195,

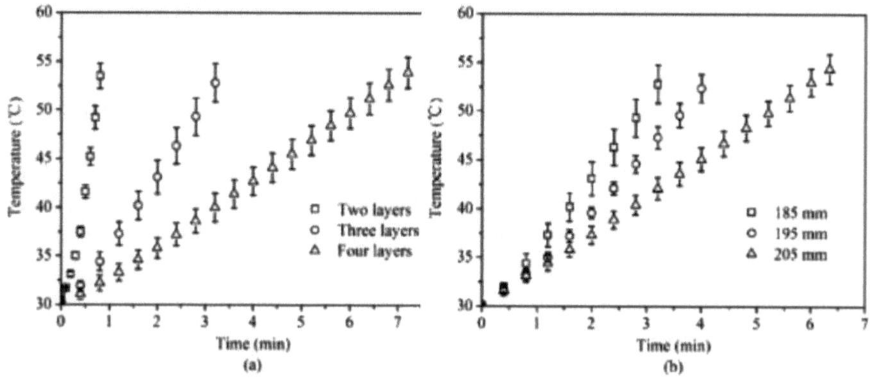

Fig. 8.24 The variation curve of the central temperature of paddy rice during radio frequency heating with sample thickness, plate spacing, and time

Table 8.10 Comparison of sample temperatures and heating uniformity after RF heat treatment for 11 minutes

	Sample temperature(°C)			Heating uniformity index(λ)		
Layer	Paddy rice	Brown rice	Rice	Paddy rice	Brown rice	Rice
Top	54.36 ± 2.73a	54.66 ± 2.08a	54.94 ± 1.78a	0.111 ± 0.043a	0.078 ± 0.026a	0.073 ± 0.012a
Medium	56.24 ± 3.34a	56.67 ± 2.31a	–	0.127 ± 0.043a	0.048 ± 0.012ab	–
Bottom	55.52 ± 2.57a	55.53 ± 1.53a	55.27 ± 1.55a	0.085 ± 0.004a	0.033 ± 0.011b	0.064 ± 0.011a

and 205 mm, the heating rates were 7.2, 5.7, and 3.8 °C/min, indicating a decline with increasing gap. An electrode gap of 195 mm, a heating rate of 5.7 °C/min, and a conveyor speed of 10.87 m/h were chosen for further studies due to improved heating uniformity.

Table 8.10 summarizes the average temperature and heating uniformity index after 11 minutes of RF treatment at room temperature. The electrode gaps were 195 mm for paddy rice, 220 mm for brown rice, and 190 mm for white rice, with conveyor speeds of 10.87, 10.31, and 12.38 m/h, respectively. The top layer lost heat to the air, resulting in the middle layer achieving the highest temperature, followed by the bottom and top layers. All samples surpassed 54 °C, ensuring 100% insect mortality. White rice exhibited better heating uniformity than brown rice, while paddy rice had the lowest. Notably, the uniformity of white rice improved compared to previous tests with stationary samples, likely due to reduced edge effects in larger-scale processing.

8.4.6.3 Effect of Radio Frequency Treatment on Agricultural Product Quality

Table 8.11 presents the physical and chemical quality data for control samples and those treated with various radio frequency (RF) methods. Nutritionally, there were no significant differences ($P > 0.05$) in fat, fatty acids, and protein content between the control and RF-treated samples. However, RF treatment significantly reduced amylose content ($P < 0.05$), resulting in softer, stickier cooked rice, which is preferred by consumers, compared to firmer rice with higher amylose.

Regarding surface color, the b^* value increased slightly, but no significant differences were observed in L^*, a^*, and b^* values ($P > 0.05$) between control and RF-treated samples. The crack and breakage rates of rice remained unchanged ($P > 0.05$) when the RF heating rate was below 5.8 °C/min, indicating that appropriate heating does not significantly affect these rates.

Table 8.11 details the sensory quality of cooked rice from untreated and RF-treated samples. Results showed no significant differences ($P > 0.05$) in aroma, appearance, structure, and taste before and after RF treatment. For samples treated at 50 °C for 26 minutes and at 52 °C for 11 minutes, slight texture changes were noted, but overall sensory quality scores for treated samples remained above 70 points, indicating "good" quality (71–80 points) according to the GB/T 15682–2008 grading system.

8.4.6.4 Key Technologies for Killing Storage Pests with Radio Frequency

In general, adult storage pests can be killed by holding them at 60 °C for around 30 seconds, with *Rhyzopertha dominica* being the most heat-resistant pest. Thus, the effective RF treatment for killing *R. dominica* can serve as a reference for other storage pests.

Effect of Grain Layer Thickness on Pest Control in Wheat

As the thickness of the grain layer increases, the mortality rate of pests also increases. When the grain layer thickness was 7 cm and 8 cm, the mortality rates were 40.7% and 79%, respectively. When the thickness increased from 8 cm to 9 cm, the mortality rate increased by 14.3%, reaching 93.3%. At a grain layer thickness of 10 cm, the mortality rate reached 100%. As the grain layer thickness increases, the RF equipment's load increases, output power rises, and current grows, significantly accelerating the temperature increase in wheat, thus raising the pest mortality rate. At 10 cm thickness, the equipment approaches its maximum output power and risks overload. Therefore, the grain layer thickness should not exceed 10 cm.

Table 8.11 Effect of different treatment weights and RF processing methods on the physicochemical quality of rice

Sensory quality (points)	Control	2.5 kg 54 °C, 6 min	2.5 kg 52 °C, 11 min	2.5 kg 50 °C, 26 min	5 kg 54 °C, 6 min	5 kg 52 °C, 11 min	5 kg 50 °C, 26 min	10 kg 54 °C, 6 min	10 kg 52 °C, 11 min	10 kg 50 °C, 26 min
Purity and richness(20)	15.67 ± 0.61a	15.12 ± 2.28a	15.00 ± 2.10a	14.44 ± 2.33a	14.98 ± 1.94a	15.58 ± 1.95a	15.24 ± 2.01a	15.56 ± 1.85a	15.56 ± 1.61a	15.60 ± 2.14a
Appearance and structure **(20)**										
Color (7)	5.20 ± 0.25	4.88 ± 0.78a	4.96 ± 1.02a	5.08 ± 0.95a	5.30 ± 0.91a	5.12 ± 1.16a	5.32 ± 0.84a	5.16 ± 0.94a	5.34 ± 1.07a	5.08 ± 0.95a
Gloss (8)	6.17 ± 0.30a	5.68 ± 1.03a	5.68 ± 1.14a	5.52 ± 1.23a	6.03 ± 1.08a	5.44 ± 0.96a	5.66 ± 0.99a	6.18 ± 0.97a	6.16 ± 0.94a	5.86 ± 0.99a
Grain integrity (5)	4.09 ± 0.20a	4.06 ± 0.98a	3.40 ± 0.82a	3.40 ± 0.87a	4.14 ± 0.73a	3.84 ± 0.99a	4.04 ± 0.97a	3.94 ± 0.92a	3.84 ± 0.81a	3.72 ± 0.72a
Palatability **(30)**										
Stickiness (10)	8.07 ± 0.27a	7.68 ± 1.31a	7.04 ± 1.65a	7.28 ± 1.88a	7.76 ± 1.01a	7.72 ± 1.39a	7.52 ± 1.48a	7.60 ± 1.29a	7.68 ± 1.28a	7.80 ± 1.47a
Elasticity (10)	8.15 ± 0.19a	7.76 ± 1.30a	7.44 ± 1.56a	7.16 ± 1.21a	7.92 ± 0.70a	7.66 ± 1.39a	7.42 ± 1.51a	7.74 ± 1.22a	7.50 ± 1.00a	8.20 ± 0.96a
Softness and hardness (10)	8.07 ± 0.24ab	7.56 ± 1.16ab	7.56 ± 1.56ab	7.00 ± 1.68b	8.22 ± 0.76a	7.85 ± 1.02ab	7.85 ± 1.73ab	7.46 ± 1.43ab	7.84 ± 1.07ab	7.69 ± 1.35ab
Taste **(25)**										
Purity and persistence (25)	18.69 ± 0.80a	18.20 ± 2.31a	18.84 ± 2.88a	18.24 ± 2.95a	18.84 ± 1.68a	18.33 ± 1.76a	18.85 ± 2.20a	18.72 ± 1.81a	18.92 ± 2.08a	18.84 ± 1.97a
Texture of cold rice**(5)**										
Cohesion, stickiness-elasticity, hardness(5)	4.09 ± 0.22a	3.64 ± 0.81ab	3.22 ± 0.94b	3.46 ± 1.32ab	4.16 ± 0.70a	4.12 ± 0.78a	3.91 ± 0.88ab	3.96 ± 0.98ab	3.78 ± 0.89ab	3.88 ± 0.92ab
Overall **(100)**	78.18 ± 2.23a	74.08 ± 4.93abc	72.97 ± 5.06bc	71.92 ± 4.96c	77.35 ± 3.91ab	76.34 ± 5.39abc	76.09 ± 5.82abc	78.09 ± 4.76a	77.06 ± 5.02ab	77.94 ± 4.70a

*Different letters in the same row indicate significant differences in mean values between different treatments ($P < 0.05$)

Effect of Different RF Frequencies on Killing *Rhyzopertha dominica*

Wheat was placed in a PP plastic box (28 cm × 19 cm × 10 cm) with an initial wheat temperature of 20 °C and a grain layer thickness of 9 cm. The distance between the RF plates was set at 10 cm. At frequencies of 6.78, 13.56, 27.12, and 40.68 MHz, the time required to achieve 100% mortality for *R. dominica* was 25 minutes, 10 minutes, 17 minutes and 53 seconds, and 7 minutes, respectively. At the same frequency, the pest mortality rate increased as the RF treatment time increased and the temperature rose. When the temperature reached 58 °C, *R. dominica* mortality reached 100%. It can be inferred that the RF frequency does not directly affect pest control but that the treatment time and the resulting temperature of the grain determine the effectiveness of pest control.

Suitable RF Treatment Time for Different Grain Types and Moisture Levels

The time required to heat corn (moisture content 11.9%), paddy rice (moisture content 11.9%), and wheat (moisture content 10.9%) to 58 °C using RF treatment was 8 minutes and 46 seconds, 8 minutes and 4 seconds, and 17 minutes and 53 seconds, respectively. The higher the moisture content of the grain, the shorter the RF treatment time needed. At a heating temperature of 58 °C, the mortality rate of *R. dominica*, *S. zeamais*, *T. castaneum*, *O. surinamensis*, and *L. serricorne* reached 100%.

8.4.7 Electrodialysis

Electrodialysis is a method that uses a direct current (DC) electric field to selectively separate ions in an electrolyte solution through ion-exchange membranes. These membranes allow only specific ions to pass: cation-exchange membranes permit cations, while anion-exchange membranes allow anions. Arranged alternately in parallel, these membranes create small chambers filled with the solution. An electric field is applied, driving ions to move and separate the solution into concentrated and diluted chambers, as illustrated with sodium chloride in Fig. 8.25.

In the electrodialysis setup, alternating cation and anion membranes create chambers where raw water enters. The applied DC electric field causes cations to pass through cation membranes while blocking anions and vice versa for anion membranes. This results in some chambers producing fresh water with low ion content, while adjacent chambers become concentrated with high ion content, effectively purifying the water.

Electrodialysis shares similarities and differences with ion exchange: ① both utilize ion-exchange resins for separation, but electrodialysis uses flat membranes, whereas ion exchange uses spherical particles. ② Ion exchange involves transfer and substitution reactions, while electrodialysis focuses on selective ion transmission

Indirect effects result from interactions among products formed by direct effects, generating compounds different from the original materials, which are also related to temperature and other conditions.

The choice of food radiation treatment depends on the preservation objectives. Since the types of food vary, the factors leading to food spoilage also differ. Based on the desired preservation period after food treatment, there are typically three methods: irradiation of aseptic processing, irradiation of pasteurization, and irradiation of storage tolerance sterilization:

1. Irradiation Aseptic Processing (Complete Sterilization): This is sufficient to reduce the microbial count to zero or a limited number, so that no spoilage microorganisms or toxins can be detected with current methods in the absence of post-processing contamination, allowing for long-term preservation. Generally, a high dose of 10–50 kGy is used, especially for meat, such as beef, where high doses may produce off-flavors; therefore, irradiation can be conducted at frozen temperatures below −30 °C. Most off-flavor formation is due to chemical reactions, as the mobility of free radicals in water decreases during freezing, preventing interactions between free radicals and meat molecules.
2. Irradiation Pasteurization (Disinfection): This is sufficient to reduce the count of specific non-spore-forming pathogenic bacteria (e.g., Salmonella) to a level that cannot be detected with current methods. This method cannot guarantee long-term storage because spore-forming bacteria may still be present in the food. It must be combined with other preservation methods, such as low temperature or reduced water activity. Additionally, if there are already large populations of microorganisms in the food, this method cannot be used since radiation cannot eliminate the toxins produced by the microorganisms. Radiation doses of 5–10 kGy are used.
3. Radiation Storage Sterilization: This method is sufficient to reduce the number of spoilage bacteria and delay the time for significant microbial proliferation (preventing reproduction). It is used to postpone the ripening period of fresh fruits and vegetables, thereby extending their shelf life. The radiation dose is <5 kGy. In China, foods that can be treated with radiation include livestock and poultry meat, aquatic products, eggs, fruits and vegetables, grains, and their products.

8.4.18 *Irradiation for Controlling Pests in Agricultural Products*

8.4.18.1 Inhibitory Effect on Hidden Stored Grain Pests

Using an electron beam at a dose rate of 25 Gy/s and static irradiation with a cobalt source at seven different doses, the lethal effect on various life stages of three hidden pests (*Sitophilus oryzae*, *Sitophilus zeamais*, and *Rhyzopertha dominica*) was examined. The lethal effect on adult insects primarily occurred 1 week after irradiation, and the time to death shortened with increasing dose, with 200–300 Gy being the most effective dose. Sensitivity to electron beam irradiation, from highest to lowest, was eggs, larvae, pupae, and adults. After irradiation of the F_0 generation, the reproductive ability of the F_1 adult generation was also affected to varying degrees. Compared to F_0 adults, F_1 adults exhibited lower reproductive ability. A dose of 200 Gy was found to be an effective dose to inhibit the reproduction of the three hidden pests.

8.4.18.2 Inhibitory Effect on Secondary Stored Grain Pests

An electron beam with dose rates of 15 Gy/s, 30 Gy/s, and 60 Gy/s was used to irradiate *Tribolium castaneum* adults at six doses: 60 Gy, 120 Gy, 240 Gy, 360 Gy, 480 Gy, and 960 Gy. The pupae, larvae, and eggs were irradiated at a dose rate of 30 Gy/s. Additionally, the same doses of gamma rays at a rate of 20 Gy/min were used to irradiate *T. castaneum* adults, pupae, larvae, and eggs. Furthermore, at a dose rate of 25.0 Gy/s, the adults, pupae, and larvae of *Oryzaephilus surinamensis* were irradiated with seven different doses: 50 Gy, 100 Gy, 200 Gy, 300 Gy, 400 Gy, 500 Gy, and 1000 Gy, and larvae were irradiated at six doses: 50 Gy, 100 Gy, 200 Gy, 300 Gy, 400 Gy, and 500 Gy. The same doses and dose rates were used to irradiate *Cryptolestes pusillus*. Sensitivity to irradiation varied among secondary pests, with the sensitivity from highest to lowest being: *T. castaneum*, *O. surinamensis*, and *C. pusillus*. The sensitivity order for different life stages was egg, larva, pupa, and adult. Electron beam irradiation exhibited a delayed lethality effect, with little mortality observed immediately after treatment. Survival rates slightly decreased in the first week and dropped significantly during the second and third weeks, stabilizing by the fourth week. A dose of 300 Gy was found to effectively inhibit the reproduction of these three secondary pests.

8.4.18.3 Inhibitory Effect on Booklice

At a dose rate of 25 Gy/s, electron beam and cobalt source static irradiation were applied to adult and nymph stages of two species of booklice. A dose of 300–400 Gy was determined to be an effective reference for controlling booklice adults and nymphs. Additionally, a dose of 50 Gy was sufficient to prevent hatching, and a 300 Gy dose of electron beam irradiation rendered adult booklice unable to produce offspring.

8.4.18.4 Pest Control in Grain Using a Conveyor Belt System

A mid-scale electron beam test was conducted using existing irradiation equipment in a grain storage facility, with 12 test batches. At doses of 300 Gy and 600 Gy, the electron beams effectively killed *Sitophilus oryzae* adults, with survival times not exceeding 14 days, and no offspring were produced. Although *Rhyzopertha dominica* adults were not killed, reproduction was inhibited, with no offspring. The electron beam also effectively killed *Tribolium castaneum* adults. Some larvae within the grain were able to develop into adults, but these adults produced no offspring.

8.4.19 Application of High-Pressure Pulsed Electric Field Technology

The high-pressure pulsed electric field food sterilization technology is a non-thermal sterilization technique. Compared to traditional food thermal sterilization methods, it has characteristics such as shorter sterilization times, lower energy consumption, and effective preservation of the nutritional components, natural colors, flavors, and aromas of food. The sterilization mechanism of high-pressure pulsed electric fields has been widely studied, demonstrating that it inactivates microorganisms in liquid media. After nearly 40 years of exploration, several representative viewpoints have emerged.

8.4.19.1 Transmembrane Potential Theory

When an external electric field is applied across the two ends of a cell, a transmembrane potential is generated. For a spherical cell with radius r in a uniform electric field E, the transmembrane potential along the direction of the electric field can be derived from the formula:

$$U(t) = 1.5rE \tag{8.56}$$

where U is the transmembrane potential in the direction of the electric field, V is the cell radius in micrometers, and E is the electric field strength in kV/mm. When the transmembrane potential reaches 1 V, the cell membrane loses its functionality.

8.4.19.2 Dielectric Breakdown Theory

This theory posits that the cell membrane can be viewed as a capacitor. Under the influence of high-voltage electric pulses, the potential difference across the membrane increases further. Due to opposing charges, they attract each other, creating compressive forces. When the transmembrane potential reaches 1 V and the compressive force exceeds the membrane's recovery force, the membrane will rupture.

8.4.19.3 Electroporation Theory

This theory suggests that high-voltage electric pulses alter the molecular structure of lipids and enlarge the openings of some protein channels, causing the cell membrane to lose its semipermeability, leading to cell swelling and death.

8.4.19.4 Cavitation Theory

When liquid food flows through a high-pressure pulsed electric field, the main gap discharge produces a powerful pulsed current, vaporizing the liquid into plasma at temperatures exceeding tens of thousands of degrees. This forms high-pressure pathways. Some gases produced, forming extremely thin "gas shells" around the sparks, transfer pressure to the liquid, generating high-speed adiabatic expansion and creating strong ultrasonic hydraulic shock waves. At the moment the discharge ends, a cavity forms within the gas shell due to the sudden drop in pressure, and the liquid refills the cavity at supersonic speed, generating a second ultrasonic refill cavity shock wave. It is this direct conversion of high-pressure pulse energy into mechanical energy that causes intense vibrations within the microbial cells of liquid food, leading to phenomena such as cell membrane rupture and thus resulting in sterilization effects.

In summary, these mechanisms suggest that the electric field's action on microorganisms is primarily manifested in two aspects: the effect of the field and the ionization effect. Many scholars both domestically and internationally have proposed various other mechanism models for the action of high-pressure pulsed electric fields in killing microorganisms, such as electromagnetic mechanism models and liposomal obstruction models. However, these mechanisms require further experimental validation. Figure 8.27 illustrates the principle of high-pressure pulsed electric field sterilization.

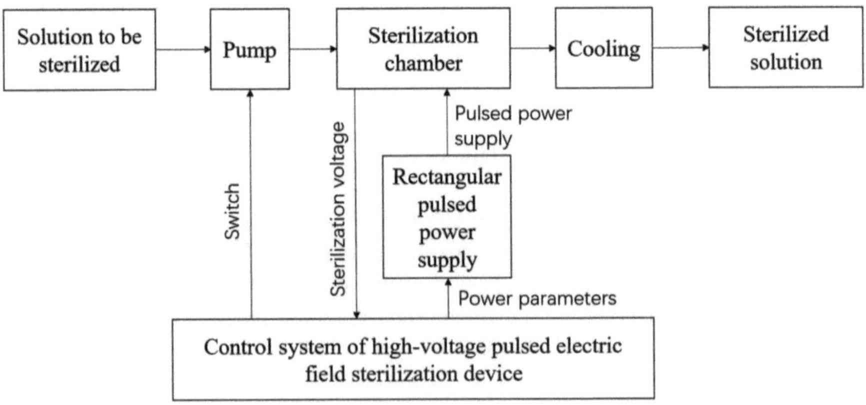

Fig. 8.27 Schematic diagram of high-voltage pulsed electric field sterilization principle

The main factors affecting high-pressure pulsed electric field sterilization are as follows:

1. Electric Field Strength: Among all factors, the electric field strength has the most significant impact on sterilization effectiveness. For instance, increasing the electric field strength significantly reduces the survival rate of target bacteria. Experiments show that when the electric field strength increases from 5 kV/cm to 25 kV/cm, the slope of the sterilization logarithmic curve doubles. If the medium's conductivity increases, the pulse frequency rises, and the pulse width decreases, the sterilization effect will decline if the number of pulses remains unchanged.
2. Number of Pulses: Increasing the number of pulses significantly enhances the sterilization effect.
3. Pulse Shape: Pulses are typically in the shape of square waves, exponential decay waves, and alternating waves, with square waves being the most effective, followed by exponential and alternating wave systems being the least effective.
4. Processing Time: The sterilization time is the total duration of each pulse discharge. As the sterilization time extends, the survival rate of target bacteria sharply decreases initially and then gradually levels off, while increasing the sterilization time beyond a certain point has little effect.

In pulsed electric field sterilization, if the food has too high resistance, the current will pass through the less resistive liquid medium without affecting the food. Some foods are composed of several substances (nonhomogeneous), with varying conductivities in different parts. The electrical characteristics of nonhomogeneous foods will change after pulse treatment, making them unsuitable for high-pressure pulsed electric field treatment. Generally, conductivity changes in the same direction as the sterilization effect. However, some studies have found that, under unchanged conditions, the medium's conductivity has little impact on sterilization efficiency.

8.4.20 Application of Pulsed Magnetic Field Sterilization Technology

Biological entities are permeable to magnetic fields because their magnetic permeability is similar to that of the vacuum under conditions of μ0. A transient magnetic field generates induced currents and high-frequency thermal effects within biological entities. Under the influence of a pulsed magnetic field, the short pulse duration results in a significant rate of change in the magnetic field, which stimulates induced currents within the cells. Research shows that the effects of pulsed magnetic fields on cells depend on the type and size of the cells; different types and sizes of cells have varying tolerances to the pulsed field strength. The mechanisms by which magnetic fields produce biological effects on cells mainly include two aspects:

1. When cells move in a magnetic field, if the motion involves cutting through magnetic field lines, it results in changes in magnetic flux, thus inducing electric currents. The magnitude, direction, and form of this current are the primary reasons for producing biological effects in cells. The larger the induced current, the more pronounced the biological effect. When cells are in a pulsed field, they can be considered stationary, with the magnetic flux passing through the cells represented as $\varphi = \mathrm{SH}(t)$, where $\mathrm{H}(t)$ is the time-varying magnetic field value, and SSS is the cross-sectional area of the magnetic field perpendicular to the cell. Due to the instantaneous appearance and disappearance of the magnetic field, a transient magnetic flux is inevitably generated within the cells, expressed as $\mathrm{d}\varphi = \mathrm{d}t$. The transient magnetic flux excites induced currents in the cells, and the density of the force that interacts with the magnetic field can disrupt the normal physiological functions of the cells. If the cell volume is large, the resulting force density will also be high, making larger cells more susceptible to death, while smaller cells are less so.
2. In a magnetic field, charged particles in the cells, especially small mass electrons and ions, are influenced by the Lorentz force, causing their trajectories to be constrained within a certain radius. The greater the magnetic field, the smaller the radius, preventing electrons and ions within the cells from transferring normally, thus affecting the cells' physiological functions. Large molecules in the cells, such as enzymes, experience distortion or deformation in their conformations due to the differing directions of movement of the different charges they carry in the magnetic field, altering the activity of the enzymes and thereby affecting the normal physiological activities of the cells.

Studies have found that alternating magnetic fields with intensities greater than 2 T possess sterilization effects, with magnetic flux densities ranging from 5 to 50 T and frequencies between 5 and 500 kHz being able to reduce the number of microorganisms by two logarithmic values. Currently, alternating magnetic fields have been used internationally to sterilize brewing condiments such as monosodium glutamate, vinegar, soy sauce, wine, and dairy products, with results showing that the quality of the sterilized foods is good and the shelf life is significantly extended.

8.5 Future Trends and Conclusions

This chapter reviews various electromagnetic properties, fundamental concepts and principles, measurement techniques, and their applications in agricultural and food processing fields. Applications include water quality testing, analysis of grains, oilseeds and their alternatives, fruits and vegetables, fish, seafood, poultry, meat substitutes, as well as aqueous solutions of sugar, starch, and salt. Given these properties and potential applications, the food industry may increasingly adopt this technology to enhance efficiency. Advances in electronics and instrumentation have led to miniaturization and improved sensitivity, paving the way for new techniques in measuring electromagnetic properties. Such data not only support quality assurance but also provide insights into the behavior of food materials under electromagnetic treatment. Future efforts should focus on developing new postharvest processes that use electromagnetic technology to maintain the quality of agricultural products and processed foods.

References

1. Hlaváčová Z. Low frequency electric properties utilization in agriculture and food treatment. Res Agric Eng. 2003;49(4):125–36.
2. Venkatesh MS, Raghavan GSV. An overview of microwave processing and dielectric properties of agri-food materials. Biosyst Eng. 2004;88(1):1–18.
3. Chen JY, Iyo C, Kawano S, Terada F. Development of calibration with sample cell compensation for determining the fat content of unhomogenised raw milk by a simple near infrared transmittance method. J Near Infrared Spectrosc. 1999;7(4):265–73.
4. Kawano S. New application of nondestructive methods for quality evaluation of fruits and vegetables in Japan (quality and its evaluation of horticultural products, for further development of horticulture in East Asia). J Jpn Soc Hortic Sci. 1998;67(6):1176–9.
5. Markx GH, Davey CL. The dielectric properties of biological cells at radiofrequencies: applications in biotechnology. Enzym Microb Technol. 1999;25(3–5):161–71.
6. Içier F, Baysal T. Dielectrical properties of food materials—1: factors affecting and industrial uses. Crit Rev Food Sci Nutr. 2004;44(6):465–71.
7. Nelson SO. Dielectric properties of grain and seed in the 1 to 50-mc range. Trans ASAE. 1965;8 (1):38–0048.
8. Feng H, Tang J, Cavalieri RP. Dielectric properties of dehydrated apples as affected by moisture and temperature. Trans ASAE. 2002;45(1):129–35.
9. Guan D, Cheng M, Wang Y, Tang J. Dielectric properties of mashed potatoes relevant to microwave and radio-frequency pasteurization and sterilization processes. J Food Sci. 2004;69 (1):FEP30-FEP37.
10. Nelson SO. Dielectric properties of agricultural products-measurements and applications. IEEE Trans Electr Insul. 1991;26(5):845–69.
11. Sipahioglu O, Barringer SA, Bircan C. The dielectric properties of meats as a function of temperature and composition. J Microw Power Electromagn Energy. 2003;38(3):161–9.
12. Datta AK, Nelson SO. Fundamental physical aspects of microwave absorption and heating in handbook of microwave technology for food applications. CHIPS; 2000.
13. Rudan-Tasic D, Klofutar C. Characteristics of vegetable oils of some Slovene manufacturers. Acta Chim Slov. 1999;46(4):511–21.
14. Sun E, Datta A, Lobo S. Composition-based prediction of dielectric properties of foods. J Microw Power Electromagn Energy. 1995;30(4):205–12.
15. Ohlsson T. Dielectric properties and microwave processing. In: Food properties and computer-aided engineering of food processing systems. Dordrecht: Springer Netherlands; 1989. p. 73–92.
16. Buffler CR. Effects of dielectric and thermal properties on the microwave heating of foods. Microw World. 1991;12(4):15–23.
17. Ahmed J, Ramaswamy HS, Raghavan VG. Dielectric properties of Indian basmati rice flour slurry. J Food Eng. 2007;80(4):1125–33.
18. Ohlsson T, Bengtsson NE, Risman PO. The frequency and temperature dependence of dielectric food data as determined by a cavity perturbation technique. J Microw Power. 1974,9(2): 129–45.

19. Bengtsson NE, Risman PO. Dielectric properties of foods at 3 GHz as determined by a cavity perturbation technique. J Microw Power. 1971;6(2):107–23.
20. Ohlsson T, Henriques M, Bengtsson NE. Dielectric properties of model meat emulsions at 900 and 2800 MHz in relation to their composition. J Food Sci. 1974;39(6):1153–6.
21. Stuchly MA, Stuchly SS. Dielectric properties of biological substances—tabulated. J Microw power. 1980;15(1):19–25.
22. Calay RK, Newborough M, Probert D, Calay PS. Predictive equations for the dielectric properties of foods. Int J Food Sci Technol. 1994;29(6):699–713.
23. Nelson SO. Frequency–and temperature–dependent permittivities of fresh fruits and vegetables from 0.01 to 1.8 GHz. Trans ASAE. 2003;46(2):567.
24. Sacilik KAMİL, Tarimci C, Colak AHMET. Dielectric properties of flaxseeds as affected by moisture content and bulk density in the radio frequency range. Biosyst Eng. 2006;93(2): 153–60.
25. Lizhi H, Toyoda K, Ihara I. Dielectric properties of edible oils and fatty acids as a function of frequency, temperature, moisture and composition. J Food Eng. 2008;88(2):151–8.
26. Stier RF. Tests to monitor quality of deep-frying fats and oils. Eur J Lipid Sci Technol. 2004;106(11):766–71.
27. Lawrence KC, Nelson SO, Bartley PG. Coaxial dielectric sensor for cereal grains. In: IMTC/98 conference proceedings. IEEE instrumentation and measurement technology conference. Where instrumentation is going (Cat. No. 98CH36222), vol. 1. IEEE; 1998. p. 541–6.
28. Kent M, Kress-Rogers E. The COST 90bis collaborative work on the dielectric properties of foods. Phys Props Foods. 1987;2:171–97.
29. Kent M. Microwave dielectric properties of fishmeal. J Microw Power. 1972;7(2):109–16.
30. Zheng M, Huang YW, Nelson SO, Bartley PG, Gates KW. Dielectric properties and thermal conductivity of marinated shrimp and channel catfish. J Food Sci. 1998;63(4):668–72.
31. Liu C, Sakai N. Dielectric properties of tuna at 2 450 MHz and 915 MHz as a function of temperature. Nippon Shokuhin Kagaku Kogaku Kaishi. 1999;46(10):652–6.
32. To, E. C, Mudgett RE, Wang DIC, Goldblith SA, Decareau RV. Dielectric properties of food materials. J Microw Power. 1974;9(4):303–15.
33. Kent M. Complex permittivity of fish meal: a general discussion of temperature density and moisture dependence. J Microw Power. 1977;12(4):341–5.
34. Meyer W, Schilz W. A microwave method for density independent determination of the moisture content of solids. J Phys D Appl Phys. 1980;13(10):1823.
35. Venkatesh MS, Raghavan GSV, Sotocinal SA. Development of a permittivity analyzer to operate at 915 and 2450MHz using cavity perturbation technique. CSAE Paper; 1998. p. 98–315.
36. Engelder DS, Buffler CR. Measuring dielectric properties of food products at microwave frequencies. Microw World. 1991;12(2):6–15.
37. Kent M, Kress-Rogers E. Microwave moisture and density measurements in particulate solids. Trans Inst Meas Control. 1986;8(3):161–8.
38. Nelson SO. Review and assessment of radio-frequency and microwave energy for stored-grain insect control. Trans ASAE. 1996;39(4):1475–84.
39. Holaday CE. An electronic method for the measurement of heat-damage in artificially dried corn. Cereal Chem. 1964;41(6):533–42.
40. Dexter ST. Separation of living and dead corn (Zea mays) kernels without germination. Agron J. 1965;57:95–6.
41. Kim YR, Morgan MT, Okos MR, Stroshine RL. Measurement and prediction of dielectric properties of biscuit dough at 27 MHz. J Microw Power Electromagn Energy. 1998;33(3): 184–94.
42. Nelson SO. Frequency dependence of the dielectric properties of wheat and the rice weevil. Iowa State University; 1972.
43. Lizhi H, Toyoda K, Ihara I. Discrimination of olive oil adulterated with vegetable oils using dielectric spectroscopy. J Food Eng. 2010;96(2):167–71.

44. Nelson SO, Guo WC, Trabelsi S, Kays SJ. Dielectric spectroscopy of watermelons for quality sensing. Meas Sci Technol. 2007;18(7):1887–92.
45. Thompson DR, Zachariah GL. Dielectric theory and bioelectrical measurements (part II. Experimental). Trans ASAE. 1971;14(2):216–0218.
46. Guo WC, Nelson SO, Trabelsi S, Kays SJ. 10–1800-MHz dielectric properties of fresh apples during storage. J Food Eng. 2007;83(4):562–9.
47. Ikediala JN, Tang J, Drake SR, Neven LG. Dielectric properties of apple cultivars and codling moth larvae. Trans ASAE. 2000;43(5):1175–84.
48. Bengtsson NE, Melin J, Remi K, Säoderlind S. Measurements of the dielectric properties of frozen and defrosted meat and fish in the frequency range 10–200 MHz. J Sci Food Agric. 1963;14(8):592–604.
49. Van Dyke D, Wang DIC, Goldblith SA. Dielectric loss factor of reconstituted ground beef: the effect of chemical composition. Food Technol. 1969;23(944):84–6.
50. Nelson SO. Dielectric properties of agricultural products and some applications. Res Agric Eng. 2008;54(2):104–12.
51. Mudgett RE. Dielectric properties of foods. In: Microwaves in the food processing industry. Orlando1: Academic; 1985. p. 5–37.
52. Guo W, Trabelsi S, Nelson SO, Jones DR. Storage effects on dielectric properties of eggs from 10 to 1800 MHz. J Food Sci. 2007;72(5):E335–40.
53. Nunes AC, Bohigas X, Tejada J. Dielectric study of milk for frequencies between 1 and 20 GHz. J Food Eng. 2006;76(2):50–255.
54. Everard CD, Fagan CC, O'donnell CP, O'callaghan DJ, Lyng JG. Dielectric properties of process cheese from 0.3 to 3 GHz. J Food Eng. 2006;75(3):415–22.
55. O'Connor JF, Synnott EC. Seasonal variation in dielectric properties of butter at 15 MHz and 4 C. Irish J Food Sci Technol. 1982:49–59.
56. Kudra TGSV, Raghavan V, Akyel C, Bosisio R, Voort FVD. Electromagnetic properties of milk and its constituents at 2.45 GHz. J Microw Power Electromagn Energy. 1992;27(4): 199–204.
57. Nelson SO, Whitney WK. Radio-frequency electric fields for stored grain insect control. Trans ASAE. 1960;3(2):133–137,144.
58. Hamid MAK, Kashyap CS, Cauwenberghe RV. Control of grain insects by microwave power. J Microw Power. 1968;3(3):126–35.
59. Shackelford RG. Studies to determine the electromagnetic spectrum effect on arthropods. Georgia Inst Technol Eng Exp Sta. Final report, project A-985, 1970.
60. Wang S, Tang J, Johnson JA, Mitcham E, Hansen JD, Hallman G, Wang Y. Dielectric properties of fruits and insect pests as related to radio frequency and microwave treatments. Biosyst Eng. 2003;85(2):201–12.
61. Johnson JA, Wang S, Tang J. Thermal death kinetics of fifth-instar Plodia interpunctella (Lepidoptera: Pyralidae). J Econ Entomol. 2003;96(2):519–24.
62. Yan R, Huang Z, Zhu H, Johnson JA, Wang S. Thermal death kinetics of adult Sitophilus oryzae and effects of heating rate on thermotolerance. J Stored Prod Res. 2014;59:231–6.

Chapter 9
Advanced Physical Technologies in Foods

Qin Ouyang

9.1 Introduction

In recent years, with the rapid advancement of technology, the food industry has undergone a profound transformation. These changes are mainly attributed to the introduction of advanced physical technologies, which have not only enhanced the understanding of food quality and safety but also significantly improved processing efficiency, driving innovative development in the food industry. Although traditional food processing and testing methods have met market demands to some extent, as consumers' requirements for food quality, nutritional content, and visual effects continue to rise, these traditional methods have gradually revealed their limitations. To meet these growing demands, cutting-edge physical technologies such as ultrasound imaging, computed tomography (CT), terahertz spectroscopy, and 4D printing have emerged and become important tools in the food industry.

Firstly, ultrasound imaging technology uses high-frequency sound waves for noninvasive detection of food, allowing inspectors to clearly observe the internal structure and quality of food [1]. This technology can effectively identify bubbles, cracks, or foreign objects in food and monitor changes during the processing in real-time, ensuring the quality and safety of the final product. The application of ultrasound imaging in food processing has significantly improved inspection efficiency and accuracy, providing a strong guarantee for food safety.

The application of computed tomography (CT) technology in the food industry is another major breakthrough. CT technology can accurately analyze the internal structure and composition of various materials by obtaining three-dimensional images [2]. The introduction of this technology has made it possible to study complex food structures, helping researchers to gain a deeper understanding of the

Q. Ouyang
School of Food and Biological Engineering, Jiangsu University, Zhenjiang, Jiangsu, China
e-mail: oyqyf@ujs.edu.cn

H. Lin et al., *Advanced Technologies for Physical Properties Measurement of Food and its Processing*, https://doi.org/10.1007/978-981-96-2347-1_9

physical and chemical properties of food. CT scanning has played an important role in food quality control and has also promoted innovation and optimization in food processing technology.

Terahertz technology uses electromagnetic radiation in the terahertz band to detect food. This technology has extremely high resolution and sensitivity, allowing for the analysis of internal components and structures without damaging the food. The application of terahertz spectroscopy in the food industry, especially in the analysis of food components and quality testing, shows broad prospects. It can help detect moisture [3], starch [4], and sugar [5] in food, and also identify potentially harmful substances [6], providing more precise detection methods for food safety.

The emergence of 4D printing technology has brought unprecedented innovation opportunities to the food industry. 4D printing can not only manufacture complex food structures based on preset designs but also achieve shape changes under different environmental conditions through intelligent material responses [7]. The introduction of this technology has enabled the production of personalized and customized food products, meeting consumer demand for personalized and diverse food products. 4D printing technology can enhance the visual appeal of food and increase its fun and interactivity while meeting nutritional needs.

However, despite the broad application prospects of these technologies in the food industry, their promotion and application still face certain challenges. For example, the high cost and complexity of the technology may limit its popularity among small and medium-sized food enterprises. Additionally, ensuring the standardized and regulated application of these advanced technologies is an urgent issue to be addressed. To fully realize the potential of these technologies, the relevant industry needs to further research and improve technical standards while strengthening training and support for technology applications.

In conclusion, the introduction of advanced physical technologies has brought unprecedented development opportunities to the food industry. These technologies have not only improved food quality, safety, and processing efficiency but also driven innovation and development in the food industry. With continuous technological progress, we can expect these technologies to play an increasingly important role in future food production and testing, further meeting consumers' ongoing pursuit of food quality and safety.

9.2 Ultrasound Imaging

9.2.1 Principles of Ultrasound Imaging

Sound waves propagate in all directions through the air medium in a vibrational state, and the frequency range of sound waves that human auditory organs can perceive is between 20 and 20,000 Hz [8]. Ultrasound refers to sound waves with a vibration frequency greater than 20,000 Hz, exceeding the upper limit of human hearing [9].

Table 9.1 Parameters of ultrasound propagation in different media

Media	Density (g/cm^3)	Ultrasonic speed (m/s)	Characteristic impedance (105R)	Test trequency (MHz)
Air	0.001293	332	0.000429	2.9
Water	0.9934	1523	1.513	2.9
Blood	1.055	1570	1.656	1.0
Soft tissue	1.016	1500	1.524	1.0
Flesh	1.074	1568	1.684	1.0
Bone	1.658	3860	5.571	1.0
Fat	0.955	1476	1.41	1.0
Panceras	1.05	1570	1.648	1.0

Ultrasound and audible sound waves are both forms of mechanical vibration modes that can reflect, refract, diffract, and scatter in a medium, typically propagating as longitudinal waves in an elastic medium [10]. The difference is that ultrasound has advantages such as high frequency, short wavelength, and good directivity [11]. Below are some characteristics of ultrasound:

Beam Directivity Ultrasound has beam directivity, which is different from general sound waves and similar to the properties of light waves, meaning it has strong directionality and can concentrate and propagate in one direction. Ultrasound emitted by a transducer is distributed in a narrow beam cylindrical shape, hence called an ultrasonic beam.

High Power When sound propagates in the air, it can cause the micro-particles in the air to vibrate reciprocally and do work on the particles. Sound wave power is a physical quantity that indicates the rate of doing work. At the same intensity, the higher the frequency of the sound wave, the greater the power it possesses. Since ultrasound has a very high frequency, compared to general sound waves, its power is very large.

Cavitation Phenomenon Ultrasound is composed of mechanical waves and propagates in a medium through the transfer of energy rather than particles [12]; in fact, the latter simply oscillate around their equilibrium positions, transferring energy from one particle to another. Oscillations propagate in all directions in the medium, thus being divided into longitudinal and transverse waves. In longitudinal waves, the oscillatory motion of particles in the transmission medium is parallel to the direction of propagation, while in transverse waves, the motion is perpendicular. Longitudinal waves (or compression waves) propagate in any medium, while transverse waves propagate only in solid media. Ultrasound is a type of longitudinal wave characterized by alternating periods of compression and rarefaction in the medium through which it propagates, leading to changes in the medium's pressure; energy is transferred from the motion of particles. This process of compression and rarefaction of particles in the medium and the subsequent collapse is known as the cavitation phenomenon [13].

Like light waves, ultrasound can form images. When ultrasound is incident on the object being tested, due to differences in the internal structural organization of the tested object, attenuation differences are formed (Table 9.1), producing different echoes, which appear as light spots of varying brightness displayed sequentially on the screen, known as the ultrasound image of the tested object, also called a sonogram. The different gray levels in the ultrasound image reflect the acoustic characteristics of different tissue structures, such as sound wave attenuation, sound speed, and acoustic impedance [14]. Below are two concepts in ultrasound images: echo and artifact.

9.2.1.1 Echo

Ultrasound passing through different body tissues will emit echoes of varying degrees. Based on the strength of the signal, it is categorized into non-echoic, hypoechoic, and hyperechoic.

1. *Non-echoic*: The area where ultrasound passes without reflection is called an anechoic dark area (shadow), which may be caused by the following situations: liquid dark area: Homogeneous liquid, with no or very little difference in acoustic impedance, does not form a reflective interface, resulting in a liquid dark area. Attenuation dark area: When ultrasound passes through some pathological tissues (such as tumors), it produces significant attenuation and no echo, resulting in an attenuation dark area. Substantial dark area: Homogeneous substances with small differences in acoustic impedance can appear as anechoic dark areas. Some transparent tissues, after pathological changes, manifest as substantial dark areas.
2. *Hypoechoic*: The echo of certain tissues (such as the liver) is originally a uniformly distributed dot-like echo. When acute inflammation occurs with exudation, its acoustic impedance is smaller than that of normal tissue, resulting in a hypoechoic area (gray shadow).
3. *Hyperechoic*: Hyperechoic can be divided into relatively strong echo, strong echo, and extremely strong echo. (1) Relatively strong echo: In parenchymal organs, the difference in acoustic impedance is large, increasing the reflective interface, enhancing local echoes, and appearing as dense bright spots or bright masses (gray-white shadow), such as cancer, myoma, and hemangioma. (2) Strong echo: The internal structure of the medium is dense, with a significant difference in acoustic impedance from adjacent soft tissues or liquids, causing strong reflection. For example, bone, stones, and calcification can appear as banded or blocky strong echo areas (white shadow). (3) Extremely strong echo: The difference in acoustic impedance with adjacent soft tissues is extremely large; almost all sound energy is reflected back, unable to transmit, resulting in an extremely strong light band.

9.2.1.2 Artifacts

Artifacts refer to image distortion or differences from the actual anatomical structure caused by the imaging system or other reasons, also known as false images. Ultrasound images often exhibit the following artifacts:

1. *Reverberation*: The phenomenon where echoes continue due to multiple reflections or scattering of sound waves after the sound source stops is called reverberation. It is caused by the sound beam vertically incident on a smooth strong reflective interface, reflecting back and forth between the probe and the interface or between interfaces, mainly in three forms: (1) Multiple reflections: Sound waves vertically incident on a strong reflective interface form multiple reflections between the probe and the interface, appearing as equidistant multiple echoes with gradually decreasing echo intensity, known as multiple reflections or multiple echoes. (2) Ring-down artifact (also known as reverberation artifact): Multiple reflections between a flat thin interface and gas. The artifact is formed as the sound energy gradually weakens during multiple reflections, causing a decrease in amplitude. (3) Mirror image artifact: Generally, appears at deeper mirror interfaces (such as the diaphragm). A lesion near the probe side of a mirror-like large interface can simultaneously appear as a symmetrical similar lesion phantom on the other side of the interface. This "reflection" or double image is the mirror image artifact.
2. *Acoustic Shadow*: Due to the presence of strong reflectors or substances causing significant sound attenuation in the front, an area where sound lines cannot reach appears behind them. In this area, the image shows a longitudinal strip-like anechoic area (dark area), a phenomenon known as acoustic shadow. This phenomenon can be used to identify the presence of stones, calcifications, and bones. The acoustic shadow caused by refraction at the edge of a round lesion is also known as edge shadow.
3. *Echo Dropout*: Large interface reflection echoes depend on the angle; when the angle between the interface and the sound beam is very small or nearly parallel, no echo is received, leading to an artifact known as echo loss or edge echo loss, which can be improved by changing the probe angle.
4. *Diffraction Effect Artifacts*: Due to the diffraction effect of ultrasound, the sound beam can bypass smaller interfaces, causing the target to lose its characteristic features, resulting in diagnostic errors. For example, stones with a diameter of less than 2 mm may lose their acoustic shadow and be missed in diagnosis.
5. *Partial Volume Effect Artifacts*: Since the sound beam is not a thin line and its scanning plane is not a very thin tomographic slice, the axial, lateral, and side resolution of the sound beam affects the displayed tomographic image, causing distortion. On the sonogram, this often appears as an image where echoes from the lesion area are superimposed with echoes from the surrounding area. For instance, small echoes may appear inside a small cyst, and a sludge-like image may form inside the gallbladder. Additionally, there are multipath reflection

artifacts, sound speed distortion and ranging errors, refractive distortion, masking invisibility, improper gain adjustment causing other artifacts, etc.

9.2.2 *Composition of Ultrasound Imaging Systems*

As shown in Fig. 9.1, an ultrasound imaging system typically consists of a transducer, ultrasound signal transmitter/receiver, data acquisition card, computer, main control circuit, and mechanical scanning device.

9.2.2.1 Ultrasound Signal Transmitter/Receiver

Ultrasound Signal Transmission Principle and Related Components

Transmission Principle

Based on the Inverse Piezoelectric Effect The basis of ultrasound signal transmission is the inverse piezoelectric effect of piezoelectric materials. When an alternating voltage is applied to both ends of a piezoelectric material, it undergoes periodic expansion and contraction. This deformation causes vibrations in the surrounding medium (such as air, liquid, or solid), generating ultrasound waves. For example, in medical ultrasound imaging, a high-frequency alternating voltage is applied to the piezoelectric crystals in the ultrasound probe, causing the crystals to vibrate and produce ultrasound waves that propagate into human tissue.

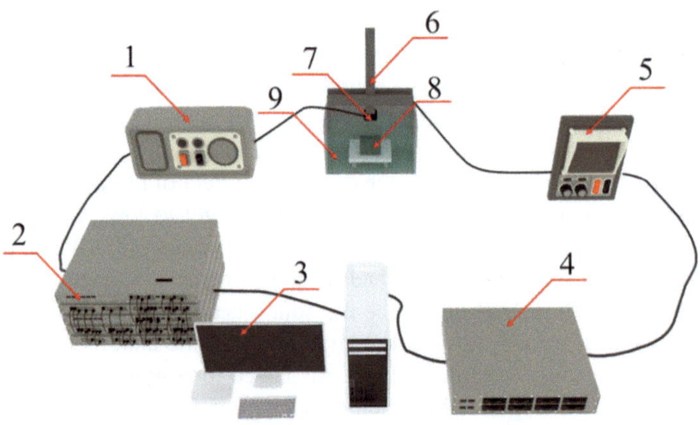

1. Ultrasonic signal transmitter/receiver 2. Data acquisition card 3. Computer 4. Motion control card 5. Main control circuit 6. 3-axis precision motor scanning mechanism 7. Ultrasonic transducer 8. Sample 9. Horizontal table

Fig. 9.1 Ultrasound imaging systems

Waveform Generation To generate an appropriate ultrasound signal waveform, the transmission circuit needs to precisely control the voltage signal applied to the piezoelectric material. Common ultrasound transmission waveforms include sine waves and pulse waves. Sine wave transmission can be used for continuous wave ultrasound imaging or Doppler ultrasound to detect blood flow velocity. Pulse waves are more commonly used in conventional ultrasound imaging, as they help avoid confusion between different echo signals and better control the propagation and reflection characteristics of ultrasound.

Transmission Related Components

Transmitting Circuit (a) Transmission Circuit Excitation Signal Source: It generates electrical signals of specific frequency and amplitude, serving as the source of ultrasound transmission. In modern ultrasound imaging systems, the excitation signal source is typically a high-precision signal generator controlled by a digital signal processor (DSP) or field-programmable gate array (FPGA). Depending on different imaging modes (such as B-mode, M-mode, color Doppler, etc.), the excitation signal source can generate signals with different frequencies, pulse widths, and repetition frequencies. (b) Power Amplifier: Since piezoelectric materials require sufficient energy to produce effective ultrasound waves, the signals generated by the excitation signal source usually need to be amplified by a power amplifier. The power amplifier must provide sufficient power while ensuring signal linearity and stability. For example, in some high-performance ultrasound imaging equipment, the power amplifier can amplify input voltages of a few volts to several hundred volts or more to meet the requirements for transmitting high-intensity ultrasound signals.

Piezoelectric Element (Transmission Part) The piezoelectric element is a key component of ultrasound transmission which works based on the inverse piezoelectric effect mentioned earlier. During ultrasound signal transmission, different types of piezoelectric elements have different performance characteristics. Common piezoelectric materials include lead zirconate titanate (PZT) ceramics, which have a high electromechanical coupling coefficient and can effectively convert electrical energy into mechanical energy, producing strong ultrasound waves. Additionally, there are new types of piezoelectric materials such as piezoelectric polymers, which, although currently not as performant as PZT ceramics in some aspects, have advantages like good flexibility and integration with flexible circuits, offering potential application value in special ultrasound imaging applications (such as wearable devices).

Ultrasound Signal Reception Principle and Related Components

Reception Principle

Based on the Positive Piezoelectric Effect When ultrasound propagates through a medium and encounters interfaces with different acoustic impedances, reflection occurs. When the reflected ultrasound acts on a piezoelectric element, due to the positive piezoelectric effect, the piezoelectric element generates an electrical signal proportional to the ultrasound pressure signal. For example, in medical ultrasound imaging, the ultrasound signals reflected from human tissues act on the piezoelectric crystals of the ultrasound probe, and the crystals produce weak electrical signals that contain information about the internal structure of the human tissues.

Signal Processing Requirements Since the reflected ultrasound signals are very weak and may be subject to noise interference, a series of signal processing steps are required. First, the signal needs to be amplified, then filtered to remove unwanted frequency components (such as environmental noise, electronic noise from the equipment, etc.), followed by demodulation and other operations to convert the received high-frequency signals into baseband signals that can reflect tissue characteristics, and finally digitized for further analysis and imaging.

Receiving Related Components

Preamplifier Since the electrical signals output from the piezoelectric element are very weak (usually in the microvolt range), the preamplifier is the first critical component in the ultrasound signal receiving chain. Its main function is to amplify the weak signals with low noise, improving the signal-to-noise ratio. The preamplifier needs to have high input impedance, low noise figure, and appropriate gain characteristics. For example, in medical ultrasound imaging equipment, the noise figure of the preamplifier is typically required to be below 1–2 dB to ensure effective amplification of weak ultrasound echo signals without introducing excessive noise.

Filter (a) Band-pass Filter: In the process of receiving ultrasound signals, a band-pass filter is used to select ultrasound signals within a specific frequency range for processing, excluding interference signals of other frequencies. For example, based on the center frequency and bandwidth of the emitted ultrasound signal, an appropriate band-pass filter is set to allow only echo signals close to the emission signal frequency to pass through. In medical ultrasound imaging, if the center frequency of the emitted ultrasound signal is 3.5 MHz and the bandwidth is 1 MHz, the band-pass filter can be set to pass signals from 3 MHz to 4 MHz. (b) Anti-aliasing Filter: Before digitizing the analog ultrasound signals, an anti-aliasing filter is essential. It prevents high-frequency signals from causing aliasing during sampling, ensuring that the digitized signals accurately reflect the characteristics of the original ultrasound signals.

Demodulator The demodulator is used to convert the received high-frequency ultrasound signals (after amplification and filtering) into baseband signals. In ultrasound imaging, common demodulation methods include envelope detection and coherent demodulation. Envelope detection is relatively simple, extracting the envelope of the ultrasound signal to obtain signals reflecting the depth and intensity of tissue reflection interfaces. Coherent demodulation can provide more information about the phase of the ultrasound signal and is important in some special ultrasound imaging techniques, such as Doppler ultrasound imaging.

Analog-to-Digital Converter (ADC) The ADC converts the processed analog ultrasound signals into digital signals for further processing by digital signal processing units, such as signal enhancement and image reconstruction. The sampling frequency and resolution of the ADC are two important parameters. The sampling frequency needs to meet the Nyquist sampling theorem to ensure accurate digitization of the ultrasound signals. For example, if the highest frequency of the ultrasound signal is 10 MHz, the sampling frequency should be at least 20 MHz. The resolution of the ADC determines the precision with which the signal amplitude can be represented, with higher resolution improving the quality of the ultrasound image.

9.2.2.2 Data Acquisition Card

Basic Functions

1. *Analog Signal Acquisition*: In ultrasound imaging systems, ultrasound signals initially exist in the form of analog signals. One important function of the data acquisition card is to acquire these analog ultrasound signals. After passing through the ultrasound transducer (transmitter/receiver component), although some preprocessing has been done, the signals remain analog. The data acquisition card receives these analog signals from the transducer and converts them into digital signals for subsequent computer processing. For example, in medical ultrasound imaging, the ultrasound signals reflected from human tissues are converted into weak analog electrical signals by the transducer, and the data acquisition card collects these signals and performs digital processing.
2. *Signal Conditioning*: During the acquisition of analog signals, the data acquisition card also conditions the signals. This includes operations such as amplification and filtering of the signals. Since ultrasound echo signals are usually very weak, amplification is needed to improve the signal-to-noise ratio. At the same time, to remove noise and interference signals, the data acquisition card filters the signals. For example, using a band-pass filter, only echo signals close to the frequency of the emitted ultrasound signal are allowed to pass, thereby improving the quality of the acquired signals.
3. *Data Conversion and Transmission*: The data acquisition card converts the collected analog signals into digital signals. This conversion process must satisfy certain sampling theorems, such as the Nyquist sampling theorem, to ensure that the digital signals can accurately reflect the original analog signals. The converted digital signals are transmitted to a computer or other data processing devices for

further analysis and imaging. Data acquisition cards typically use high-speed interfaces (such as PCI, USB, etc.) to transmit data to the host to meet real-time imaging requirements.

Working Principle

1. *Analog Input Channels*: The data acquisition card has one or more analog input channels, which are used to connect ultrasound transducers or other analog signal sources. When ultrasound signals enter the data acquisition card through these channels, they first encounter the analog front-end circuit. The analog front-end circuit includes components such as amplifiers and filters, responsible for the preliminary conditioning of the input analog signals.
2. *Analog-to-Digital Conversion (ADC)*: After the analog signals are conditioned, they enter the analog-to-digital conversion module. The ADC is one of the core components of the data acquisition card. It converts the analog signals into digital signals based on the set sampling frequency. The choice of sampling frequency should be determined according to the characteristics of the ultrasound signals, generally more than twice the highest frequency of the ultrasound signals (Nyquist sampling theorem). For example, if the highest frequency of the ultrasound signal is 10 MHz, then the sampling frequency should be at least 20 MHz. The resolution of the ADC is also important as it determines the accuracy with which the digital signal can represent the amplitude of the analog signal. Higher resolution can improve the quality of ultrasound images.
3. *Digital Signal Processing and Transmission*: The converted digital signals may undergo some simple digital signal processing within the data acquisition card, such as data buffering and format conversion. Then, these digital signals are transmitted to a computer or other data processing devices through the output interface of the data acquisition card (such as PCI or USB interface). In the computer, specialized software will further process these digital signals, such as image reconstruction and display operations.

9.2.2.3 Motion Control Card

Basic Functions

1. *Position Control*: The motion control card can precisely control the position of the ultrasound probe. In three-dimensional ultrasound imaging, the probe needs to scan at different spatial positions, and the motion control card ensures that the probe accurately moves to the specified position by sending commands. For example, in medical ultrasound examinations, when performing three-dimensional ultrasound imaging of a fetus, the motion control card controls the probe to move along a predetermined trajectory on the surface of the mother's abdomen to obtain ultrasound image data from different angles and depths. It can

achieve various motion modes, such as linear motion and circular motion. For some special ultrasound imaging applications, such as intravascular ultrasound imaging (IVUS), the motion control card can control the probe to linearly advance within the blood vessel while maintaining stable rotation of the probe, thereby comprehensively scanning the inner wall of the blood vessel.

2. *Speed Control*: In addition to position control, the motion control card can also precisely control the movement speed of the probe. In the ultrasound imaging process, different imaging requirements have different demands on the probe's movement speed. For example, in rapid screening applications, the probe may need to move at a faster speed to cover a larger inspection area; whereas in cases requiring high-resolution imaging, the probe's movement speed will slow down to collect data more meticulously. The motion control card can make the probe move at a constant speed or according to a specific speed curve based on the set parameters.
3. *Direction Control*: It can control the movement direction of the ultrasound probe. In complex imaging scenarios, such as ultrasound imaging of irregularly shaped organs like joints, the motion control card can make the probe scan in specific directions, such as tilting and rotating, to obtain comprehensive and accurate image data.

Working Principle

1. *Command Reception and Parsing*: The motion control card first receives commands from the host (such as a computer). These commands can be issued by programs written in high-level programming languages (such as C, C++, etc.). The microprocessor inside the motion control card will parse the received commands to determine the specific motion operations to be executed, such as whether to perform translational or rotational motion, and the parameters of distance, speed, and direction.
2. *Motor Drive Control*: In most cases, the motion control card achieves the movement of the ultrasound probe by controlling the motor. It sends signals to the motor driver based on the parsed commands. The motor driver converts the control signals into current or voltage signals that the motor can recognize, thereby driving the motor to rotate. For high-precision ultrasound imaging systems, stepper motors or servo motors may be used, and the motion control card can precisely control the steps (for stepper motors) or angles (for servo motors) of these motors to achieve precise probe movement.
3. *Feedback Regulation*: To ensure the accuracy of the probe's movement, the motion control card also features a feedback regulation mechanism. It receives feedback information from sensors, such as position sensors and speed sensors. For example, during the probe's movement, the position sensor detects the current position of the probe in real-time and feeds this information back to the motion control card. The motion control card compares the feedback information with the preset target position, and if there is any deviation, it adjusts the control

signal to the motor to bring the probe back to the correct position or maintain the correct motion trajectory.

9.2.2.4 Three-Axis Precision Motor Scanning Mechanism

Basic Structure and Composition

1. *Motors for Three Axes*: (a) X-Axis Motor: Responsible for controlling the movement of the ultrasound probe in the horizontal direction (usually defined as the x-axis direction). In medical ultrasound imaging, for example, when scanning the human abdomen, the x-axis motor can move the probe laterally along the abdomen to cover different areas. The x-axis motor typically has a high-precision rotating shaft and a smooth transmission mechanism, enabling precise position control. (b) Y-Axis Motor: Mainly controls the movement of the ultrasound probe in the horizontal direction perpendicular to the x-axis (y-axis direction). This axial movement, combined with the x-axis movement, allows for scanning of planar areas. In industrial ultrasound inspection, for planar workpieces, the y-axis motor can move the probe orderly over the workpiece surface to detect internal defects. The performance characteristics of the y-axis motor are similar to those of the x-axis motor, focusing on precision and stability. (c) Z-Axis Motor: Used to control the movement of the ultrasound probe in the vertical direction (z-axis direction). In three-dimensional ultrasound imaging, the z-axis motor can move the probe closer to or further from the target object or tissue. For example, when performing three-dimensional imaging of human organs, the z-axis motor can adjust the distance between the probe and the organ surface to obtain ultrasound images at different depth levels. The z-axis motor needs to have good axial positioning accuracy to ensure accurate vertical movement of the probe.
2. *Transmission Mechanism*: (a) Lead Screw Transmission: The lead screw is a common transmission component. It converts the rotational motion of the motor into linear motion through the interaction of the lead screw and nut. In the three-axis scanning mechanism of an ultrasound imaging system, lead screw transmission offers high transmission accuracy, achieving micron-level or even higher positioning precision. Moreover, lead screw transmission can withstand significant loads, suitable for the weight of the ultrasound probe and its auxiliary equipment. (b) Synchronous Belt Transmission: Synchronous belt transmission has advantages such as high transmission efficiency and low noise. In some ultrasound imaging scanning mechanisms with high-speed requirements and relatively light loads, synchronous belt transmission can be used to connect the motor and moving parts. It ensures synchronized movement between multiple axes, which is very helpful for achieving precise planar scanning (such as the x-y plane). (c) Linear Guide: Linear guides provide precise guidance for moving parts, ensuring linear movement of the ultrasound probe in each axis direction. They feature low friction and high rigidity, reducing wobble and deviation during

movement. In the three-axis scanning mechanism, linear guides work in conjunction with motors and transmission mechanisms to enhance the overall motion precision and stability of the scanning mechanism. (d) Support Structure: The support structure provides physical support for the entire three-axis scanning mechanism, ensuring its stability. It is usually made of metal materials such as aluminum alloy or stainless steel, with sufficient strength and rigidity. When designing the support structure, it is important to consider reducing vibration transmission to avoid affecting the quality of ultrasound imaging. For example, using vibration isolation design, adding vibration isolation pads between the support structure and external equipment or base, reduces interference from external vibration sources on the scanning mechanism.

Working Principle

1. *Motor Drive and Motion Conversion*: Each axis motor performs rotational motion based on the received control signals. As previously mentioned, through methods such as lead screw transmission or synchronous belt transmission, the motor's rotational motion is converted into linear motion of the ultrasound probe in the corresponding axis direction. For example, when the x-axis motor receives a command to move forward a certain distance, the motor rotates, and through lead screw transmission, the nut drives the ultrasound probe to move linearly in the x-axis direction.
2. *Three-Axis Coordinated Motion*: During three-dimensional scanning, the motors for the three axes need to work in coordination. The control system (such as a motion control card) sends coordinated instructions to make the x-, y-, and z-axis motors move according to a predetermined trajectory and speed. For example, when performing ultrasound imaging of a complex-shaped object, it may be necessary to first perform planar scanning in the x-y plane and then gradually move the probe in the z-axis direction to obtain image data at different depths. This coordinated motion requires precise control algorithms and real-time feedback regulation mechanisms.
3. *Feedback Control Mechanism*: To ensure the precision of the ultrasound probe's movement, a three-axis precision motor scanning mechanism employs feedback control. Each axis is equipped with position sensors (such as grating scales, encoders, etc.). These sensors detect the probe's position in real-time along the respective axis and feed the position information back to the control system. The control system compares the feedback position with the preset target position, and if there is a deviation, it adjusts the motor's drive signal to bring the probe back to the correct position or move along the correct trajectory. For example, in the x-axis direction, if the probe's actual position is ahead of the target position, the control system will reduce the x-axis motor's speed, slowing down the probe until it reaches the correct position.

9.2.2.5 Transducer

Basic Principles

Piezoelectric Effect

1. *Piezoelectric Effect*: When pressure is applied to certain crystals (such as quartz, piezoelectric ceramics, etc.), charges are generated on the crystal's surface, a phenomenon known as the direct piezoelectric effect. In ultrasound transducers, the direct piezoelectric effect is used to convert received ultrasound pressure signals into electrical signals. For instance, when ultrasound waves reach the piezoelectric material surface of the transducer, the material generates charges proportional to the pressure due to the applied pressure, which can be collected and further processed.
2. *Inverse Piezoelectric Effect*: Conversely, when an electric field is applied to piezoelectric materials, the material deforms, which is the inverse piezoelectric effect. In ultrasound imaging systems, by applying high-frequency electrical signals to the piezoelectric material in the transducer, the inverse piezoelectric effect is utilized to make the material vibrate at high frequencies, thereby emitting ultrasound waves.

Electromechanical Coupling Coefficient

The electromechanical coupling coefficient is an important parameter that measures the efficiency of converting electrical energy to mechanical energy (inverse piezoelectric effect) and mechanical energy to electrical energy (direct piezoelectric effect) in the transducer's piezoelectric material. A higher electromechanical coupling coefficient means less energy loss during the conversion process, enhancing the transducer's performance. Different piezoelectric materials have different electromechanical coupling coefficients; for example, PZT (lead zirconate titanate) ceramics have a high electromechanical coupling coefficient and are one of the commonly used materials in ultrasound transducers.

Structural Composition

1. *Piezoelectric Elements*: (1) Material Selection: As mentioned earlier, common piezoelectric materials include quartz crystals and piezoelectric ceramics. Quartz crystals have good stability and a high-quality factor, but their electromechanical coupling coefficient is relatively low. Piezoelectric ceramics (such as PZT) have a higher electromechanical coupling coefficient, are relatively cheaper, and can produce larger vibration amplitudes, making them widely used in most ultrasound transducers. Additionally, some new piezoelectric materials, such as piezoelectric polymers, are being researched and developed. They have advantages like good flexibility and are expected to be used in special application scenarios (such as

wearable ultrasound devices). (2) Shape and Size: The shape and size of piezoelectric elements significantly affect the transducer's performance. Common shapes include circular and rectangular. When designing a transducer, the size of the piezoelectric elements needs to be determined based on different application requirements. For example, in high-frequency ultrasound transducers, to achieve higher resolution, the size of the piezoelectric elements is usually smaller, as smaller sizes can produce higher frequency ultrasound waves. In low-frequency ultrasound transducers, used for penetrating deeper tissues or objects, the size of the piezoelectric elements is relatively larger.

2. *Matching Layer*: (a) Function: The matching layer is located between the piezoelectric element and the propagation medium (such as human tissue or the object being detected). Due to the significant difference in acoustic impedance between the piezoelectric element and the propagation medium, without a matching layer, severe reflection of ultrasound waves would occur at their interface, leading to energy loss and reduced imaging quality. The function of the matching layer is to reduce this acoustic impedance difference and improve the transmission efficiency of ultrasound waves. (b) Design Requirements: The thickness of the matching layer is generally one-quarter of the wavelength of the ultrasound wave in that material. Its acoustic impedance value needs to be calculated and determined based on the acoustic impedance of the piezoelectric element and the propagation medium to achieve optimal matching. For example, in medical ultrasound transducers, for media like human tissue with an acoustic impedance of about 1.5×10^6 kg/(m$^2 \cdot$ s), materials with suitable acoustic impedance need to be selected to make the matching layer, and its thickness must be precisely controlled.
3. *Backing Material*: (a) Function: The backing material is located on the back of the piezoelectric element. Its main function is to absorb the ultrasound waves propagating backward from the piezoelectric element, preventing multiple reflections of ultrasound within the transducer, thereby reducing clutter interference and improving the resolution and contrast of imaging. (b) Characteristics: The backing material needs to have a high acoustic attenuation coefficient to effectively absorb ultrasound energy. Common backing materials include composite materials such as epoxy resin with tungsten powder. By adjusting the content of tungsten powder, the acoustic attenuation properties of the backing material can be altered.
4. *Casing*: (a) Protective Function: The casing serves to protect the components inside the transducer, preventing damage from the external environment. In medical ultrasound transducers, the casing also needs to meet biocompatibility requirements to ensure safety when in contact with the human body. (b) Material and Design: Casing materials are usually selected from plastics or metals. When designing the casing, the usage environment of the transducer and ease of operation need to be considered. For example, the casing of a handheld ultrasound transducer typically has an ergonomic shape to facilitate holding and operation by doctors.

9.2.2.6 Three-Axis Precision Motor Scanning Mechanism

Basic Structure and Composition

1. *Motors for Three Axes*: (a) X-axis Motor: Responsible for controlling the movement of the ultrasound probe in the horizontal direction (usually defined as the x-axis direction). In medical ultrasound imaging, for example, when scanning the human abdomen, the x-axis motor can move the probe laterally along the abdomen to cover different areas. The x-axis motor typically has a high-precision rotating shaft and a smooth transmission mechanism, enabling precise position control. (b) Y-axis Motor: Mainly controls the movement of the ultrasound probe in the horizontal direction perpendicular to the x-axis (y-axis direction). This axial movement, combined with the x-axis movement, allows for scanning of planar areas. In industrial ultrasound inspection, for flat workpieces, the y-axis motor can move the probe orderly over the workpiece surface to detect internal defects. The performance characteristics of the y-axis motor are similar to those of the x-axis motor, focusing on precision and stability. (c) Z-axis Motor: Used to control the movement of the ultrasound probe in the vertical direction (z-axis direction). In three-dimensional ultrasound imaging, the z-axis motor can move the probe closer to or farther from the target object or tissue. For example, when performing three-dimensional imaging of human organs, the z-axis motor can adjust the distance between the probe and the organ surface to obtain ultrasound images at different depth levels. The z-axis motor needs to have good axial positioning accuracy to ensure accurate vertical movement of the probe.
2. *Transmission Mechanism*: (a) Lead Screw Transmission: The lead screw is a common transmission component. It converts the rotational motion of the motor into linear motion through the cooperation of the lead screw and nut. In the three-axis scanning mechanism of an ultrasound imaging system, lead screw transmission has high transmission accuracy, capable of achieving micron-level or even higher positioning accuracy. Moreover, lead screw transmission can withstand large loads, suitable for the weight of the ultrasound probe and its auxiliary equipment. (b) Synchronous Belt Transmission: Synchronous belt transmission has advantages such as high transmission efficiency and low noise. In some ultrasound imaging scanning mechanisms with high-speed requirements and relatively light loads, synchronous belt transmission can be used to connect the motor and moving parts. It ensures synchronous movement between multiple axes, which is very helpful for achieving precise planar scanning (such as the x-y plane). (c) Linear Guide: The linear guide provides precise guidance for moving parts, ensuring linear motion of the ultrasound probe in each axis direction. It features low friction and high rigidity, reducing wobble and deviation during movement. In the three-axis scanning mechanism, the linear guide works in conjunction with the motor and transmission mechanism to improve the overall motion accuracy and stability of the scanning mechanism. (d) Support Structure: The support structure provides physical support for the entire three-axis scanning

mechanism, ensuring its stability. It is usually made of metal materials (such as aluminum alloy or stainless steel) with sufficient strength and rigidity. When designing the support structure, it is necessary to consider reducing vibration transmission to avoid affecting the quality of ultrasound imaging, for example, using vibration isolation design, adding vibration isolation pads between the support structure and external equipment or base to reduce interference from external vibration sources on the scanning mechanism.

Working Principle

1. *Motor Drive and Motion Conversion*: Each axis motor performs rotational motion according to the received control signals. As previously mentioned, through methods such as lead screw transmission or synchronous belt transmission, the rotational motion of the motor is converted into linear motion of the ultrasound probe in the corresponding axis direction. For example, when the x-axis motor receives a command to move forward a certain distance, the motor rotates, and through screw transmission, the nut drives the ultrasonic probe to perform linear motion in the x-axis direction.
2. *Three-Axis Coordinated Motion*: During three-dimensional scanning, the motors in the three axes need to work in coordination. By sending coordinated commands through a control system (such as a motion control card), the X-, Y-, and Z-axis motors move according to a predetermined trajectory and speed. For example, when performing ultrasonic imaging on an object with a complex shape, it may be necessary to first perform a planar scan on the x-y plane and then gradually move the probe in the z-axis direction to obtain image data at different depths. This coordinated motion requires precise control algorithms and real-time feedback adjustment mechanisms.
3. *Feedback Control Mechanism*: To ensure the accuracy of the ultrasonic probe's movement, the three-axis precision motor scanning mechanism adopts feedback control. Each axis is equipped with position sensors (such as grating scales, encoders, etc.). These sensors detect the probe's position in the corresponding axis in real time and feed the position information back to the control system. The control system compares the feedback position with the preset target position, and if there is a deviation, it adjusts the motor's drive signal to bring the probe back to the correct position or move along the correct trajectory. For example, in the x-axis direction, if the probe's actual position is ahead of the target position, the control system will reduce the x-axis motor's speed, slowing down the probe until it reaches the correct position.

9.2.3 Application of Ultrasonic Imaging in Food

The thawing process of beef is crucial as it significantly affects the quality, safety, and nutritional integrity of the beef. Current time-based thawing control methods often lead to under-thawing or over-thawing, affecting the meat's quality, color, nutritional content, and safety. Chen et al. [15] used ultrasonic signals and ultrasonic velocity methods to quantify beef thawing. Thermal imaging of beef cross-sections was used for quantitative evaluation of thawing degree, and ultrasonic signals and ultrasonic velocity were collected and preprocessed at various thawing stages to develop a regression model. The study results showed that the model, combining ultrasonic signals and ultrasonic velocity, demonstrated high accuracy and excellent fitting ability in predicting the degree of beef thawing (Fig. 9.2).

During the processing of potato, various physicochemical parameters such as density, porosity, water activity, and textural properties change. In this sense, the textural properties of dehydrated potato snacks are one of the main quality parameters. However, instrumental techniques for evaluating textural attributes are destructive and very time-consuming and can only be applied to a very small batch of the total yield. Sanchez-Jimenez et al. [16] used air-coupled ultrasound as a fast, non-destructive, and noninvasive technique to monitor the drying of potato chips. As shown in Fig. 9.3, the PCR model for modulus of elasticity and relaxation capacity achieved high correlation coefficients. The results indicate that air-coupled

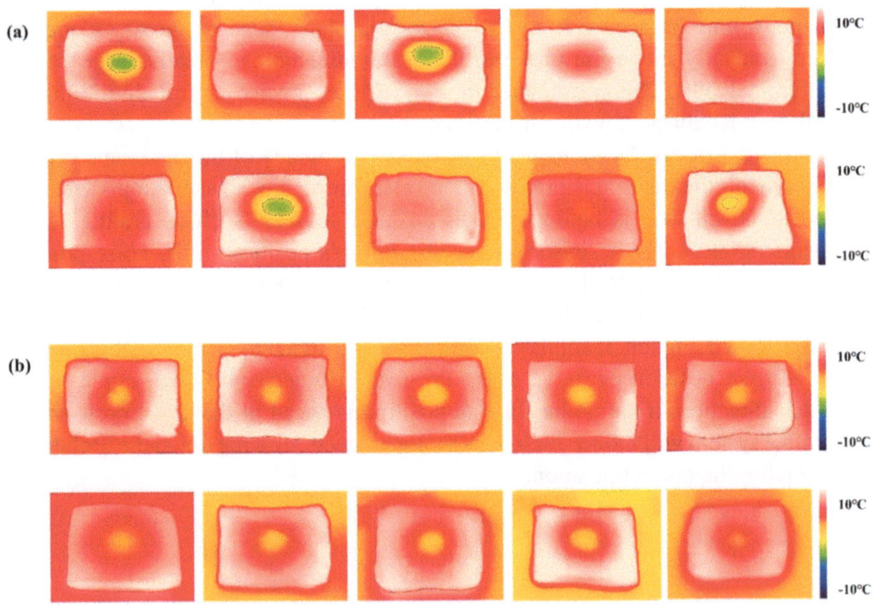

Fig. 9.2 (**a**) Thawing endpoints determined by conventional time-control methods; (**b**) thawing endpoints determined by predictive modeling methods [15]

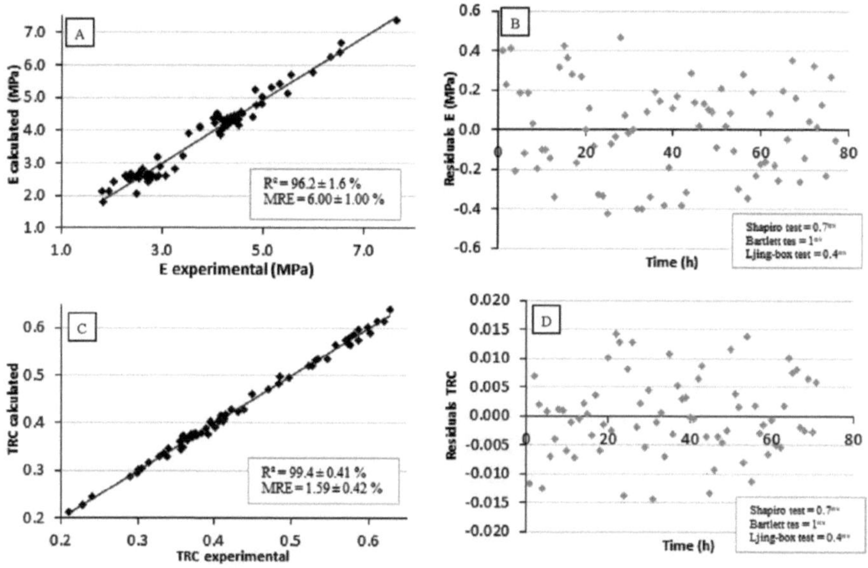

Fig. 9.3 Principal component analysis (**a**) and residual analysis (**b**) for modulus of elasticity; principal component analysis (**c**) and residual analysis (**d**) for total relaxation capacity [16]

ultrasound technology can accurately describe the textural characteristics of potato slices during the drying process.

Mechanically separated meat is a method of obtaining raw materials by mechanically separating soft tissue residues from bones, which remain after cutting and stamping the carcasses of poultry, pork, and beef. It is widely used as an ingredient in various meat products. Using mechanically separated meat as an ingredient in meat products has a significant impact on its quality, consumer satisfaction, and safety. Wieja et al. [17] proposed an innovative ultrasonic analysis method for identifying and studying mechanically separated meat samples by utilizing the differences in ultrasonic velocity in different types of meat. Figure 9.4 shows a highly significant correlation between ultrasound speed and the protein, fat, sodium, calcium, and density content of the studied meat. The developed ultrasonic method's applicability in identifying various meats and determining protein, fat, sodium, and density content was demonstrated.

The slight cracking of cotton seeds is a key factor affecting the germination rate of cotton, as foam acid or water enters the cotton seeds through foam acid or water. A new method for detecting slightly cracked cotton seeds using air-coupled ultrasound and sound-image encoding with a lightweight Vision Transformer (ViT) was proposed by Zhang et al. [11]. The experimental results show that the average overall recognition accuracy for slightly cracked cotton seeds among regular cotton seeds is 90.7%, indicating that the proposed method is reliable for the task of detecting slightly cracked cotton seeds.

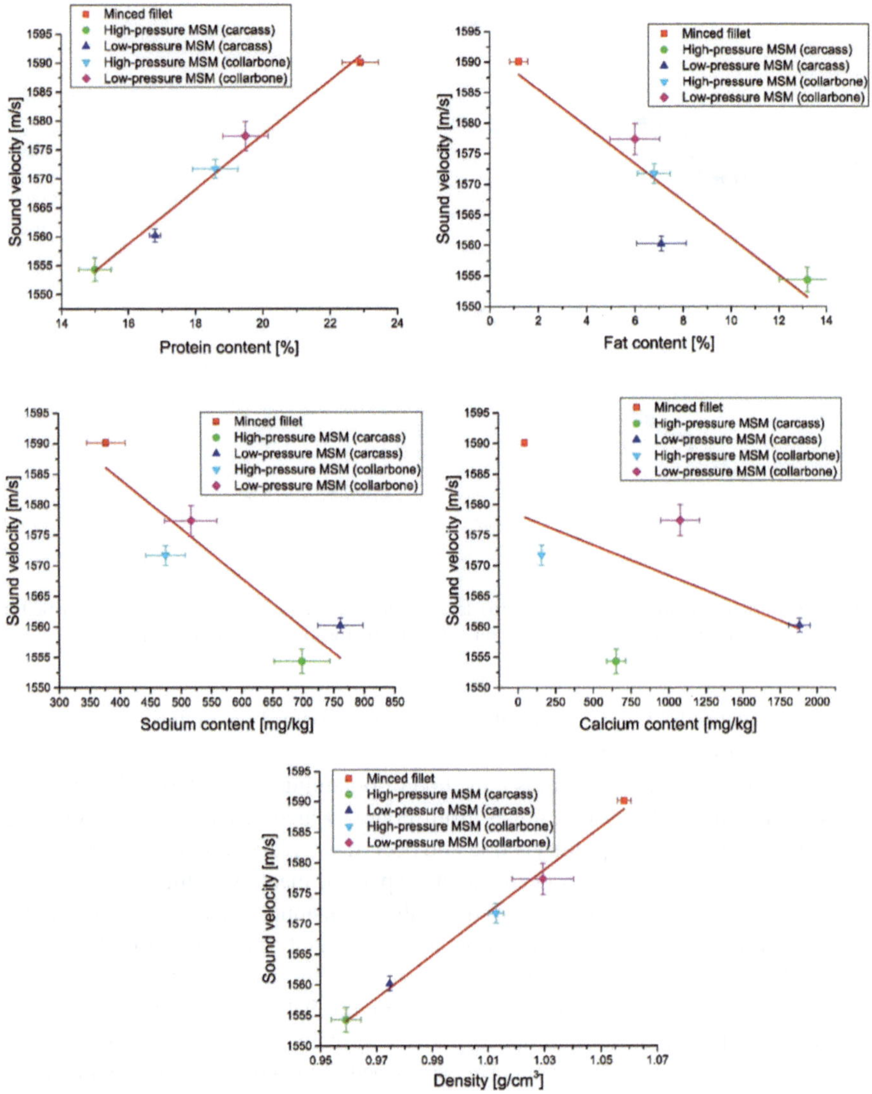

Fig. 9.4 Correlation of ultrasound speed with (**a**) protein, (**b**) fat, (**c**) sodium, (**d**) calcium, and (**e**) density [17]

9.3 Computerized Tomography

9.3.1 Principles of Computerized Tomography

Computerized tomography (CT) scanner is an important non-destructive testing technology, a product of the development of physics and computer science. Based on the principle of interaction between rays and matter, it obtains digital images of the object being inspected through projection reconstruction methods. Computerized tomography acquires a series of projection data about the object from multiple perspectives, and through image reconstruction technology, it can obtain tomographic images of the object [18].

Compared with conventional imaging examination methods, CT has the following three advantages.

1. *Rich image information*: The cross-sectional images obtained by CT have accurate layer thickness and clear images. Additionally, the cross-sectional images obtained by CT scanning can be reconstructed through computer software processing to obtain the multidirectional (such as coronal, sagittal) sectional images required for diagnosis [19]. CT tomographic images are digital images that can be directly processed digitally, facilitating numerical analysis and two-dimensional and three-dimensional display in various forms, and are also convenient for storage and transmission.
2. *High density resolution*: Generally, the density resolution of CT is about 20 times higher than that of conventional X-rays. The reasons are the following: (1) the ray beam of CT is strictly collimated when passing through the object to reach the detector, resulting in less scattered rays; (2) CT machines use high-sensitivity, high-efficiency receivers; (3) According to detection needs, computer software can be used to control the gray scale, adjusting it to a range suitable for human visual observation.
3. *Capable of quantitative analysis*: CT can accurately measure the radiation absorption attenuation values of various tissues, allowing for quantitative analysis through various calculations. CT imaging utilizes the differences in attenuation characteristics of emitted rays (such as X rays, γ rays, etc.) in different tissues for imaging. In a homogeneous object, the attenuation of rays is proportional to the distance traveled within the object. If the proportional constant is μ, the travel distance of the rays is dX, and the intensity of the X-rays after passing through the object is dI, then:

$$dI = I_0 e^{-ux} dX \tag{9.1}$$

By performing indefinite integration on the above equation, the total path length (the thickness of the object through which the X-rays pass) is d, and then the above equation can be written as:

$$I = I_0 e^{-\mu d} \tag{9.2}$$

In the formula, I is the intensity of the rays after passing through the object, I_0 is the intensity of the incident rays, e is Euler's constant (2.718), μ is the linear absorption coefficient, and d is the thickness of the object. This is the law of intensity attenuation when rays pass through a uniform object. Because in practical situations, the attenuation of rays is also related to the density and atomic number of the material, i.e., the greater the density and the higher the atomic number, the greater the attenuation of the rays, the above formula can be rewritten as:

$$I = I_0 e^{-\mu_m d_m} \tag{9.3}$$

In the formula, $\mu_{m=}\ \mu/\rho$ is the mass absorption coefficient, and $d_m = \rho d$ is the mass thickness.

In actual scanning, when rays pass through the tissue of the object being measured, the density at various locations is often nonuniform. For ease of analysis, the target through which the rays pass is divided into several small blocks, with each block having a length of l (assuming at this time that the absorption coefficient μ within the small block is a constant), for each small block, the following equation can be used:

$$I_1 = I_0 e^{-\mu_1 1} \tag{9.4}$$

The intensity for *n* small blocks is:

$$I_n = I_0 e^{-l(\mu_1 + \mu_2 + L + \mu_n)} \tag{9.5}$$

Generally speaking, the μ-*value* of soft tissue is close to that of water, with the μ value of muscle being about 5% higher than that of water, and the μ value of fat being about 10% lower than that of water. The μ value of bone is approximately twice that of water. In practical applications, using the absorption coefficient is not very convenient. The inventor of CT, Hounsfield, defined a new scale for the absorption attenuation coefficient, dividing the change in the X-ray linear attenuation coefficient between air and dense bone into 2000 units, called the CT value. To commemorate Hounsfield's immortal contribution, this scale unit was named HU. Nowadays, in applications, HU is used as the measurement unit for CT values. In the practical application of CT, the absorption attenuation values of various tissues, including air, are compared with water, with dense bone set as the upper limit of +1000 and air as the lower limit of −1000, thus creating a relative absorption coefficient, which can be expressed by the following formula:

$$\text{CT value} = \frac{\mu_{\text{tissue}} - \mu_{\text{water}}}{\text{water}} \times a \tag{9.6}$$

In the formula, a is the amplification factor, generally taken as 1000, and the CT value of water is 0. The CT values for different tissues of organisms are shown in Table 9.2.

Table 9.2 CT values for different tissues of organisms

Organization	CT	Organization	CT
Air	−1000	Blood	10–80
Fat	−100	Skeleton	200–1000
Water	0	Soft tissue	50–100

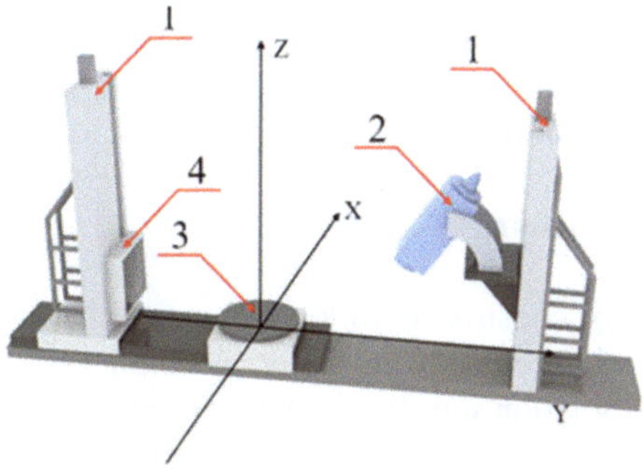

Fig. 9.5 Simple CT system

1. Vertical translation table 2. Radiation source 3. Rotating table 4. Detector

9.3.2 Computerized Tomography System

A simple computerized tomography system is shown in Fig. 9.5, mainly including components such as a vertical translation stage, a radiation source, a rotating stage, and detectors.

9.3.2.1 Vertical Platform

Basic Structure

It usually consists of high-precision guide rails, sliders, a drive motor, a transmission mechanism (such as a screw nut mechanism or belt drive), and a carrying platform. The guide rail provides precise linear motion guidance, and the slider cooperates with the guide rail to ensure the smooth movement of the translation stage. The drive motor provides power for the movement of the translation stage, and the transmission mechanism converts the rotational motion of the motor into linear motion of the slider along the guide rail. The carrying platform is used to install the radiation source or other related equipment.

Working Principle

When the drive motor receives a control signal, it starts to rotate, driving the slider to move vertically along the guide rail through the transmission mechanism. In the CT system, the movement of the vertical translation stage can change the relative position between the radiation source and the object being detected in the vertical direction, thereby achieving scanning of different height levels of the object being detected. For example, in multilayer spiral CT scanning, the precise movement of the vertical translation stage, in coordination with the rotation of the rotating stage, can perform continuous tomographic scanning of different height parts of the human body.

9.3.2.2 Radiation Source

Basic Structure

It mainly consists of core components that generate radiation (such as the X-ray tube), a high-voltage generator, a cooling system, etc. The X-ray tube is the key part of the radiation source, containing a cathode and an anode inside. The cathode emits electrons, and the anode attracts electrons under high voltage and decelerates them suddenly, thereby producing X-rays. The high-voltage generator provides high voltage to the X-ray tube to accelerate electrons to produce X-rays with sufficient energy. The cooling system is used to dissipate the heat generated by the X-ray tube during operation, as a large amount of energy is converted into heat during the radiation generation process. Common cooling methods include air cooling and water cooling.

Working Principle

The high-voltage generator applies high voltage between the cathode and anode of the X-ray tube, causing the cathode filament to heat up and emit electrons. These electrons, under the influence of the high-voltage electric field, fly at high speed towards the anode target. When high-speed electrons strike the anode target, most of the kinetic energy is converted into heat energy, and a small portion (about 1%) is converted into X-ray energy. The generated X-rays pass through a specific collimator, forming a beam with a certain shape and intensity distribution, directed at the object being inspected. After the rays penetrate the object, due to the different absorption and attenuation levels of the object's internal tissues or structures, the rays carrying the internal structural information of the object reach the detector.

9.3.2.3 Rotating Table

Basic Structure

It consists of a rotating platform, motor, rotating shaft, bearings, and support structure. The rotating platform is the part where the object to be inspected is placed. The motor is connected to the rotating shaft through a transmission device (such as gear or belt transmission), driving the rotating platform to perform circular motion around the rotating shaft. Bearings are used to support the rotating shaft, reducing friction during the rotation process and ensuring smooth rotation. The support structure provides stable physical support for the entire rotating table, ensuring no shaking or deviation occurs during rotation.

Working Principle

After receiving instructions from the control system, the motor starts to rotate, driving the rotating shaft through the transmission device, which in turn makes the rotating platform rotate uniformly around the rotating shaft. During the CT scan, the object to be inspected is placed on the rotating table. As the rotating table turns, the rays emitted by the radiation source penetrate the object from different angles and reach the detector. By collecting and processing the penetration information of rays from multiple angles, the internal three-dimensional structure image of the object can be reconstructed. For example, in a complete CT scan, the rotating table may rotate 360° or 180° plus a fan angle (depending on the specific scanning mode) to obtain comprehensive information about the object.

9.3.2.4 Detector

Basic Structure

It is composed of multiple detection units (such as units composed of scintillation crystals and photodiodes or gas detector units) arranged in an array form. The scintillation crystal detector is a more common type, where the scintillation crystal produces visible light photons upon receiving X-ray photons, and the photodiode converts the visible light photons into electrical signals. These detection units are tightly arranged in a two-dimensional array to cover a certain area, capable of simultaneously receiving rays that have passed through the object being inspected. The detector is also equipped with signal acquisition circuits, amplification circuits, and data transmission circuits, used to collect, amplify, and transmit the weak electrical signals generated by the detection units.

Working Principle

When the rays penetrate the object and reach the detector, the photons in the rays interact with the detection units of the detector. For scintillation crystal detectors, X-ray photons interact with the scintillation crystal to produce visible light photons, which are detected by the photodiode and converted into electrical signals. The intensity of these electrical signals is related to the intensity of the received rays. Due to the different attenuation of rays by different parts of the object's interior, the detection units at different positions of the detector receive rays of varying intensities, thus generating electrical signals of different intensities. These electrical signals are collected by the signal acquisition circuit, amplified by the amplification circuit, and then transmitted to the computer system through the data transmission circuit for further processing and image reconstruction, ultimately obtaining the internal image information of the object.

9.3.3 Application of Computerized Tomography Technology in Food

Carrots are a vegetable crop with significant economic value. When carrots contain excessive undesirable fibrous tissue, it can affect human health. Donis-González [20] and others proposed a method to investigate the presence of unwanted fibrous tissue in carrots through computerized tomography (CT) images. Figure 9.6 shows a reconstructed 3D image obtained from a set of 2D CT images, which can spatially visualize the presence of unwanted fibrous tissue and its impact on the entire carrot. This study indicates CT has great potential for non-destructive visualization of the internal quality attributes of freshly processed carrots.

Chestnuts are popular with consumers as a supplement to traditional foods. Besides external factors such as color, shape, size, and visible surface cracks, consumers are also concerned about inherent internal defects in fresh chestnuts,

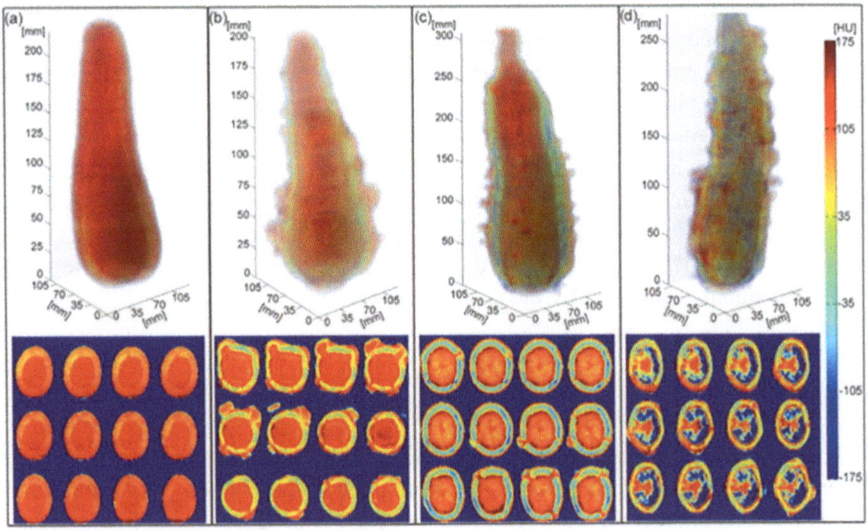

Fig. 9.6 3D CT reconstruction of processed carrots (top) and its corresponding example set of 2D CT images (bottom) for (**a**) fiberless, (**b**) gelatinous-fibrous, (**c**) mixed (gelatinous and woody) fibrous, and (**d**) woody-fibrous carrots [20]

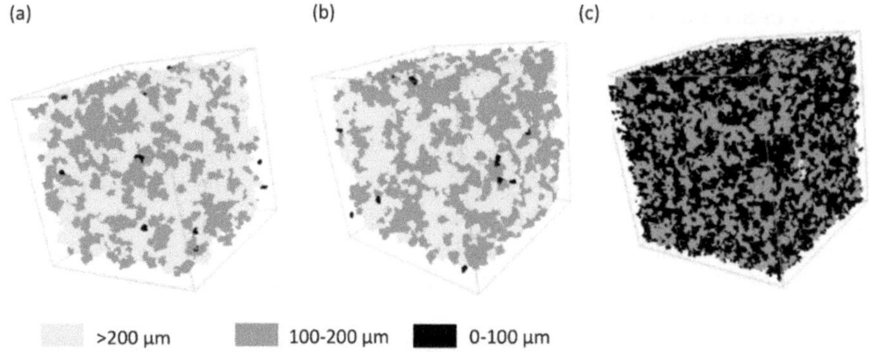

Fig. 9.7 3D visualization of μCT-based technique ice crystals at three different freezing rates of (**a**) 0.9 °C/min, (**b**) 1.4 °C/min and (**c**) 3.3 °C/min [23]

such as pest damage, mold, and calcification. Wang et al. [21] combined X-ray CT technology with image processing methods to develop a method for identifying internal defects in chestnuts. The results showed that the recognition rates for healthy, slightly defective, and severely defective chestnuts were as high as 92.9%, 83.6%, and 93.7%, respectively. This study indicates that the designed detection process can effectively identify chestnut defects and can thus be used for non-destructive monitoring of chestnut quality during postharvest storage.

To maintain the quality of fresh pear fruits after harvest and provide high-quality fruits year-round, a controlled supply chain and long-term storage are employed. However, during storage, internal diseases may occur due to poor storage conditions, which may not cause externally visible symptoms. Van De Looverbosch et al. [22] combined support vector machines with feature extraction algorithms and X-ray computed tomography for non-destructive detection of internal diseases in pear fruits. The classifier was able to distinguish between defective and sound fruits, with classification accuracy ranging from 90.2% to 95.1%. Additionally, low false positive and false negative rates were achieved, ranging from 0.0% to 6.7% and 5.7% to 13.3%, respectively.

The freezing rate determines the size distribution and morphology of ice crystals. Ice crystal characteristics can directly affect the microstructure of food, causing irreversible cell and tissue damage, leading to quality loss. Mulot et al. [23] used an innovative method to directly image and quantify ice crystals in frozen minced beef samples through X-ray micro-computed tomography. The results obtained using μCT showed smaller, rounder, and more numerous ice crystals with faster freezing rates. This method is applicable to frozen and freeze-dried sample collection (Fig. 9.7).

9.4 Terahertz Technology

9.4.1 Principles of Terahertz Technology

Terahertz waves (THz waves), also known as terahertz rays (THz rays), were officially named in the late 1980s. Before that, scientists collectively referred to them as far-infrared rays. Terahertz waves refer to electromagnetic waves with frequencies in the range of 0.1 THz to 10 THz, with wavelengths approximately in the range of 0.03–3 mm. The band of terahertz is located between microwaves and infrared [24]. Far-infrared technology has achieved many results and has been industrialized. However, research results and data involving the terahertz band are very few. The electromagnetic energy of THz band photons is approximately 1–10 meV, which is roughly equivalent to the energy transitions between molecular rotational energy levels [25]. Weak interactions between molecules (such as the extension of DNA hydrogen bonds), the backbone vibrations of nucleic acid macromolecules (conformational bending), the rotational and vibrational transitions of dipoles, and the low-frequency vibrational absorption frequencies of crystal lattices all correspond to specific frequency ranges in the THz band [26]. Most polar molecules, such as water molecules and ammonia molecules, strongly absorb THz radiation, and the transitions between the vibrational and rotational energy levels of many organic macromolecules (DNA, proteins, etc.) also fall within the THz band range [27, 28]. The THz spectrum of substances (including emission, reflection, and transmission spectra) contains rich physical and chemical information, and studying the absorption and dispersion characteristics of the spectrum can be used for the detection of various biological samples.

Compared to X-rays, THz photon energy is extremely low (10^7 to 10^8 times weaker than X-ray photons), and THz radiation does not cause photodamage or photoionization in biological tissues, making it particularly suitable for in vivo examination of biological tissues. Similar to radio waves, THz waves can penetrate most dielectric materials (such as plastics, ceramics, clothing, cardboard, wood, fat, bones, ice, various powders, dry food, etc.) [29]. THz imaging technology can achieve better contrast than X-ray technology [30]. Like light waves, THz waves can propagate in the air and are also reflected by metals, solid objects, and the human body. Due to these unique properties, THz wave detection technology has shown great advantages in areas such as chemical and biomass detection, label-free gene detection, non-destructive food testing, agricultural product analysis, and quality control.

One of the advantages of THz imaging is its sensitivity to phase, which means it can achieve material identification and component measurement. Since different materials have different refractive indices, using THz phase information, it is possible to perform three-dimensional imaging of objects to identify internal defects and texture of materials. At the same time, THz imaging can achieve higher contrast than X-ray imaging. Utilizing the strong absorption properties of water for THz radiation, it can be used to measure the moisture content in substances, such as determining the freshness of food inside packaging by measuring moisture. THz detection of meat products has found that lean meat absorbs THz radiation strongly, while fat almost does not absorb THz radiation, allowing for quality inspection of meat products using this characteristic. Based on the low interference and nonionizing characteristics, THz imaging has the potential to become a powerful tool in fields such as safe in vivo medical imaging and non-destructive testing of agricultural products (such as quality assessment of seeds [31], variety identification [32], and internal quality inspection of fruits [33]).

THz imaging systems primarily perform transmission and reflection imaging of samples, obtaining image information of samples for analysis and processing. A THz imaging system mainly consists of a light source, optical system, THz detector, and computer system. Depending on the imaging method, current THz imaging methods mainly include time-domain pulse THz imaging (TPI) and continuous THz wave (CW) imaging [34]. Compared to time-domain pulse THz imaging (TPI), continuous wave imaging has the advantage of achieving high contrast, and the imaging system is simple and inexpensive, facilitating miniaturization [35]. Currently, continuous wave THz radiation is primarily generated using quantum cascade terahertz lasers. To achieve a large dynamic range and high spatial resolution, TPI imaging is required. Although TPI imaging systems are complex, they can obtain more useful information. With the development of THz generation, detection, and transmission equipment, TPI imaging systems will gradually become miniaturized in the future. THz wave coherent tomography technology (T-CT) can obtain the three-dimensional distribution of the complex refractive index of the object under test in the THz band, thus allowing the detection of the material's three-dimensional structural information, which has broad application prospects in the field of non-destructive testing.

9.4.2 Terahertz Imaging System

The basic principle of terahertz electromagnetic wave imaging is as follows: using terahertz electromagnetic waves with a known waveform as imaging rays, the intensity and phase of the terahertz electromagnetic waves that pass through (or reflect from) the imaging sample contain the spatial distribution of the sample's complex dielectric constant. By recording the two-dimensional information of the transmitted terahertz electromagnetic wave's intensity and phase and through appropriate digital processing and spectral analysis, a three-dimensional image of the sample's terahertz electromagnetic wave can be obtained. A notable feature of terahertz wave imaging technology is the large amount of information; each image source corresponds to a terahertz time-domain spectrum, and by performing a Fourier transform on the time-domain spectrum, the terahertz frequency response spectrum at each point can be obtained. Generally, without the sample, the THz waveform is first obtained using sampling as reference information; then the sample is added to derive the waveform information of that pixel. Traditional transmission-

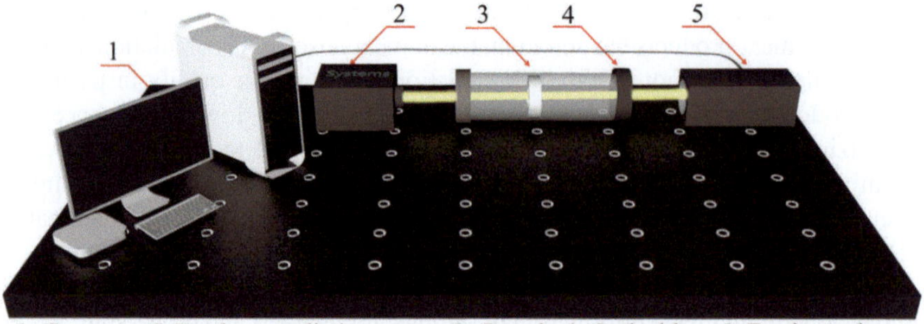

1. Computer 2. Terahertz radiation source 3. Sample 4. Optical lens 5. Terahertz detector

Fig. 9.8 Terahertz imaging system

type terahertz imaging can provide information for each pixel, including normalized peak amplitude information, normalized pulse delay information, pulse width information, etc., extracting various information depending on the specific situation. After Fourier transformation, we obtain frequency domain information, which includes (1) amplitude variation information at a certain frequency, (2) phase variation information at a certain frequency, (3) intensity integral information over a certain frequency range, and (4) ln(a/n) value at a certain frequency (considering both amplitude and phase) as well as transmittance, time delay, absorption rate, refractive index, etc.

The terahertz imaging system, as shown in Fig. 9.8, generally includes a computer, terahertz radiation source, sample, optical lens, and terahertz detector, among others.

9.4.2.1 Computer

Basic Structure

In terms of hardware, it consists of a central processing unit (CPU), memory, hard drive, motherboard, graphics card (which may play a role in visualizing terahertz-related data), and input-output interfaces (such as USB interfaces for connecting terahertz system devices for data transmission, etc.). The central processing unit is the core computing component of the computer, responsible for executing various instructions; memory is used to temporarily store running programs and data; and the hard drive is used for long-term storage of system software, applications, and data. In terms of software, it is equipped with an operating system (such as Windows, Linux, etc.), as well as software specifically for terahertz data acquisition control, processing, and analysis. This software may include user interfaces for setting parameters of the terahertz radiation source and detector, controlling the data acquisition process, and data processing algorithms (such as Fourier transform for converting time-domain signals to frequency-domain signals, etc.).

Working Principle

In the terahertz system, the computer sends control instructions to the terahertz radiation source through interfaces, setting parameters such as the output frequency and power of the radiation source. During data acquisition, the computer controls the terahertz detector to start collecting data and receives electrical signals from the detector (representing information after the interaction of terahertz waves with the sample). The software within the computer processes these electrical signals, for example, by converting analog signals to digital signals through analog-to-digital conversion, and then using various algorithms for data processing, such as noise removal and signal correction. Finally, the computer visualizes the processed terahertz data, displaying it to the user in image form (such as terahertz frequency band spectrum images, terahertz imaging of samples, etc.), and can also store the data for subsequent analysis.

9.4.2.2 Terahertz Radiation Source

Basic Structure

Radiation Sources Based on Electronic Technology For example, backward wave oscillators (BWO), which consist of components such as an electron gun, slow-wave structure, and collector. The electron gun emits an electron beam, which interacts with the radio frequency field in the slow-wave structure, generating terahertz waves through special electron motion and energy conversion mechanisms, and finally, the electrons are collected by the collector. Gunn diodes, made from semiconductor materials such as N-type gallium arsenide (GaAs), generate terahertz waves through high-frequency oscillations of electrons under the influence of high-field domains formed within the semiconductor when an appropriate voltage is applied across the ends due to the Gunn effect.

Radiation Sources Based on Optical Technology Terahertz quantum cascade laser (THz-QCL) has an internal structure composed of a cascade of multiple quantum wells. By injecting current, electrons transition between quantum wells, achieving population inversion through quantum confinement effects and band structure, thereby emitting laser in the terahertz frequency range. Photoconductive antenna radiation sources are composed of photoconductive materials (such as low-temperature grown gallium arsenide) and metal electrodes. When an ultrashort laser pulse irradiates the photoconductive material, electrons in the material are excited, forming a transient current under the influence of an external electric field, which then radiates terahertz waves.

Working Principle

Radiation Sources Based on Electronics Conversion of electrical energy into electromagnetic radiation energy in the terahertz frequency band by utilizing the kinematic properties of electrons in special structures or materials, such as acceleration, oscillation and other processes of electrons. For example, in a return-wave tube, an electron beam interacts with a radio-frequency field in a slow-wave structure to produce terahertz waves through energy exchange.

Optical-Based Radiation Source mainly utilizes optical principles such as photoelectric effect or quantum leap. Such as photoconductive antenna radiation source, ultrashort laser pulse excitation photoconductive material in the electron, in the external electric field to produce transient current, according to Maxwell's system of equations, the change of the current will be radiated electromagnetic wave, its frequency in the terahertz band; terahertz quantum cascade laser is through the electron in the quantum well of the quantum leap to achieve the number of particles inverted, resulting in the output of the terahertz laser by the stimulated radiation.

9.4.2.3 Optical Lenses

Basic Structure

Optical lenses are optical components made from materials (such as silicon, germanium, etc.) with good transmittance in the terahertz band. Their surfaces are usually precisely machined into spherical or aspherical shapes to achieve focusing or collimating functions for terahertz waves. Depending on their function, there are converging lenses (thicker in the middle, thinner at the edges) and diverging lenses (thinner in the middle, thicker at the edges). To reduce reflection losses, the lens surface may also be coated with anti-reflective coatings.

Working Principle

For converging lenses, when terahertz waves are incident on the lens surface, they refract at the interface between the lens material and the surrounding medium (such as air) according to the law of refraction. The shape and material properties of the lens cause terahertz waves incident from different directions to converge at the lens's focal point after refraction. Diverging lenses, on the other hand, cause terahertz waves to diverge away from the optical axis after passing through the lens. In terahertz systems, optical lenses can be used to converge terahertz waves emitted by a terahertz radiation source onto a sample to enhance the interaction between the terahertz waves and the sample; they can also be used to converge terahertz waves transmitted through the sample onto a terahertz detector to improve detection efficiency.

9.4.2.4 Terahertz Detectors

Basic Structure

Detectors Based on Thermal Effects For example, pyroelectric detectors are composed of pyroelectric crystal materials (such as triglycine sulfate (TGS)) and electrodes. The crystal is encapsulated in a suitable housing, and the electrodes are used to extract electrical signals. (2) Detectors based on photon effects: Terahertz Schottky diode detectors consist of a Schottky barrier structure (formed by metal and semiconductor contact) and are usually equipped with an antenna structure to receive terahertz waves. Superconducting detectors, such as superconducting tunnel junction (STJ) detectors, have a tunnel junction structure made of superconducting materials and operate in low-temperature environments.

Working Principle

Detectors Based on Thermal Effects For example, pyroelectric detectors, when a terahertz wave is irradiated onto a pyroelectric crystal, the energy of the terahertz wave is absorbed by the crystal, causing the temperature of the crystal to change. Due to the pyroelectric effect, the polarization strength of the crystal changes with temperature, resulting in charge accumulation on the electrodes and the formation of an electrical signal related to the intensity of the terahertz wave.

Photon Effect-Based Detectors such as terahertz Schottky diode detectors, where the terahertz wave is received by the antenna structure and interacts with the electrons in the Schottky barrier. The electric field of the terahertz wave causes a change in the state of motion of the electrons in the barrier region, resulting in a nonlinear current-voltage characteristic, which is detected by detecting the change in this characteristic. Superconducting detectors (e.g., superconducting tunnel junction detectors) utilize the special physical properties in the superconducting state. When a terahertz wave irradiates a superconducting tunnel junction, it destroys the Cooper pairs in the superconducting state and produces quasiparticles, which causes changes in the electrical properties (e.g., currents, etc.) of the tunnel junction, and detects terahertz waves based on such changes.

9.4.3 Applications of Terahertz Technology in Food

The processing of fish is complex and tedious, and during processing, fish bones or various foreign objects can easily remain. Traditional manual and visual inspection methods are difficult to detect these foreign objects, posing a serious threat to human health. Hu [36] and others explored the feasibility of using terahertz imaging and spectroscopy for rapid, non-destructive detection of endogenous foreign objects (fish

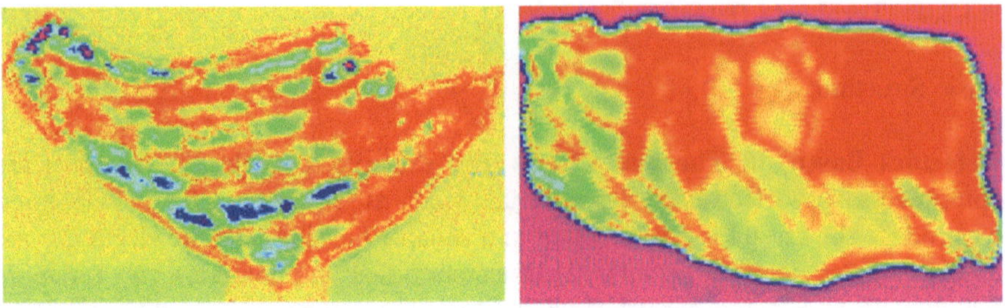

(a) (b)

Fig. 9.9 Terahertz compressed image after CARS feature selection [36]

bones) and exogenous foreign objects (metal, plastic, wooden toothpicks) in fish. Figure 9.9 shows terahertz images of foreign objects in fish before and after CARS feature selection, which reduces interference from a large amount of redundant information, and the boundary information of foreign objects is clear and continuous. The features extracted by CARS obtained the optimal SVM model with an accuracy of 99.56%. Meanwhile, the terahertz images of samples at 0.1–4.0 THz were visualized and analyzed. The study found that the contours of foreign objects in the fish body are clear and the shape and size of foreign objects in the fish body can be well distinguished by terahertz images.

Wheat is one of the main food sources for humans and is important for human life. Impurities such as wheat husks, wheat straw, wheat leaves, wheat grains, weeds, and ladybugs can easily mix with wheat during harvesting, lowering the quality of the harvest and affecting the quality of wheat to some extent. Shen [37] and others used terahertz spectral imaging to study the spectral characteristics of wheat, wheat husks, wheat straw, wheat leaves, wheat grains, weeds, and ladybugs in the 0.2–1.6 THz range and obtained the corresponding frequency domain spectra using Fourier transform. They designed a new Wheat-V2 convolutional neural network (CNN) to extract data and information related to spectral imaging features. The results showed that the designed Wheat-V2 model could effectively identify impurities in wheat images, with recognition accuracies of 97.56% and 98.58% for the Top_1 and Top_5 of the validation set, respectively.

Wheat is highly adaptable, widely used, and has many varieties. Different wheat varieties contain different nutrients and taste different, and they are sold at different prices in the market. According to protein content, wheat can be divided into strong gluten wheat, medium gluten wheat, and weak gluten wheat. Strong gluten wheat is usually used for making bread. Medium gluten wheat can be used for making Chinese buns and noodles, while weak gluten wheat is suitable for making cakes and cookies. Considering the numerous wheat varieties, there may be adulterated wheat in the domestic and international markets to seek illegal profits. Chen [38] and others used terahertz time-domain spectroscopy (THz-TDS) technology combined with convolution to distinguish 12 wheat varieties (strong gluten wheat, medium gluten wheat, weak gluten wheat), achieving calibration set accuracy and prediction

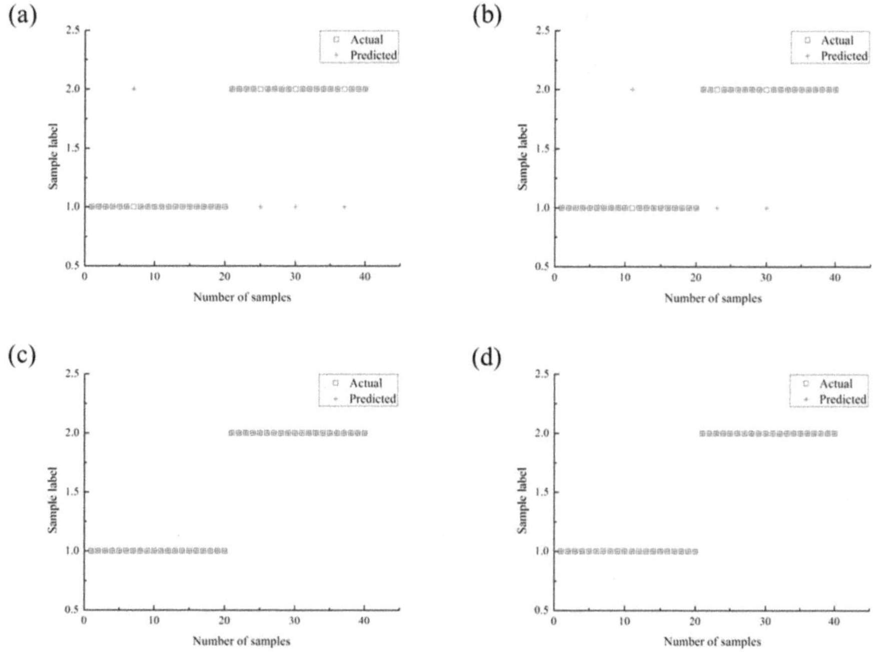

Fig. 9.10 Determination of the detection limit of AFB_1 in soybean oil by the BPNN combined with t-SNE method. (**a**) 1 μg/kg AFB_1; (**b**) 2 μg/kg AFB_1; (**c**) 10 μg/kg AFB_1; (**d**) 20 μg/kg AFB_1 [39]

set accuracy of 98.7% and 97.8%, respectively, with an error recognition rate of only 2.2%.

Aflatoxin B_1 (AFB_1), as a metabolite of *Aspergillus flavus* and *Aspergillus parasiticus*, is well-known for its toxicity, carcinogenicity, and mutagenicity, posing serious harm to humans. Soybean oil, a widely consumed dietary product made from soybeans, is considered inevitably contaminated by aflatoxin. Liu [39] and others studied the feasibility of using terahertz spectroscopy for the rapid determination of AFB_1 in soybean oil. Using BPNN combined with t-SNE, excellent prediction results were obtained, with a prediction coefficient (Rp) of 0.9948 and a root mean square error of prediction (RMSEP) of 0.7124 μg/kg. Additionally, terahertz spectroscopy was proven to detect AFB_1 in soybean oil at a concentration of 1 μg/kg (accuracy over 90%). The conclusion is that the combination of terahertz spectroscopy and chemometrics will be a promising technique for the rapid determination of AFB_1 concentration in soybean oil (Fig. 9.10).

9.5 4D Printing Technology

9.5.1 Principles of 4D Printing Technology

3D printing (three-dimensional printing) refers to the technology of generating three-dimensional entities based on a three-dimensional digital model, through specific forming equipment (3D printers), using digital model files, and employing adhesive materials such as engineering plastics, photosensitive resins, and powdered metals for continuous physical layering. It is the main implementation method of "additive manufacturing" technology [40], and has wide applications in fields such as electronics, energy, biomedicine, and aviation.

4D printing (four-dimensional printing) was initially defined as 4D printing = 3D printing + time [41], as shown in Fig. 9.11. In this, the shape, properties, or functions of a 3D printed structure can change over time, undergoing morphological changes such as expansion, bending, and sharpening. With further research, 4D printing is defined as the targeted evolution of 3D printed structures in terms of shape, performance, and function. 4D printing can achieve self-assembly, multifunctionality, and self-repair, with time-relatedness, printer independence, and predictability [42]. Therefore, 4D printing can manufacture dynamic structures with adjustable shapes, characteristics, or functions. The realization of dynamic structures is closely related to smart materials and their combinations in 3D space, and how to achieve this requires mathematical modeling of the distribution of various materials in the structure. In 4D printed structures, there are at least two stable states, and the structure can transition from one state to another under corresponding stimuli.

9.5.2 4D Printing System

The elements for realizing 4D printing include 3D printing equipment, stimuli or triggers, smart (stimuli-responsive) materials, interaction mechanisms, and mathematical modeling [43].

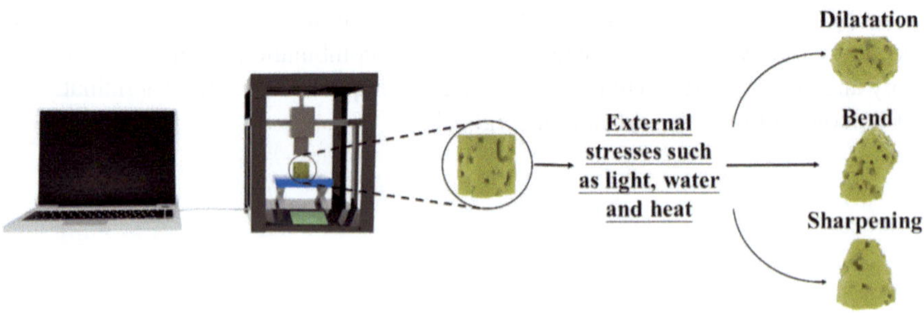

Fig. 9.11 4D printing process

9.5.2.1 3D Printing Equipment

4D printed structures are created by combining multiple materials into a single one-time printed structure with appropriate distribution. Differences in material properties, such as expansion rate and thermal expansion coefficient, will lead to the desired shape transformation behavior of the printed structure under external stimuli. Therefore, 3D printing is essential for manufacturing multi-material structures with simple geometric shapes. 3D printing equipment that can be used for 4D printing includes fused deposition modeling (FDM) 3D printers based on material jetting technology, stereolithography appearance (SLA) 3D printers based on photopolymerization technology, selective laser sintering (SLS), and selective laser melting (SLM) 3D printers based on powder fusion technology [44].

9.5.2.2 Stimuli or Triggers

Stimuli or triggers are external stresses used to induce changes in shape/properties/functions of 4D printed structures. So far, the stimuli used in 4D printing by researchers include water, heat, a combination of heat and light, and a combination of water and heat. The choice of stimuli depends on specific application requirements and also determines the type of smart materials used in the 4D printed structure.

9.5.2.3 Smart (Stimuli-Responsive) Materials

Stimuli-responsive materials are the most critical component in 4D printing, possessing characteristics such as self-sensing, decision-making, reactivity, shape memory, adaptability, multifunctionality, and self-repair [45]. From the perspective of structural shape, stimuli-responsive materials can be divided into shape-changing materials and shape-memory materials. Shape-changing materials refer to materials whose shape changes immediately upon external stimuli and immediately returns to the original state once the stimuli are removed, such as volumetric linear expansion or contraction materials; shape-memory materials refer to materials that change to a temporary shape upon an external stimulus or action and return to the initial shape upon another suitable stimulus or action. Shape-memory materials can be further divided into shape memory alloys (SMA), shape memory polymers (SMP), shape memory hydrogels (SMH), shape memory ceramics (SMC), shape memory gel (SMG), etc. [46]

9.5.2.4 Interaction Mechanism

In certain cases, simply exposing smart materials to stimuli cannot directly achieve the desired shape of 4D printed structures. The stimuli need to be applied in a specific sequence at the appropriate time, which is known as the interaction mechanism. Taking the constrained thermal mechanism as an example, the stimulus is heat, and the smart material has a shape memory effect. The implementation of the constrained thermal mechanism mainly includes four steps: first, the structure deforms under external load at high temperature; second, the temperature is lowered while maintaining the external load; third, the structure is unloaded at low temperature and reaches the desired shape; and fourth, the original shape can be restored by reheating the structure.

9.5.2.5 Mathematical Modeling

Mathematical modeling is necessary for the material distribution and structure required for the designed shape, properties, or functional changes. It is essential to develop theoretical and numerical models among the four core elements: material structure, desired final shape, material properties, and stimulus properties.

9.5.3 Application of 4D Printing Technology in Food

As a protein-rich food, surimi is suitable for use as an “ink” matrix because its viscosity and gel characteristics are relatively close to the bonding and supporting capabilities required for 3D printing. However, most reports on 3D food focus on how to improve the printability of surimi, with little attention to extending the shelf life of printed food. Although the freshness of the print changes during storage, it cannot be perceived due to the color of the material itself. This leads to the destruction of the printed shape traditionally used to determine freshness, which is detrimental to the application of printed functional foods. Li [47] and others prepared 4D printed color-changing materials to address the difficulty in discerning freshness changes of printed functional surimi while maintaining the printed shape. As shown in Fig. 9.12, the 4D printed material undergoes significant color changes with the freshness of surimi. The results show that the prepared 4D color-changing material can simultaneously achieve enhanced printing effects, preservation, and non-destructive freshness monitoring, promoting the application of 3D printing technology in food processing.

Some fruits, such as mangosteen, lychee, and kiwi, have a thick peel that conceals their edible contents inside, making internal spoilage invisible. This often leads to fruit waste. Smart labels can respond to changes in gases, temperature, and microbial load in the environment where the food is placed, indicating its freshness and

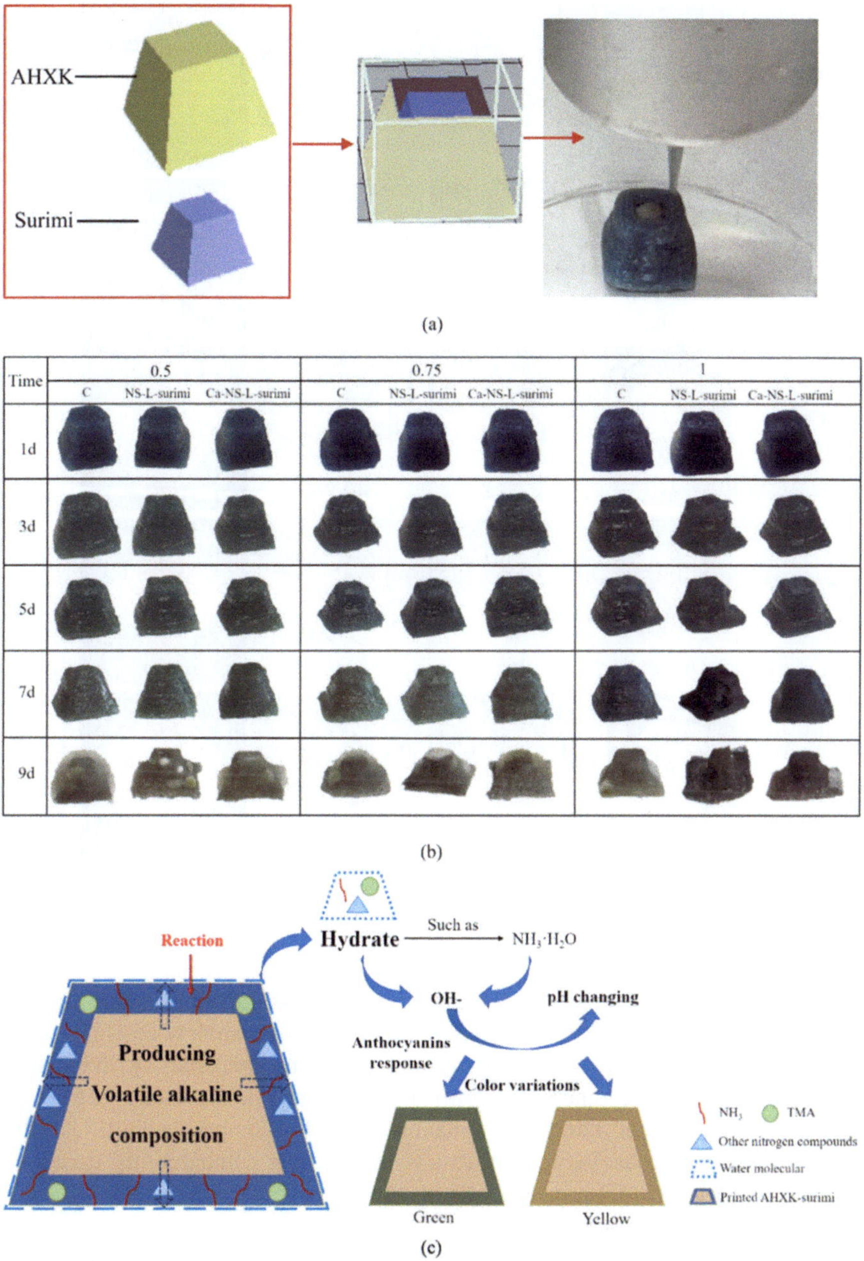

Fig. 9.12 Freshness monitoring of minced fish by 4D-printed color-changing materials Annotation (**a**) Procedure; (**b**) Actual color change [47]

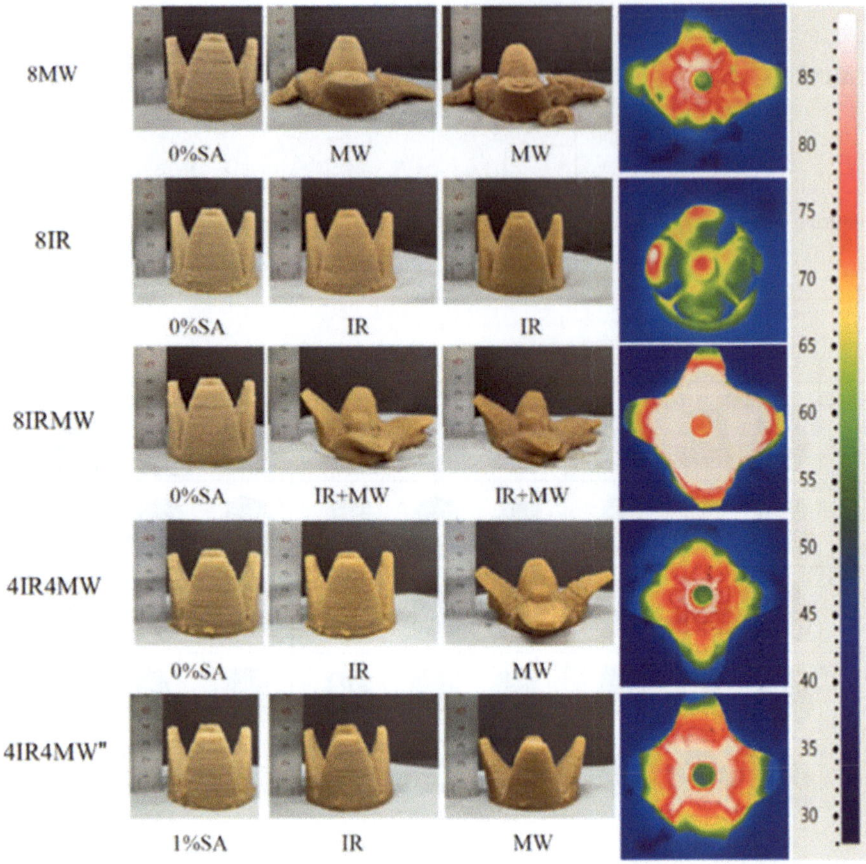

Fig. 9.13 4D-printed foods with different SA contents after IR and/or microwave heat distribution treatments. (MW microwave heating, IR infrared heating) [49]

providing early warnings. Among the available freshness indicators, pH-responsive labels are efficient and economical alternatives because they are low-cost and easy to prepare. Traditional pH-responsive markers are films prepared by the casting method. Teng [48] and others developed pH-responsive labels using methacrylic anhydride (MA) grafted konjac glucomannan (KGM) combined with 4D printing and photocuring technology. The results show that compared to traditional pH-responsive labels, 4D printed labels respond faster and more clearly to quality changes in kiwi during storage.

Chestnuts are gluten-free and low in fat, which gives chestnut flour great commercial potential. Gluten provides elasticity and moisture, giving products a better texture. Therefore, the lack of gluten makes products harder, limiting the application of chestnut flour in 3D printing. Lv [49] and others used sodium alginate (SA) and chestnut flour for 4D printing under microwave and infrared heating conditions. As shown in Fig. 9.13, the puffing effect during microwave heating causes directional deformation of the 3D printed product, while 1.0% SA content contributes to more ideal deformation. Infrared heating for 4 min, followed by microwave heating for 4 min, reduces deformation and avoids collapse. Microwave and infrared heating help with flavor, texture, and swallowability.

References

1. Bariya AR, Rathod NB, Patel AS, Nayak JKB, et al. Recent developments in ultrasound approach for preservation of animal origin foods. Ultrason Sonochem. 2023;101:106676.
2. Yu Y, Jia C, Wang J, Pi F, et al. Characterizing the internal structure of Chinese steamed bread during storage for quality evaluation using X-ray computer tomography. Sensors (Basel, Switzerland). 2023;23(21):8804.
3. Ren Y, Lei T, Sun D-W. In-situ indirect measurements of real-time moisture contents during microwave vacuum drying of beef and carrot slices using terahertz time-domain spectroscopy. Food Chem. 2023;418:135943.
4. Nakajima S, Shiraga K, Suzuki T, Kondo N, et al. Quantification of starch content in germinating mung bean seedlings by terahertz spectroscopy. Food Chem. 2019;294:203–8.
5. Suhandy D, Suzuki T, Ogawa Y, Kondo N, et al. A quantitative study for determination of glucose concentration using attenuated total reflectance terahertz (ATR-THz) spectroscopy. Eng Agric Environ Food. 2012;5(3):90–5.
6. Baek SH, Kang JH, Hwang YH, Ok KM, et al. Detection of methomyl, a carbamate insecticide, in food matrices using terahertz time-domain spectroscopy. J Infrared Millim Terahertz Waves. 2016;37(5):486–97.
7. Cao Q, Chen W, Zhong Y, Ma X, et al. Biomedical applications of deformable hydrogel microrobots. Micromachines. 2023;14(10):1824.
8. Fu J, Deng Z, Liu C, Liu C, et al. Intelligent, flexible artificial throats with sound emitting, detecting, and recognizing abilities. Sensors (Basel, Switzerland). 2024;24(5):1493.
9. Chen Q, Zhang C, Zhao J, Ouyang Q. Recent advances in emerging imaging techniques for non-destructive detection of food quality and safety. TrAC Trends Anal Chem. 2013;52:261–74.
10. Moyano DB, Paraiso DA, González-Lezcano RA. Possible effects on health of ultrasound exposure, risk factors in the work environment and occupational safety review. Healthcare (Basel, Switzerland). 2022;10(3):423.
11. Zhang C, Huang W, Liang X, He X, et al. Slight crack identification of cottonseed using air-coupled ultrasound with sound to image encoding. Front Plant Sci. 2022;13:956636.
12. Cano-Lamadrid M, Martínez-Zamora L, Castillejo N, Artés-Hernández F. From pomegranate byproducts waste to worth: a review of extraction techniques and potential applications for their revalorization. Food Secur. 2022;11(17):2596.
13. Gallo M, Ferrara L, Naviglio D. Application of ultrasound in food science and technology: a perspective. Food Secur. 2018;7(10):164.
14. Liao YY, Yang KC, Lee MJ, Huang KC, et al. Multifeature analysis of an ultrasound quantitative diagnostic index for classifying nonalcoholic fatty liver disease. Sci Rep. 2016;6:35083.
15. Chen H, Sun Q, Xuan L, Lin Z, et al. Ultrasonic technology for predicting beef thawing degree and endpoint. J Food Eng. 2024;383:112236.
16. Sanchez-Jimenez V, Collazos-Escobar GA, González-Mohino A, Gomez Alvarez-Arenas TE, et al. Non-invasive monitoring of potato drying by means of air-coupled ultrasound. Food Control. 2023;148:109653.

17. Wieja K, Kiełczyński P, Szymański P, Szalewski M, et al. Identification and investigation of mechanically separated meat (MSM) with an innovative ultrasonic method. Food Chem. 2021;348:128907.
18. Li X, Luo S. A compressed sensing-based iterative algorithm for CT reconstruction and its possible application to phase contrast imaging. Biomed Eng Online. 2011;10:73.
19. Shen L. Implementation of CT image segmentation based on an image segmentation algorithm. Appl Bionics Biomech. 2022;2022:2047537.
20. Donis-González IR, Guyer DE, Chen R, Pease A. Evaluation of undesirable fibrous tissue in processing carrots using Computed Tomography (CT) and structural fiber biochemistry. J Food Eng. 2015;153:108–16.
21. Wang J, Lu Z, Xiao X, Xu M, et al. Non-destructive determination of internal defects in chestnut (Castanea mollissima) during postharvest storage using X-ray computed tomography. Postharvest Biol Technol. 2023;196:112185.
22. Van De Looverbosch T, Rahman Bhuiyan MH, Verboven P, Dierick M, et al. Nondestructive internal quality inspection of pear fruit by X-ray CT using machine learning. Food Control. 2020;113:107170.
23. Mulot V, Fatou-Toutie N, Benkhelifa H, Pathier D, et al. Investigating the effect of freezing operating conditions on microstructure of frozen minced beef using an innovative X-ray micro-computed tomography method. J Food Eng. 2019;262:13–21.
24. Wu X, Liang X, Wang Y, Wu B, et al. Non-destructive techniques for the analysis and evaluation of meat quality and safety: a review. Food Secur. 2022;11(22):3713.
25. Zhang J, Yang Y, Feng X, Xu H, et al. Identification of bacterial blight resistant rice seeds using terahertz imaging and hyperspectral imaging combined with convolutional neural network. Front Plant Sci. 2020;11:821.
26. Li K, Li D, Zhang Y. Terahertz spectral properties of 5-substituted uracils. Sensors (Basel, Switzerland). 2021;21(24):8292.
27. Lin YS, Huang ST, Hsu SS, Tang KY, et al. Sensitivity enhancement and probiotic detection of microfluidic chips based on terahertz radiation combined with metamaterial technology. Micromachines. 2022;13(6):904.
28. Liu W, Liu C, Chen F, Yang J, et al. Discrimination of transgenic soybean seeds by terahertz spectroscopy. Sci Rep. 2016;6:35799.
29. Jiang Y, Ge H, Lian F, Zhang Y, et al. Early detection of germinated wheat grains using terahertz image and chemometrics. Sci Rep. 2016;6:21299.
30. Vohra N, Bowman T, Diaz PM, Rajaram N, et al. Pulsed terahertz reflection imaging of tumors in a spontaneous model of breast cancer. Biomed Phys Eng Express. 2018;4(6):5025.
31. Jiang W, Wang J, Lin R, Chen R, et al. Machine learning-based non-destructive terahertz detection of seed quality in peanut. Food Chem: X. 2024;23:101675.
32. Luo H, Zhu J, Xu W, Cui M. Identification of soybean varieties by terahertz spectroscopy and integrated learning method. Optik. 2019;184:177–84.
33. Hu J, Qiao P, Yang L, Lv H, et al. Research on nondestructive detection of pine nut quality based on terahertz imaging. Infrared Phys Technol. 2023;134:104798.
34. Yu L, Hao L, Meiqiong T, Jiaoqi H, et al. The medical application of terahertz technology in non-invasive detection of cells and tissues: opportunities and challenges. RSC Adv. 2019;9(17): 9354–63.
35. Zhang R, Zhang L, Wu T, Zuo S, et al. Contrast-enhanced continuous-terahertz-wave imaging based on superparamagnetic iron oxide nanoparticles for biomedical applications. Opt Express. 2016;24(8):7915–21.
36. Hu J, Zhan C, Shi H, Qiao P, et al. Rapid non-destructive detection of foreign bodies in fish based on terahertz imaging and spectroscopy. Infrared Phys Technol. 2023;131:104448.
37. Shen Y, Yin Y, Li B, Zhao C, et al. Detection of impurities in wheat using terahertz spectral imaging and convolutional neural networks. Comput Electron Agric. 2021;181:105931.
38. Chen F, Shen Y, Li G, Ai M, et al. Classification of wheat grain varieties using terahertz spectroscopy and convolutional neural network. J Food Compos Anal. 2024;129:106060.

39. Liu W, Zhao P, Wu C, Liu C, et al. Rapid determination of aflatoxin B1 concentration in soybean oil using terahertz spectroscopy with chemometric methods. Food Chem. 2019;293: 213–9.
40. Mai Z, Liu D, Chen Z, Lin D, et al. Fabrication and application of photocatalytic composites and water treatment facility based on 3D printing technology. polymers. 2021;13(13):2196.
41. Fahmy AR, Derossi A, Jekle M. Four-dimensional (4D) printing of dynamic foods-definitions, considerations, and current scientific status. Food Secur. 2023;12(18):3410.
42. Shokrani H, Shokrani A, Seidi F, Mashayekhi M, et al. Polysaccharide-based biomaterials in a journey from 3D to 4D printing. Bioeng Trans Med. 2023;8(4):e10503.
43. Shie MY, Shen YF, Astuti SD, Lee AK, et al. Review of polymeric materials in 4D printing biomedical applications. Polymers. 2019;11(11):1864.
44. Shuo Z. Research on 4D printing of bionic intelligent responsive hydrogels and their drug-controlled release characteristics [D], 2022.
45. Wang M, Li W, Tang G, Garciamendez-Mijares CE, et al. Engineering (bio)materials through shrinkage and expansion. Adv Healthc Mater. 2021;10(14):e2100380.
46. Wang Y, Wang Y, Mushtaq RT, Wei Q. Advancements in soft robotics: a comprehensive review on actuation methods, materials, and applications. Polymers. 2024;16(8):1087.
47. Li G, Zhan J, Huang J, Xu E, et al. Enhanced fresh-keeping capacity of printed surimi by Ca2+-nano starch-lutein and its nondestructive freshness monitoring based on 4D printed anthocyanin. Int J Biol Macromol. 2023;252:126543.
48. Teng X, Zhang M, Adhikari B, Ma Y. 4D printed pH-responsive labels of methacrylic anhydride grafted konjac glucomannan for detecting quality changes in respiring climacteric fruits. Food Hydrocoll. 2024;149:109543.
49. Lv Y, Wang B, Cheng Y, Lv W, et al. Printing characteristics and microwave infrared-induced 4D printing of chestnut powder composite paste. J Food Eng. 2024;382:112197.

MIX
Papier aus verantwortungsvollen Quellen
Paper from responsible sources
FSC® C105338

If you have any concerns about our products,
you can contact us on
ProductSafety@springernature.com

In case Publisher is established outside the EU,
the EU authorized representative is:
Springer Nature Customer Service Center GmbH
Europaplatz 3, 69115 Heidelberg, Germany

Printed by Libri Plureos GmbH
in Hamburg, Germany